INTRODUCTION TO MICROPROCESSORS

INTRODUCTION TO MICROPROCESSORS

By

Manish Varshney
M.Sc., M.Tech.
HOD (CS/IT), SSVIT
Bareilly (U.P)

Jyoti Agarwal
B.E., M.E.
Sr. Lecturer, SRMSCET
Bareilly (U.P)

UNIVERSITY SCIENCE PRESS
(An Imprint of Laxmi Publications Pvt. Ltd.)
An ISO 9001:2015 Company
BENGALURU • CHENNAI • GUWAHATI • HYDERABAD
KOCHI • KOLKATA • LUCKNOW • RANCHI
NEW DELHI

Printed and bound in India
First Edition : 2016; *Edition* : 2025
ISBN **:** 978-93-81159-78-1

Branches

✆	Bengaluru	080-26 75 69 30
✆	Chennai	044-24 34 47 26
✆	Guwahati	0361-254 36 69
✆	Hyderabad	040-27 55 53 83
✆	Kochi	0484-405 13 03
✆	Kolkata	033-40 04 77 79
✆	Lucknow	0522-430 36 13
✆	Ranchi	0651-224 24 64

Published in India by

UNIVERSITY SCIENCE PRESS
(An Imprint of Laxmi Publications Pvt. Ltd.)
An ISO 9001:2015 Company
113, GOLDEN HOUSE, GURUDWARA ROAD, DARYAGANJ,
NEW DELHI - 110002, INDIA
Corporate Office : B-33, Sector-5, Noida,
Gautambuddha Nagar (U.P.) - 201 301, INDIA
Telephone : +91-120-44 46 900
www.laxmipublications.com
info@laxmipublications.com

PREFACE

This textbook has been written as per the new syllabus of the subject "Introduction to Microprocessor" of B.Tech degree of U.P. Technical University. This book is offered to third year students of Electronics and Communication Engg. This book is divided into 13 chapters and all the topics are illustrated and clear diagrams and simple language is used throughout the text to facilitate easy understanding of the concept.

The following is the brief description of the topics that are covered in each chapter:

Chapter 1 introduces the number system and gives the concept of various digital devices. Various number systems, their conversions and computer data format and code format are given in this chapter which is very necessary to revise before starting microprocessor. In this chapter, we have also discussed various logic circuits like flip flop, counters etc. in brief.

Chapter 2 introduces the microprocessor evaluation. This chapter includes types of computer, various generations of microprocessor.

Chapter 3 gives the basic concepts of microprocessor. This includes the concept of address bus, data bus, control bus and registers, various data transfer schemes etc. This also includes about the instruction cycle.

Chapter 4 discusses the architectural advancements of microprocessors, typical microprocessor development schemes. This gives the concept of RISC and CISC architecture, pipeline concept, cache memory and multi core processor design concept.

Chapter 5 gives the outlines of 8-bit microprocessor with its pin configuration, internal architecture and various timing signals are introduced. The machine cycles of simple instructions are also discussed.

Chapter 6 outlines the basics of instruction set of 8085 processor and various addressing modes. The instruction formats, opcodes, mnemonics, number of bytes are cleared illustrated. We have also introduced in this chapter, how to write the assembly language program and also explained with examples.

Chapter 7 deals with 16-bit microprocessor 8086 architecture. This chapter includes register organization, memory segmentation, internal architecture of 8086, operating modes, physical memory organization, general bus operations etc.

Chapter 8 discusses the basics of instruction set of 8086 processor. In this various groups of instructions and interrupt of 8086 processor is also discusses.

Chapter 9 gives the Assembly language programming techniques of 8085. This includes Looping, counting and indexing concept. This also covers stack and subroutine concept.

Chapter 10 gives the Assembly language programming techniques of 8086. This chapter also includes various Assembly language programs based on 8086.

Chapter 11 discusses the basics of interfacing with microprocessor. The various interfacing devices of 8237, 8255, 8253/8254 and 8259 are discussed in detail.

Chapter 12 discusses the concepts of DAC and ADC

Chapter 13 discusses the concepts of Advanced Microprocessors.

The language of the book is very lucid for the students of Engineering.

—Authors

ACKNOWLEDGEMENTS

It is pleasant to acknowledge the help we have received from different individuals. We are particularly indebted to our Chairman respected Shri Anupam Kapoor ji for his constant encouragement, support and motivation for such a good task. We are also thankful to Mr. Saurabh Mehrotra, Secretary of SSVIT Trust, Bareilly, and our Principal Prof. Anil Kumar for their important suggestions and support provided in this endeavour.

We would also like to thank all the faculty members, family members, friends and well wishers whose blessings, knowledge, advice and interaction have made this book a possible venture of our life.

We are also grateful to all our family members for their constant motivation and support.

—**Authors**

CONTENTS

1

Computer Number Formats and Digital Devices

Number system is used to represent information in quantitative form. Some of the common number systems are binary, octal, decimal and hexadecimal. A number system of base (also called radix) *r* is a system, which has *r* distinct symbols for *r* digits. A string of these symbolic digits represents a number. To determine the value that a number represents, we multiply the number by its place value that is an integer power of *r* depending on the place it is located and then find the sum of weighted digits.

The number system we are all familiar with is the **decimal** number system, or **base 10** number system. Computers, on the other hand, use the **binary** number system, which is **base 2**. Other number systems, notably **hexadecimal** (base 16) and, to a lesser degree, **octal** (base 8) are also used in the study and programming of computers, but only because they provide a more compact notation for representing binary values. Internally the computer *only knows the binary system*. As will be shown, *hexadecimal is shorthand for binary*.

1.1 CLASSIFICATION OF NUMBER SYSTEM

The following are the types of Number System:

1. **Decimal Number System:** It is a well known number system with base 10. The numbers included in the Decimal Number System are 0, 1, 2, 3, 4, 5, 6, 7, 8, 9.
2. **Binary Number System:** It is a well known number system with base 2. The numbers present in the Binary Number System are 0, 1.
3. **Octal Number System:** It is a well known number system with base 8. The numbers included in the Octal Number System are 0, 1, 2, 3, 4, 5, 6, 7.
4. **Hexa Decimal Number System:** It is a well known number system with base 16. The numbers included in the Hexa Decimal Number System are: 0, 1, 2, 3, 4, 5, 6, 7, A, B, C, D, E, F.

1.2 DECIMAL NUMBER SYSTEM

Decimal Number System is the most commonly used number system, to the base ten. Decimal numbers do not necessarily contain a decimal point; 563, 5.63, and –563 are all decimal numbers. Other systems are mainly used in computing and include the binary number system, octal number system, and hexadecimal number system.

The decimals 0.3, 0.51, and 0.023 can be expressed as the decimal fractions 3/10, 51/100, and 23/1,000. They are all terminating decimals. These fractions can equally be expressed as the percentages 30%, 51%, and 2.3%.

1.2.1 Decimal to Binary Conversion

Divide the number repeatedly by 2 (Answer being a whole number and a remainder, The same as Modula-2 Div and Mod commands). The First remainder is the LSB (Least Significant Bit or Digit, the digit on the right of the number) of the Binary number. The next remainder is the next digit, and so on until the value is 0.

We write the remainders from bottom to top.

Example 1. Convert $(145)_{10}$ into base 2.

- 145 ÷ 2 =
  - 72
  - Rem 1
- 72 ÷ 2 =
  - 36
  - Rem 0
- 36 ÷ 2 =
  - 18
  - Rem 0
- 18 ÷ 2 =
  - 9
  - Rem 0
- 9 ÷ 2 =
  - 4
  - Rem 1
- 4 ÷ 2 =
  - 2
  - Rem 0
- 2 ÷ 2 =
  - 1
  - Rem 0
- 1 ÷ 2 =
  - 0
  - Rem 1

So 145 = 10010001

Example 2. Convert 147_{10} to binary

$$
\begin{aligned}
147 \div 2 &= 73 \ldots 1\\
73 \div 2 &= 36 \ldots 1\\
36 \div 2 &= 18 \ldots 0\\
18 \div 2 &= 9 \ldots 0\\
9 \div 2 &= 4 \ldots 1\\
4 \div 2 &= 2 \ldots 0\\
2 \div 2 &= 1 \ldots 0\\
1 \div 2 &= 0 \ldots 1
\end{aligned}
$$

Reading the remainders from bottom to top, we have: $147_{10} = 10010011_2$.

Converting Decimal Fractions to Binary

A good way to organize the problem is to use a table, with columns **old**, **bit**, and **new**. We start with the number we're planning to convert in the top row **old** entry.

Now, we work across each row, and multiply the **old** entry by 2 to get a new value. Now we take the integer part of the new value and put it in the **bit** entry, and put the fraction part in the **new** entry. If the **new** entry is non-zero, we copy it into the **old** entry in the next row down and continue (if it is 0, we're done).

The result is the contents of the **bit** column, with the most significant bit in the top row.

Example. Convert $(0.375)_{10}$ to $(?)_2$

Old	Bit	New	Notes
0.375	0	0.75	Multiply 0.375 by 2, get 0.75
0.75	1	0.5	Multiply 0.75 by 2, get 1.5
0.5	1	0	Multiply 0.5 by 2, get 1.0

So the result is 0.011.

1.2.2 Decimal to Octal Conversion

To convert decimal to octal is slightly more difficult. The typical method to convert from decimal to octal is repeated division by 8. While we may also use repeated subtraction by the weighted position value, it is more difficult for large decimal numbers.

Repeated Division by 8 Method

For this method, divide the decimal number by 8, and write the remainder on the side as the least significant digit. This process is continued by dividing the quotient by 8 and writing the remainder until the quotient is 0. When performing the division, the remainders which will represent the octal equivalent of the decimal number are written beginning at the least significant digit (right) and each new digit is written to the next more significant digit (the left) of the previous digit.

Example 1. *Convert $(44978)_{10}$ to Octal number.*

Solution.

Division	Quotient	Remainder	Octal Number
44978 / 8	5622	2	2
5622 / 8	702	6	62
702 / 8	87	6	662
87 / 8	10	7	7662
10 / 8	1	2	27662
1 / 8	0	1	127662

Example 2. *Convert $(25)_{10}$ to octal number system.*

Solution.

The division problem will be setup as 2510 ÷ 8

2510 ÷ 8 = 3

2510 – 24 = Remainder 1 ----> 1 (LSD)

3 ÷ 8 = 0

3 – 0 = Remainder 3 ----> 3 (MSD)

Now, write the number from MSD to LSD as $(31)_8$.

Example 3. *Convert $(258)_{10}$ to octal.*

The division problem will be setup as 25810 ÷ 8

$25810 \div 8 = 32$
$18 - 16$ = Remainder 2 ----> 2 (LSD)
$32 \div 8 = 4$
$32 - 32$ = Remainder 0 ----> 0
$4 \div 8 = 0$
$4 - 0$ = Remainder 4 ----> 4 (MSD)

Now, write the number from MSD to LSD. You should end up with $(402)_8$.

Converting Decimal Fractions to Octal

To convert a decimal fraction to octal, you must multiply the fraction by 8. Then, you need to extract the number(s) that appear to the left of the decimal point. The first number that is extracted will be the MSD and will follow the decimal point. The last number that is extracted will be the LSD. Continue to desired accuracy. This type of problem may result in results containing a mixed decimal number, which requires the whole number and the fraction to become split in order to achieve its equivalent octal conversion.

Example. Convert $(0.175)_{10}$ to Octal.

The multiplication problem will be setup as

0.017510
× 8
0.1400 ----> 0 (MSD)

0.1400
× 8
1.1200 ----> 1

0.1200
× 8
0.9600 ----> 0

0.9600
× 8
7.6800 ----> 7

0.6800
× 8
5.4400 ----> 5

0.4400
× 8
3.5200 ----> 3

0.5200
× 8
4.1600 ----> 4

0.1600
× 8
1.2800 ----> 1 (LSD)

Writing from MSD to LSD, you should end with $(.01075341)_8$.

1.2.3 Decimal to Hexa Decimal Conversión

The following are the steps which we have to follow for converting a decimal number into equivalent Hexa Decimal Number.

1. Divide the decimal number by 16. Treat the division as an integer division.
2. Write down the remainder (in hexadecimal).
3. Divide the result again by 16. Treat the division as an integer division.
4. Repeat steps 2 and 3 until result is 0.
5. The hex value is the digit sequence of the remainders from the last to first.

Example 1. Convert the number **1128** Decimal to Hexadecimal.

Notes	Division	Result	Remainder (in Hexadecimal)
Start by dividing the number by 16. In this case, 1128 divided by 16 is 70.5. So the integer division result is 70 (throw out anything after the decimal point). The remainder is (70.5 - 70) multiplied with 16; or (0.5 times 16), which is 8.	1128 / 16	70	8
Then, divide the result again by 16 (the number 70 on the DIVISION column comes from the previous RESULT). In this case, 70/16=4.375. So the integer division result is 4 (throw out anything after the decimal point). The remainder is (0.375 multiplied with 16, which is 6.	70 / 16	4	6
Repeat. Note here that 4/16=0.25. So the integer division result is 0. The remainder is (0.25-0) multiplied with 16, which is 4.	4 / 16	0	4
Stop because the result is already 0 (0 divided by 16 will always be 0)			
Well, here is the answer. These numbers come from the REMAINDER column values (read from bottom to top)			468

Example 2. Convert the number **256** Decimal to Hexadecimal.

Division	Result	Remainder (in Hex)
256 / 16	16	0
16 / 16	1	0
1 / 16	0	1
Answer		100

Example 3. Convert the number **921** Decimal to Hexadecimal.

Division	Result	Remainder (in Hex)
921 / 16	57	9
57 / 16	3	9
3 / 16	0	3
Answer		399

Example 4. Convert the number **188** Decimal to Hexadecimal.

Division	Result	Remainder (in Hex)
188 / 16	11	C (12 decimal)
11 / 16	0	B (11 decimal)
Answer		**BC**

Note that here; the answer would not be 1112, but BC. Remember to write down the remainder in hex, not decimal.

Example 5. Convert the number **590** Decimal to Hexadecimal.

Division	Result	Remainder (in Hex)
590 / 16	36	E (14 decimal)
36 / 16	2	4 (4 decimal)
2 / 16	0	2 (2 decimal)
Answer		**24E**

Converting Decimal Fractions to Hexadecimal

To convert a decimal fraction to Hexadecimal, you must multiply the fraction by 16. Then, you need to extract the number(s) that appear to the left of the decimal point. The first number that is extracted will be the MSD and will follow the decimal point. The last number that is extracted will be the LSD. Continue to desired accuracy. This type of problem may result in results containing a mixed decimal number, which requires the whole number and the fraction to become split in order to achieve its equivalent octal conversion.

Example. *Convert* 0.695_{10} *to Hexadecimal number system.*

Solution:

```
                                 695
                              ×   16
                              ------
                               4.170
                               6.950
MSD ←—— B16 ←——————           11.120
                              ×   16
                              ------
                                 720
                               1.200
                              ------
        I16 ←——————            1.920
                              ×   16
                              ------
                               5.520
                               9.200
                              ------
        E16 ←——————           14.720
                              ×   16
                              ------
                               4.320
                               7.200
                              ------
LSD ←—— B16 ←——————           11.520
```

So the solution is $B1EB_{16}$

1.3 BINARY NUMBER SYSTEM

The basic parts of a computer are the central processing unit (CPU), memory, a keyboard or other input device and a screen or other output device. Sounds simple, doesn't it? But, how does the computer know how to add and subtract, and how can its memory remember the answers it computes? We know that the computer doesn't have a real brain inside. It fact, it is made up mostly of plastic, metal and silicon. Yet, a computer acts in many ways as though it does have a real brain.

To find the answer, we must take a close look at how we understand numbers. We have ten digits in our number system: 0, 1, 2, 3, 4, 5, 6, 7, 8, 9. Digit is a fancy word for a single number. It's interesting that digit also means a finger or toe. A number system based on ten is called a decimal system.

Computers don't use the ten digits of the decimal system for counting and arithmetic. Their CPU and memory are made up of millions of tiny switches that can be either ON or OFF. Two digits, 0 and 1, can be used to stand for the two states of ON and OFF. So, we can see that computers could work with a number system based on two digits. This type of system is called a binary numbering system. How does this type of number system work? First, let's look more closely at our own decimal system. As you have learned in your arithmetic classes, our decimal system is based on place, or location. That is, the place of each digit tells you the value of that digit. For example, the number 17 has a 7 in the one's place and a 1 in the ten's place. In other words, 1 ten plus 7 ones equals 17. The number 138 has a 1 in the hundred's place, a 3 in the ten's place, and an 8 in the one's place. Written in numerals this is $(1 \times 100) + (3 \times 10) + (8 \times 1) = 138$.

The binary system works in exactly the same way, except that its place value is based on the number two. In the binary system, we have the one's place, the two's place, the four's place, the eight's place, the sixteen's place, and so on. Each place in the number represents two times (2X's) the place to its right.

Here's a comparison of decimal and binary numbers:

Decimal	Binary	Decimal	Binary
0	0	6	110
2	10	7	111
3	11	8	1000
4	100	9	1001
5	101	10	1010

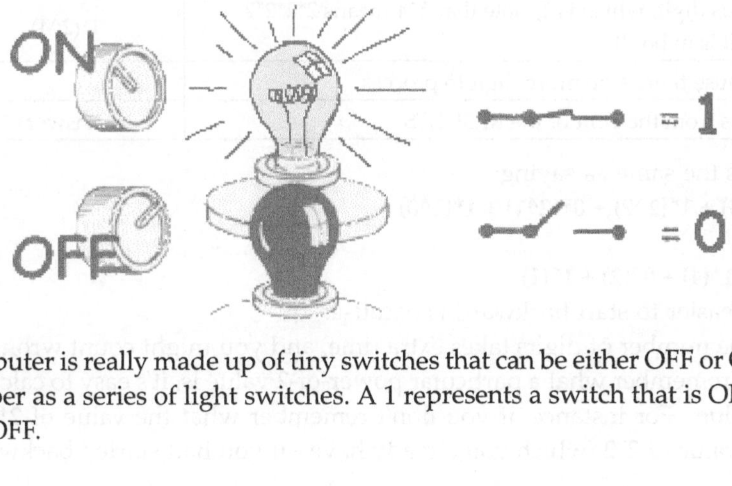

Since the computer is really made up of tiny switches that can be either OFF or ON, you can look at a binary number as a series of light switches. A 1 represents a switch that is ON, and a 0 means a switch that is OFF.

As you can see, numbers can become rather longin the binary system. For example, to show the number 10, we need four light switches, or four places. However, the real switches inside a computer are tiny and they are able to turn on and off very rapidly. The binary number system suits a computer extremely well.

1.3.1 Binary to Decimal Conversion

To convert binary into decimal is very simple and can be done as shown below:

Say we want to convert the 8 bit value **10011101** into a decimal value, we can use a formula like that below:

128	64	32	16	8	4	2	1
1	0	0	1	1	1	0	1

As you can see we have placed the numbers 1, 2, 4, 8, 16, 32, 64, 128 (powers of two) in reverse numerical order and then written the binary value below. To convert you simply take a value from the top row wherever there is a 1 below and add the values together, for instance in our example we would have **128 + 16 + 8 + 4 + 1 = 157.** For a 16 bit value you would use the decimal values 1, 2, 4, 8, 16, 32, 64, 128, 256, 512, 1024, 2048, 4096, 8192, 16384, 32768 (powers of two) for the conversion.

Because we know binary is base 2 then the above could be written as:

$1*2^7 + 0*2^6 + 0*2^5 + 1*2^4 + 1*2^3 + 1*2^2 + 0*2^1 + 1*2^0 = 157.$

Example 1. Convert Binary **11101** to Decimal

Notes	**Multiplication**	**Result**
Start from the last digit, which is 1, multiply that digit with 2^0, note that the power of 0 of any number is always 1 11101 (current digit is in bold)	1*(2^0)	1
Process the previous digit, which is 0, multiply that digit with the increasing power of 2 11101 (current digit is in bold)	0*(2^1)	0
Process the previous digit, which is 1, note that 2^2 means 2*2 11101 (current digit is in bold)	1*(2^2)	4
Process the previous digit, which is 1, note that 2^3 means 2*2*2 11101 (current digit is in bold)	1*(2^3)	8
Process the previous digit, which is 1, note that 2^4 means 2*2*2*2 11101 (current digit is in bold)	1*(2^4)	16
Here, we stop because there's no more digit to process		
This number comes from the **sum** of the RESULTS	Answer	29

Basically, this is the same as saying:

1*(2^4) + 1*(2^3) + 1*(2^2) + 0*(2^1) + 1*(2^0)

Or

1*(16) + 1*(8) + 1*(4) + 0*(2) + 1*(1)

The reason it's easier to start backward is because:

- Counting the number of digits takes extra time, and you might count wrongly.
- If you don't remember what a particular power-of-2 value is, it's easy to calculate it from the previous value. For instance, if you don't remember what the value of 2*2*2 is, then just double the value of 2*2 (which you already have - if you had started backward).

Example 2. Convert Binary **1010** to Decimal

Multiplication	Result
0*(2^0)	0
1*(2^1)	2
0*(2^2)	0
1*(2^3)	8
Answer	**10**

Is constructing a table like above required? No, it just depends on your preference. Some people are visual, and the table might help. Without a table, it's also easy. If you want to be a speed counter, just remember that the value of the multiplier is always the double of the previous one.

1, 2, 4, 8, 16, 32, 64, 128, 256, 512, ...

Power of 2s	Result
2^0	1
2^1 = 2	2
2^2 = 2*2	4
2^3 = 2*2*2	8
2^4 = 2*2*2*2	16

Example 3. Convert Binary **1010001** to Decimal.

Again, I'm starting backward here:

(1*1) + (0*2) + (0*4) + (0*8) + (1*16) + (0*32) + (1*64) = 1 + 0 + 0 + 0 + 16 + 0 + 64 = 81

Converting Binary Fractions to Decimal

Remember your binary place value. Starting at the decimal point and moving right you have 1/2 place, then 1/4, 1/8, 1/16, and 1/32. If there is a 1 in a particular place, simply add that fraction to the result to get the result.

So 0.11 (binary) is 1/2 + 1/4 or 0.75 decimal. 0.101 binary is 1/2 + 1/8 = 0.625 decimal.

The thing that makes binary easy is to do the same thing in binary you do in base 10.

1.3.2 Binary to Octal Conversion

Converting binary to octal is also a simple process. Break the binary digits into groups of three starting from the binary point and convert each group into its appropriate octal digit. For whole numbers, it may be necessary to add a zero as the MSB in order to complete a grouping of three bits. Note that this does not change the value of the binary number. Similarly, when representing fractions, it may be necessary to add a trailing zero in the LSB in order to form a complete grouping of three.

Example 1. Converting $(010111)_2$ to Octal

111 = 7 (LSB)

010 = 2 (MSB)

Thus, $(010111)_2 = (27)_8$

Example 2. Converting $(0.110101)_2$ to Octal

110 = 6 (MSB)

101 = 5 (LSB)

Thus, $(0.110101)_2 = (0.65)_8$

1.3.3 Binary to Hexadecimal Conversion

Converting binary to hexadecimal is another simple process. Break the binary digits into groups of four starting from the binary point and convert each group into its appropriate hexadecimal digit. For whole numbers, it may be necessary to add a zero as the MSB in order to complete a grouping of four bits. Note that this addition does not change the value of the binary number. Similarly, when representing fractions, it may be necessary to add a trailing zero in the LSB in order to form a complete grouping of four.

Example 1. *Convert $(1010111)_2$ to hexadecimal.*

Solution:

0111 = 7 (LSB)

0101 = 5 (MSB)

Thus $(1010111)_2 = (57)_{16}$

Example 2. *Convert $(0.00111111)_2$ to hexadecimal.*

Solution:

0011 = 3 (MSB)

1111 = F (LSB)

Thus, $(0.00111111)_2 = (0.3F)_{16}$

1.4 BINARY ARITHMETIC

Binary arithmetic is fundamental to all digital computers and most other digital systems.

In this context we will discuss the Concept of Binary Addition and Binary Subtraction, Binary Multiplication and Binary Division.

1.4.1 Binary Addition

In particular, addition is the most important binary arithmetic process because it can be used to perform all other arithmetic operations such as subtraction, multiplication and division. Thus it is important to fully understand binary addition.

The following Table shows the four basic rules for binary addition.

Addition Rules		
A+B	**Sum**	**Carry**
0+0	0	0
0+1	1	0
1+0	1	0
1+1	0	1

Example 1.

Decimal	Binary
5	101
+3	+011
8	1000

Example 2.

Decimal	Binary
74	1001010
+19	+ 10011
93	1011101

1.4.2 Binary Subtraction

The process of binary subtraction may be viewed as the addition of a negative number. For example, 9–4 may be viewed as 9 + (–4). However to determine the negative representation of a binary number, one must become familiar with 1's and 2's complement.

1's complement of a binary number

The 1's complement of binary number is found by changing all the 1's to 0s and vice versa as illustrated by the examples below:

Number	1's Complement
10001	01110
101001	010110

2's complement of a binary number

The 2's complement of binary number is found by adding 1 to the 1's complement representation as illustrated by the examples below:

Number	1's Complement	2's Complement
10001	01110	01110 +1 ------ 01111 ------
101001	010110	010110 +1 ------ 010111 ------

1.4.2.1 Subtracting Using 1's Complement

For subtracting a smaller number from a larger number, the 1's complement method is as follows:

1. Determine the 1's complement of the smaller number.
2. Add the 1's complement to the larger number.
3. Remove the final carry and add it to the result. This step is called the end-around carry.

Example. *11001–10011*

Solution:

Result from Step1: 01100

Result from Step2: 100101

Result from Step3: 00110

To verify, note that 25 – 19 = 6

For subtracting a larger number from a smaller number, the 1's complement method is as follows:

1. Determine the 1's complement of the larger number.
2. Add the 1's complement to the smaller number.
3. There is no carry. The result has the opposite sign from the answer and is the 1's complement of the answer.
4. Change the sign and take the 1's complement of the result to get the final answer.

Example. *1001 – 1101*

Solution:

Result from Step1: 0010
Result from Step2: 1011
Result from Step3: -0100
To verify, note that 9 – 13 = – 4

1.4.2.2 Subtracting Using 2's Complement

For subtracting a smaller number from a larger number, the 2's complement method is as follows:

1. Determine the 2's complement of the smaller number.
2. Add the 2's complement to the larger number.
3. Discard the final carry (there is always one in this case)

Example. *11001 – 10011*

Solution:

Result from Step1: 01101
Result from Step2: 100110
Result from Step3: 00110
Again, to verify, note that 25 – 19 = 6

For subtracting a larger number from a smaller number, the 2's complement method is as follows:

1. Determine the 2's complement of the larger number.
2. Add the 2's complement to the smaller number.
3. There is no carry from the left-most column. The result is in 2's complement form and is negative.
4. Change the sign and take the 2's complement of the result to get the final answer.

Example. *1001 – 1101*

Solution:

Result from Step1: 0011
Result from Step2: 1100
Result from Step3: – 0100
Again to verify, note that 9 – 13 = – 4

1.4.3 Binary Multiplication

Multiplication in the binary system works the same way as in the decimal system:

- 1*1=1
- 1*0=0
- 0*1=0

Example.

```
   101
*   11
------
```

```
 101
1010
----
1111
```

Note that multiplying by two is extremely easy. To multiply by two, just add a 0 on the end.

1.4.4 Binary Division

Binary division follows the same rules as in decimal division. For the sake of simplicity, throw away the remainder.

Example:

111011/11

```
   10011 r 10
  --------
11)111011
  - 11
  -----
    101
   - 11
   ----
    101
     11
   ----
     10
```

1.5 OCTAL NUMBER SYSTEM

Although this was once a popular number base, especially in the Digital Equipment Corporation PDP/8 and other old computer systems, it is rarely used today. The Octal system is based on the binary system with a 3-bit boundary. The Octal Number System:

- uses base 8
- includes only the digits 0 through 7 (any other digit would make the number an invalid octal number)

The weighted values for each position are as follows:

8^5	8^4	8^3	8^2	8^1	8^0
32768	4096	512	64	8	1

1.5.1 Octal to Decimal Conversion

To convert from Octal to Decimal, multiply the value in each position by its Octal weight and add each value,

Example 1. *Convert* $(127662)_8$ *into Decimal Form*

Solution:

1*8^5	2*8^4	7*8^3	6*8^2	6*8^1	2*8^0
1*32768	2*4096	7*512	6*64	6*8	2*1
32768	8192	3584	384	48	2

Answer is 32768 + 8192 + 3584 + 384 + 48 + 2 = 44978

Example 2. *Convert* $(237.04)_8$ *to decimal form.*
Solution:

Weights	8^2	8^1	8^0	8^{-1}	8^{-2}	
Weight Value	64	8	1	0.125	0.015625	
Octal Number	2	3	7	0	4	
Decimal Value	128	24	7	0	0.0625	Total $(159.0625)_{10}$

1.5.2 Octal to Binary Conversion

The primary application of octal numbers is representing binary numbers, as it is easier to read large numbers in octal form than in binary form. Because each octal digit can be represented by a three-bit binary number (see following Table), it is very easy to convert from octal to binary. Simply replace each octal digit with the appropriate three-bit binary number as indicated in the examples below.

The following table shows the octal digit and the equivalent Binary digit.

Octal Digit	Binary Digit
0	000
1	001
2	010
3	011
4	100
5	101
6	110
7	111

Example 1. *Convert* $(13)_8$ *into equivalent binary number.*
Solution:

$$(13)_8 = (001011)_2$$

Example 2. *Convert* $(37.12)_8$ *into equivalent Binary number.*
Solution:

$$(37.12)_8 = (011111.\ 001010)_2$$

Example 3. *Convert* $(127662)_8$ *into equivalent binary number.*
Solution: This yields the binary number 001010111110110010 or 00 1010 1111 1011 0010 in our more readable format.

1	2	7	6	6	2
001	010	111	110	110	010

1.5.3 Octal to Hexadecimal Conversion

This is intermediate Conversion. The following steps are to be followed to implement the Octal to Hexadecimal Conversion.

Step No 1: Convert Octal to Binary or Decimal No.

Step No 2: Convert Decimal/Binary to Hexadecimal No.

Example. *Convert* $(53.7)_8$ *into equivalent Hexadecimal number system.*

Solution:

The first step to convert the 537 into equivalent binary form is as follows:

5	3.	7_8
101	011.	111_2

Now regroup the binary digits into groups of four and add zeros where needed to complete groups; then convert the binary to hexadecimal.

0010	1011.	1110_2
2	B.	E_{16}

So the solution of 53.7_8 is equal to $2B.E_{16}$

1.6 HEXADECIMAL NUMBER SYSTEM

Just like the octal number system, the hexadecimal (or base-sixteen) number system provides a convenient way to express binary numbers. The following table shows the weighting for the hexadecimal number system up to 3 decimal places before and 2 places after the *hexadecimal* point. Based on the trend in previous number systems, the methods used to convert hexadecimal to decimal and vice versa should be intuitive.

Weights	16^2	16^1	16^0	.	16^{-1}	16^{-2}

The following table lists the equivalent decimal, binary and hexadecimal representations for the decimal numbers ranging from 0 to 15.

Decimal	Binary	Hexadecimal
0	0000	0
1	0001	1
2	0010	2
3	0011	3
4	0100	4
5	0101	5
6	0110	6
7	0111	7
8	1000	8
9	1001	9
10	1010	A
11	1011	B
12	1100	C
13	1101	D
14	1110	E
15	1111	F

1.6.1 Hexadecimal to Decimal Conversion

The following steps are to be followed to convert Hexadecimal into decimal number system.

1. Get the last digit of the hex number; call this digit the **current Digit**.
2. Make a variable, let's call it **power**. Set the value to 0.
3. Multiply the **current digit** with (16^**power**), store the result.
4. Increment **power** by 1.
5. Set the **current Digit** to the previous digit of the hex number.
6. Repeat from step 3 until all digits have been multiplied.
7. Sum the result of step 3 to get the answer number.

Example 1. *Convert the number 1128 Hexadecimal to Decimal*

Solution:

Multiplication	Result	Notes
8 × (16^0)	8	Start from the last digit of the number. In this case, the number is 1128. The last digit of that number is 8. Note that the power of 0 of any number is always 1
2 × (16^1)	32	Process the previous digit, which is 2. Multiply that number with an increasing power of 16.
1 × (16^2)	256	Process the previous digit, which is 1, note that 16^2 means 16 x 16
1 × (16^3)	4096	Process the previous digit, which is 1, note that 16^3 means 16 x 16 x 16
		Here, we stop because there's no more digit to process
Answer	4392	This number comes from the **sum** of the all Results (8+32+256+4096)=4392

Once discerned, notice that the above process is essentially performing this calculation:
1 × (16^3) + 1 × (16^2) + 2 × (16^1) + 8 × (16^0)

Example 2. *Convert the number 589 Hexadecimal to Decimal*

Solution:

Multiplication	Result
9 x (16^0)	9
8 x (16^1)	128
5 x (16^2)	1280
Answer	**1417**

If you want to be a speed counter, it's beneficial to memorize the values of the smaller powers of 16s, such as in this table

Power of 16s	Result
16^0	1
16^1 = 16	16
16^2 = 16x16	256
16^3 = 16x16x16	4096
16^4 = 16x16x16x16	65536

Example 3. *Convert the number 1531 Hexadecimal to Decimal*
Solution:

Multiplication	Result
1 x 1	1
3 x 16	48
5 x 256	1280
1 x 4096	4096
Answer	**5425**

Example 4. *Convert the number 8F Hexadecimal to Decimal*
Solution:

Division	Result
F x 1	15
8 x 16	128
Answer	**143**

Example 5. *Convert the number A0 Hexadecimal to Decimal*
Solution:

Division	Result
0 x 1	0
A x 16	160
Answer	**160**

Example 6. *Convert the number 12 Hexadecimal to Decimal*
Solution:

Division	Result
2 x 1	2
1 x 16	16
Answer	**18**

Example 7. *Convert the number 35432 Hexadecimal to Decimal*
Solution:

$2x(16^0) + 3x(16^1) + 4x(16^2) + 5x(16^3) + 3x(16^4)$

$= 2 + 3x16 + 4*256 + 5*4096 + 3*65536$

$= 2 + 48 + 1024 + 20480 + 196608 = 218162$

1.6.2 Hexadecimal to Binary Conversion

To convert hexadecimal to binary, at each hexadecimal digit we associate the 4 bits binary number using the table of follows:

0 0000
1 0001
2 0010
3 0011
4 0100
5 0101
6 0110
7 0111
8 1000
9 1001
A 1010
B 1011
C 1100
D 1101
E 1110
F 1111

Example 1. *Convert* $(F8)_{16}$ *into equivalent binary form.*

Solution:

Write down the binary for F first, then the binary for 8.

F 8
1111 1000

So, the answer is 11111000.

Example 2. *Convert hex number 1A to binary.*

Solution:

1 A
0001 1010

So, the answer is 00011010. (Note: Once you got the answer, you can ignore the zeros at the beginning, so this can be also written as 11010.)

Example Table. The following table represents the conversion between hexadecimal to binary conversion.

hex	D	C	C	
bin	1101	1100	1100	
hex	1	0	0	3
bin	0001	0000	0000	0011
hex	F	3	A	2
bin	1111	0011	1010	0010

1.6.3 Hexadecimal to Octal Conversion

1. Convert Octal (Hexadecimal) to Binary first.
2. (*a*) Regroup the binary number in 3 bits, a group ***starts from the LSB*** if Octal is required.
   (*b*) Regroup the binary number in 4 bits, a group starts from the LSB if Hexadecimal is required.

Example. Convert $5A8_{16}$ to Octal

$5A8_{16}$ = 0101 1010 1000 (Binary)
= 2650 (Octal)

1.7 BINARY CODES

The usual way of expressing a decimal number in terms of a binary number is known as *pure binary coding*. A number of other techniques can be used to represent a decimal number. These are summarized below.

1.7.1 8421 BCD Code

In the 8421 Binary Coded Decimal (BCD) representation each decimal digit is converted to its 4-bit pure binary equivalent.

For example: $\mathbf{57_{dec} = 0101\ 0111_{bcd}}$

Addition is analogous to decimal addition with normal binary addition taking place from right to left. For example,

6	0110	BCD for 6	42	0100 0010	BCD for 42
+3	0011	BCD for 3	+27	0010 0111	BCD for 27
	1001	BCD for 9		0110 1001	BCD for 69

Where the result of any addition exceeds 9(1001) then six (0110) must be added to the sum to account for the six invalid BCD codes that are available with a 4-bit number. This is illustrated in the example below:

8	1001	BCD for 8
+7	0111	BCD for 7
	1111	exceeds 9 (1001) so
	0110	add six (0110)
	0001 0101	BCD for 15

Note that in the last example the 1 that was carried forward from the first group of 4 bits has made a new 4-bit number and so represents the "1" in "15".

In the examples above the BCD numbers are split at every 4-bit boundary to make reading them easier. This is not necessary when writing a BCD number down.

This coding is an example of a binary coded (each decimal number maps to four bits) weighted (each bit represents a number, 1, 2, 4, etc.) code.

1.7.2 4221 BCD Code

The 4221 BCD code is another binary coded decimal code where each bit is weighted by 4, 2, 2 and 1 respectively. Unlike BCD coding there are no invalid representations. The decimal numbers 0 to 9 have the following 4221 equivalents

Decimal	4221	1's complement
0	0000	1111
1	0001	1110
2	0010	1101

3	0011	1100
4	1000	0111
5	0111	1000
6	1100	0011
7	1101	0010
8	1110	0001
9	1111	0000

The 1's complement of a 4221 representation is important in decimal arithmetic. In forming the code remember the following rules

- Below decimal 5 use the right-most bit representing 2 first
- Above decimal 5 use the left-most bit representing 2 first
- Decimal 5 = 2+2+1 and not 4+1

1.7.3 Gray Code

Gray coding is an important code and is used for its speed, it is also relatively free from errors. In pure binary coding or 8421 BCD the counting from 7 (0111) to 8 (1000) requires 4 bits to be changed simultaneously. If this does not happen then various numbers could be momentarily generated during the transition thereby creating spurious numbers which could be read.

Gray coding avoids this since only one bit changes between subsequent numbers. To construct the code there are two simple rules. First start with all 0s and then proceed by changing the least significant bit (lsb) which will bring about a new state.

The first 16 Gray coded numbers are indicated below.

Decimal	Gray Code
0	0000
1	0001
2	0011
3	0010
4	0110
5	0111
6	0101
7	0100
8	1100
9	1101
10	1111
11	1110
12	1010
13	1011
14	1001
15	1000

1.8 COMPUTER DATA FORMATS

All data stored and processed by digital computers exist as binary (base 2) numbers. In the binary system, the only numbers are 0s and 1s. **This means that all computer data (numbers, words/text, pictures, sounds, and video) consist of some combination of 0s and 1s.**

Nonetheless, it is important to understand the way in which these various types of data are stored, processed and delivered by computers. Only by understanding these various digital formats can you put them to good use.

1.8.1 ASCII Code

It is a very well-known fact that computers can manage internally only 0s (zeros) and 1s (ones). This is true, and by means of sequences of 0s and 1s the computer can express any numerical value as its binary translation, which is a very simple mathematical operation.

Nevertheless, there is no such evident way to represent letters and other non-numeric characters with 0s and 1s. Therefore, in order to do that, computers use *ASCII tables*, which are tables or lists that contain all the letters in the roman alphabet plus some additional characters. In these tables each character is always represented by the same order number. For example, the ASCII code for the capital letter "A" is always represented by the order number 65, which is easily representable using 0s and 1s in binary: 65 expressed as a binary number is 1000001.

The standard ASCII table defines 128 character codes (from 0 to 127), of which, the first 32 are control codes (non-printable), and the remaining 96 character codes are representable characters:

<table>
<tr><td>*</td><td>0</td><td>1</td><td>2</td><td>3</td><td>4</td><td>5</td><td>6</td><td>7</td><td>8</td><td>9</td><td>A</td><td>B</td><td>C</td><td>D</td><td>E</td><td>F</td></tr>
<tr><td>0</td><td>NUL</td><td>SOH</td><td>STX</td><td>ETX</td><td>EOT</td><td>ENQ</td><td>ACK</td><td>BEL</td><td>BS</td><td>TAB</td><td>LF</td><td>VT</td><td>FF</td><td>CR</td><td>SO</td><td>SI</td></tr>
<tr><td>1</td><td>DLE</td><td>DC1</td><td>DC2</td><td>DC3</td><td>DC4</td><td>NAK</td><td>SYN</td><td>ETB</td><td>CAN</td><td>EM</td><td>SUB</td><td>ESC</td><td>FS</td><td>GS</td><td>RS</td><td>US</td></tr>
<tr><td>2</td><td></td><td>!</td><td>"</td><td>#</td><td>$</td><td>%</td><td>&</td><td>'</td><td>(</td><td>)</td><td>*</td><td>+</td><td>,</td><td>-</td><td>.</td><td>/</td></tr>
<tr><td>3</td><td>0</td><td>1</td><td>2</td><td>3</td><td>4</td><td>5</td><td>6</td><td>7</td><td>8</td><td>9</td><td>:</td><td>;</td><td><</td><td>=</td><td>></td><td>?</td></tr>
<tr><td>4</td><td>@</td><td>A</td><td>B</td><td>C</td><td>D</td><td>E</td><td>F</td><td>G</td><td>H</td><td>I</td><td>J</td><td>K</td><td>L</td><td>M</td><td>N</td><td>O</td></tr>
<tr><td>5</td><td>P</td><td>Q</td><td>R</td><td>S</td><td>T</td><td>U</td><td>V</td><td>W</td><td>X</td><td>Y</td><td>Z</td><td>[</td><td>\</td><td>]</td><td>^</td><td>_</td></tr>
<tr><td>6</td><td>'</td><td>a</td><td>b</td><td>c</td><td>d</td><td>e</td><td>f</td><td>G</td><td>h</td><td>i</td><td>j</td><td>k</td><td>l</td><td>m</td><td>n</td><td>o</td></tr>
<tr><td>7</td><td>p</td><td>q</td><td>r</td><td>s</td><td>t</td><td>u</td><td>v</td><td>w</td><td>x</td><td>y</td><td>z</td><td>{</td><td>|</td><td>}</td><td>~</td><td></td></tr>
</table>

* This panel is organized to be easily read in hexadecimal: row numbers represent the first digit and the column numbers represent the second one. For example, the "A" character is located at the 4th row and the 1st column, for that it would be represented in hexadecimal as 0×41 (65).

Because most systems nowadays work with 8bit bytes, which can represent 256 different values, in addition to the 128 standard ASCII codes there are other 128 that are known as *extended ASCII*, which are platform- and locale-dependent. So there is more than one extended ASCII character set.

1.8.2 EBCDIC

EBCDIC (Extended Binary Coded Decimal Interchange Code) is a binary code for alphabetic and numeric characters that IBM developed for its larger operating systems. It is the code for text files that is used in IBM's OS/390 operating system for its S/390 servers that thousands of corporations use for their legacy applications and databases. In an EBCDIC file, each alphabetic or numeric character is represented with an 8-bit binary number (a string of eight 0's or 1's). 256 possible characters (letters of the alphabet, numerals, and special characters) are defined.

IBM's PC and workstation operating systems do not use IBM's proprietary EBCDIC. Instead, they use the industry standard code for text, ASCII. Conversion programs allow different operating systems to change a file from one code to another.

1.9 DIGITAL DEVICES

Digital device is a device that computes, especially a programmable electronic machine that performs high-speed mathematical or logical operations or that assembles, stores, correlates, or otherwise processes information.

1.9.1 Buffer

In computing, a **buffer** is a region of memory used to temporarily hold data while it is being moved from one place to another. Typically, the data is stored in a buffer as it is retrieved from an input device (such as a keyboard) or just before it is sent to an output device (such as a printer). However, a buffer may be used when moving data between processes within a computer. This is comparable to buffers in telecommunication. Buffers can be implemented in either hardware or software, but the vast majority of buffers are implemented in software. Buffers are typically used when there is a difference between the rate at which data is received and the rate at which it can be processed, or in the case that these rates are variable, for example in a printer spooler.

1.9.1.1 Applications

Buffers are often used in conjunction with I/O to hardware, such as disk drives, sending or receiving data to or from a network, or playing sound on a speaker. A line to a rollercoaster in an amusement park shares many similarities. People who ride the coaster come in at an unknown and often variable pace, but the roller coaster will be able to load people in bursts (as a coaster arrives and is loaded). The queue area acts as a buffer: a temporary space where those wishing to ride wait until the ride is available. Buffers are usually used in a FIFO (first in, first out) method, outputting data in the order it arrived.

1.9.1.2 Telecommunication Buffer

A buffer routine or storage medium used in telecommunications compensates for a difference in rate of flow of data, or time of occurrence of events, when transferring data from one device to another.

Buffers are used for many purposes, such as

- interconnecting two digital circuits operating at different rates,
- holding data for use at a later time,
- allowing timing corrections to be made on a data stream,
- collecting binary data bits into groups that can then be operated on as a unit,
- Delaying the transit time of a signal in order to allow other operations to occur.

1.9.1.3 Difference between Buffer and Cache

The terms "buffer" and "cache" are not mutually exclusive and the functions are frequently combined; however, there is a difference in intent. A buffer is a temporary memory location, that is traditionally used because CPU instructions cannot directly address data stored in peripheral devices. Thus, addressable memory is used as intermediate stage.

Additionally such a buffer may be feasible when a large block of data is assembled or disassembled (as required by a storage device), or when data may be delivered in a different order than that in which it is produced. Also a whole buffer of data is usually transferred sequentially (for example to hard disk), so buffering itself sometimes increases transfer performance. These benefits are present even if the buffered data are written to the buffer once and read from the buffer once.

A cache also increases transfer performance. A part of the increase similarly comes from the possibility that multiple small transfers will combine into one large block. But the main performance-gain occurs because there is a good chance that the same datum will be read from cache multiple times, or that written data will soon be read. A cache's sole purpose is to reduce accesses to the underlying slower storage. Cache is also usually an abstraction layer that is designed to be invisible from the perspective of neighboring layers.

1.9.2 Flip-Flops

In digital circuits, a **flip-flop** is a term referring to an electronic circuit (a bistable multivibrator) that has two stable states and thereby is capable of serving as one bit of memory. Today, the term *flip-flop* has come to mostly denote *non-transparent* (*clocked* or *edge-triggered*) devices, while the simpler *transparent* ones are often referred to as latches; however, as this distinction is quite new, the two words are sometimes used interchangeably (see history).

A flip-flop is usually controlled by one or two control signals and/or a gate or clock signal. The output often includes the complement as well as the normal output. As flip-flops are implemented electronically, they require power and ground connections.

1.9.2.1 Implementation

Flip-flops can be either simple (transparent) or clocked. Simple flip-flops can be built around a pair of cross-coupled *inverting* elements: vacuum tubes, bipolar transistors, field effect transistors, inverters, and inverting logic gates have all been used in practical circuits—perhaps augmented by some gating mechanism (an enable/disable input). The more advanced clocked (or non-transparent) devices are specially designed for synchronous (time-discrete) systems; such devices therefore ignore its inputs except **at** the transition of a dedicated clock signal (known as clocking, pulsing, or strobing). This causes the flip-flop to either change or retain its output signal based upon the values of the input signals at the transition. Some flip-flops change output on the rising edge of the clock, others on the falling edge.

Clocked flip-flops are typically implemented as master-slave devices where two basic flip-flops (plus some additional logic) collaborate to make it insensitive to spikes and noise between the short clock transitions; they nevertheless also often include asynchronous *clear* or *set* inputs which may be used to change the current output independent of the clock.

1.9.2.2 Classification of Flip-Flops

Flip-flops can be further divided into types that have found common applicability in both asynchronous and clocked sequential systems: the **SR** ("set-reset"), **D** ("data" or "delay"), **T** ("toggle"), and **JK** types are the common ones; all of which may be synthesized from (most) other types by a few logic gates. The behaviour of a particular type can be described by what is termed the characteristic equation, which derives the "next" (*i.e.*, after the next clock pulse) output, Q_{next}, in terms of the input signal(s) and/or the current output, Q.

(a) SR Flip-flops

The fundamental latch is the simple *SR flip-flop* (also commonly known as *RS flip-flop*), where S and R stand for *set* and *reset*, respectively. It can be constructed from a pair of cross-coupled NAND or NOR logic gates. The stored bit is present on the output marked Q.

Normally, in storage mode, the S and R inputs are both low, and feedback maintains the Q and Q outputs in a constant state, with Q the complement of Q. If S is pulsed high while R is held low, then the Q output is forced high, and stays high even after S returns low; similarly, if R is pulsed high while S is held low, then the Q output is forced low, and stays low even after R returns low.

The next-state equation of the SR flip-flop is:

$$Q_{next} = S + \bar{R}\,Q$$

where Q is the current state. Q_{next} becomes Q (the stored value) at clock edge.

The Block Diagram of SR flip-flops is as follows:

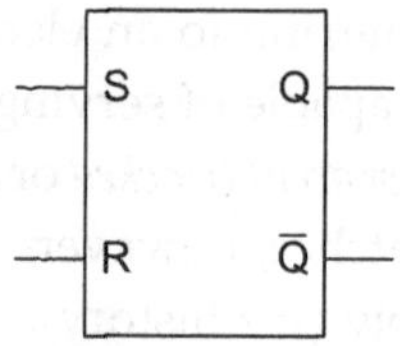

The symbol for an SR latch

The Table of SR Flip-Flop is as follows:

SR Flip-Flop operation (Built With Nor Gates)								
Characteristic table			Excitation table					
S	R	Action	Q(t)	Q(t+1)	S	R	Action
0	0	Keep state	0	0	0	X	No change
0	1	Q = 0	1	0	0	1	reset
1	0	Q = 1	0	1	1	0	set
1	1	Unstable combination	1	1	X	0	race condition

('X' denotes a Don't care condition; meaning the signal is irrelevant)

(b) T Flip-flops

If the T input is high, the T flip-flop changes state ("toggles") whenever the clock input is strobed. If the T input is low, the flip-flop holds the previous value. This behaviour is described by the characteristic equation:

$$Q_{next} = T \oplus Q = \bar{T}Q + \bar{T}Q \text{ Expanding the XOR operator}$$

And can be described in a truth table:

T Flip-Flop operation							
Characteristic table				Excitation table			
T	Q	Q_{next}	Comment	Q	Q_{next}	T	Comment
0	0	0	hold state (no clk)	0	0	0	No change
0	1	1	hold state (no clk)	1	1	0	No change
1	0	1	toggle	0	1	1	Complement
1	1	0	toggle	1	0	1	Complement

When T is held high, the toggle flip-flop divides the clock frequency by two; that is, if clock frequency is 4 MHz, the output frequency obtained from the flip-flop will be 2 MHz This "divide by" feature has application in various types of digital counters. A T flip-flop can also be built using a JK flip-flop (J & K pins are connected together and act as T) or D flip-flop (T input and $Q_{previous}$ is connected to the D input through an XOR gate).

The Circuit Symbol of T-Flip-flop is as follows:

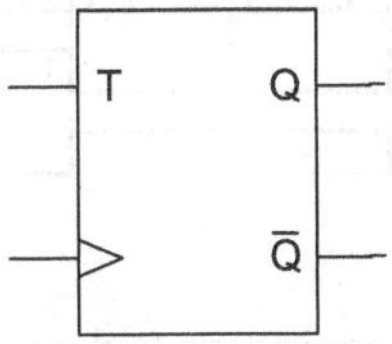

A circuit symbol for a T-type flip-flop, where > is the clock input, T is the toggle input and Q is the stored data output is shown above.

(c) JK Flip-flop

The JK flip-flop augments the behaviour of the SR flip-flop (J=Set, K=Reset) by interpreting the S = R = 1 condition as a "flip" or toggle command. Specifically, the combination J = 1, K = 0 is a command to set the flip-flop; the combination J = 0, K = 1 is a command to reset the flip-flop; and the combination J = K = 1 is a command to toggle the flip-flop, *i.e.*, change its output to the logical complement of its current value. Setting J = K = 0 does NOT result in a D flip-flop, but rather, will hold the current state. To synthesize a D flip-flop, simply set K equal to the complement of J. The JK flip-flop is therefore a universal flip-flop, because it can be configured to work as an SR flip-flop, a D flip-flop, or a T flip-flop.

NOTE: The flip-flop is positive-edge triggered (rising clock pulse) as seen in the timing diagram.

The characteristic equation of the JK flip-flop is:

$$Q_{next} = J\bar{Q} + \bar{K}Q$$

And the corresponding truth table is:

JK Flip Flop operation								
Characteristic table				**Excitation table**				
J	**K**	**Q_{next}**	**Comment**	**Q**	**Q_{next}**	**J**	**K**	**Comment**
0	0	Q_{prev}	hold state	0	0	0	X	No change
0	1	1	reset	0	1	1	X	Set
1	0	1	set	1	0	X	1	Reset
1	1	Q_{prev}	toggle	1	1	X	0	No change

A circuit symbol for a **negative edge triggered** JK flip-flop is represented as follows:

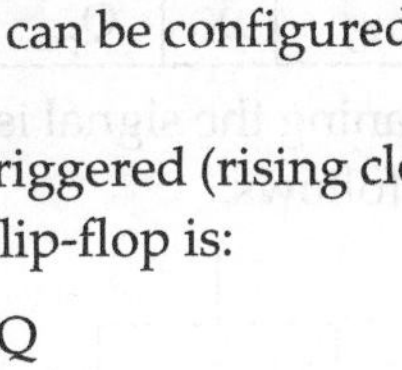

where > is the clock input, J and K are data inputs, Q is the stored data output, and Q' is the inverse of Q.

JK flip-flop timing diagram is represented as follows:

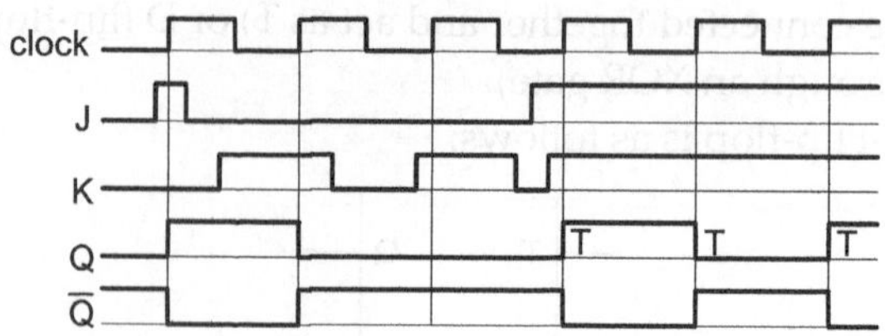

(d) D Flip-flop

The D flip-flop is the most common flip-flop in use today.

The Q output always takes on the state of the D input at the moment of a positive edge (or negative edge if the clock input is active low). It is called the **D** flip-flop for this reason, since the output takes the value of the **D** input or *Data* input, and *Delays* it by one clock count. The D flip-flop can be interpreted as a primitive memory cell, zero-order hold, or delay line. Whenever the clock pulses, the value of Q_{next} is D and Q_{prev} otherwise.

Truth table of D flip-flop is as follows:

Clock	D	Q	Q_{prev}
Rising edge	0	0	X
Rising edge	1	1	X
Non-Rising	X	Q_{prev}	

('X' denotes a *Don't care* condition, meaning the signal is irrelevant)

The Circuit Symbol of D flip-flop is as follows:

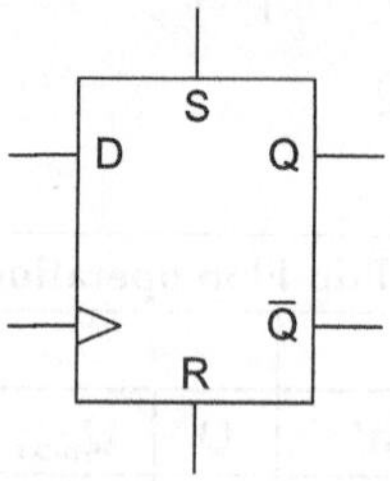

These flip-flops are very useful, as they form the basis for shift registers, which are an essential part of many electronic devices. The advantage of the D flip-flop over the D-type latch is that it "captures" the signal at the moment the clock goes high, and subsequent changes of the data line do not influence Q until the next rising clock edge. An exception is that some flip-flops have a "reset" signal input, which will reset Q (to zero), and may be either asynchronous or synchronous with the clock.

The above circuit shifts the contents of the register to the right, one bit position on each active transition of the clock. The input X is shifted into the leftmost bit position.

(e) Master-slave (Pulse-triggered) D Flip-flop

A master-slave D flip-flop is created by connecting two gated D latches in series, and inverting the *enable* input to one of them. It is called master-slave because the second latch in the series only changes in response to a change in the first (master) latch.

The term *pulse-triggered* means that data is entered on the rising edge of the clock pulse, but the output does not reflect the change until the falling edge of the clock pulse.

For a positive-edge triggered master–slave D flip-flop, when the clock signal is low (logical 0) the "enable" seen by the first or "master" D latch (the inverted clock signal) is high (logical 1). This allows the "master" latch to store the input value when the clock signal transitions from low to high. As the clock signal goes high (0 to 1) the inverted "enable" of the first latch goes low (1 to 0) and the value seen at the input to the master latch is "locked". Nearly simultaneously, the twice inverted "enable" of the second or "slave" D latch transitions from low to high (0 to 1) with the clock signal. This allows the signal captured at the rising edge of the clock by the now "locked" master latch to pass through the "slave" latch. When the clock signal returns to low (1 to 0), the output of the "slave" latch is "locked", and the value seen at the last rising edge of the clock is held while the "master" latch begins to accept new values in preparation for the next rising clock edge.

The Circuit Diagram of a Master-Slave D flip-flop is as follows:

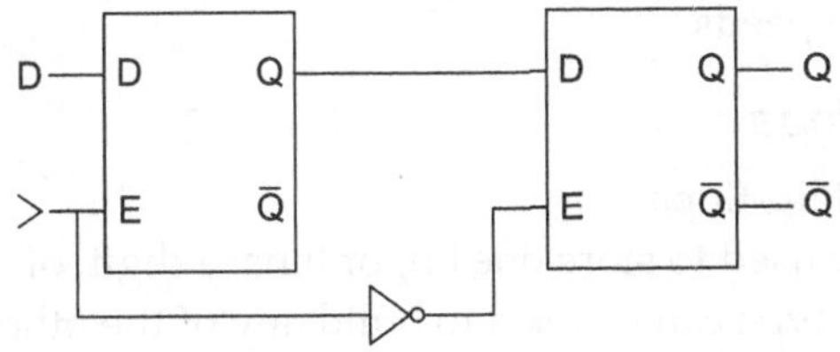

A master-slave D flip-flop responds on the negative edge of the *enable* input (usually a clock).

An implementation of a master–slave D flip-flop that is triggered on the positive edge of the clock is represented as follows:

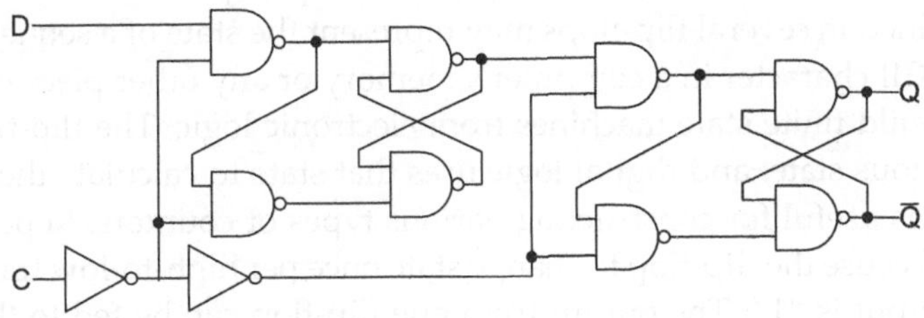

By removing the leftmost inverter in the above circuit, a D-type flip flop that strobes on the *falling edge* of a clock signal can be obtained. This has a truth table like this:

D	Q	>	Q_{next}
0	X	Falling	0
1	X	Falling	1

Most D-type flip-flops in ICs have the capability to be set and reset, much like an SR flip-flop. Usually, the illegal S = R = 1 condition is resolved in D-type flip-flops.

A positive-edge-triggered D flip-flop is represented as follows:

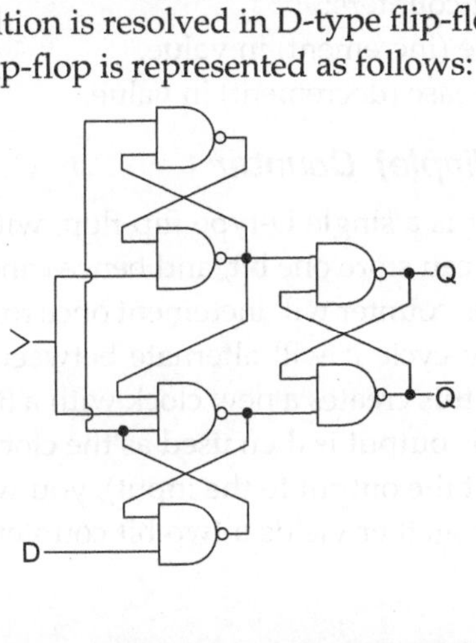

(f) Edge-triggered D flip-flop

A more efficient way to make a D flip-flop is not so easy to understand, but it works the same way. While the master–slave D flip-flop is also triggered on the edge of a clock, its components are each triggered by clock levels. The "edge-triggered D flip-flop" does not have the master–slave properties.

Edge Triggered D flip flops are often implemented in integrated high speed operations using dynamic logic. This means that the digital output is stored on parasitic device capacitance while the device is not transitioning. This design of dynamic flip flops also enables simple resetting since the reset operation can be performed by simply discharging one or more internal nodes. A common dynamic flip flop variety is the True Single Phase Clock (TSPC) which performs the flip flop operation with little power and at high speeds.

1.9.2.3 Uses of Flip-flops

The following are the uses of Flip-flops:

- A single flip-flop can be used to store one bit, or binary digit, of data.
- Any one of the flip-flop type can be used to build any of the others.
- Many logic synthesis tools will not use any other type than D flip-flop and D latch.
- Level sensitive latches cause problems with Static Timing Analysis (STA) tools and Design for Test (DFT). Therefore, their usage is often discouraged.
- Many FPGA devices contain only edge-triggered D flip-flops
- The data contained in several flip-flops may represent the state of a sequencer, the value of a counter, an ASCII character in a computer's memory or any other piece of information.
- One use is to build finite state machines from electronic logic. The flip-flops remember the machine's previous state, and digital logic uses that state to calculate the next state.
- The T flip-flop is useful for constructing various types of counters. Repeated signals to the clock input will cause the flip-flop to change state once per high-to-low transition of the clock input, if its T input is "1". The output from one flip-flop can be fed to the clock input of a second and so on. The final output of the circuit, considered as the array of outputs of all the individual flip-flops, is a count, in binary, of the number of cycles of the first clock input, up to a maximum of $2^n - 1$, where n is the number of flip-flops used.

1.9.3 Counters

In digital logic and computing, a **counter** is a device which stores (and sometimes displays) the number of times a particular event or process has occurred, often in relationship to a clock signal. In practice, there are two types of counters:

- Up counters, which increase (increment) in value.
- Down counters, which decrease (decrement) in value.

1.9.3.1 Asynchronous (Ripple) Counter

An asynchronous (ripple) counter is a single D-type flip-flop, with its D (data) input fed from its own inverted output. This circuit can store one bit, and hence can count from zero to one before it overflows (starts over from 0). This counter will increment once for every clock cycle and takes two clock cycles to overflow, so every cycle it will alternate between a transition from 0 to 1 and a transition from 1 to 0. Notice that this creates a new clock with a 50% duty cycle at exactly half the frequency of the input clock. If this output is then used as the clock signal for a similarly arranged D flip-flop (remembering to invert the output to the input), you will get another 1 bit counter that counts half as fast. Putting them together yields a two bit counter:

Cycle	Q1	Q0	(Q1:Q0)dec
0	0	0	0
1	0	1	1
2	1	0	2
3	1	1	3
4	0	0	0

You can continue to add additional flip-flops, always inverting the output to its own input, and using the output from the previous flip-flop as the clock signal. The result is called a ripple counter, which can count to $2^n - 1$ where n is the number of bits (flip-flop stages) in the counter. Ripple counters suffer from unstable outputs as the overflows "ripple" from stage to stage, but they do find frequent application as dividers for clock signals, where the instantaneous count is unimportant, but the division ratio overall is. (To clarify this, a 1-bit counter is exactly equivalent to a divide by two circuit; the output frequency is exactly half that of the input when fed with a regular train of clock pulses).

The use of flip-flop outputs as clocks leads to timing skew between the count data bits, making this ripple technique incompatible with normal synchronous circuit design styles.

Asynchronous counter created from JK flip-flops is represented as follows:

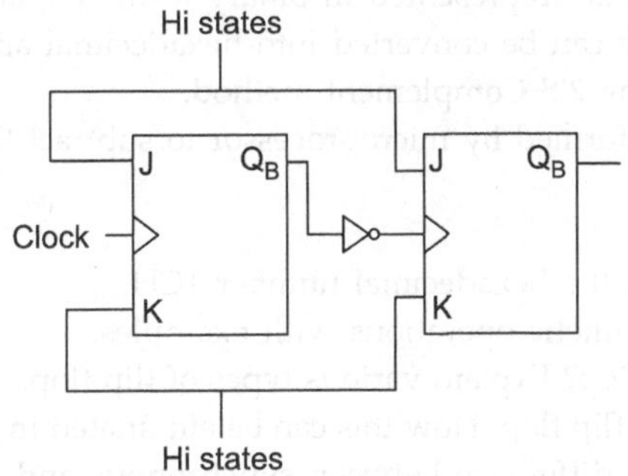

1.9.3.2 Synchronous Counter

A *synchronous counter*, in contrast to an *asynchronous counter*, is one whose output bits change state simultaneously, with no ripple. The only way we can build such a counter circuit from J-K flip-flops is to connect all the clock inputs together, so that each and every flip-flop receives the exact same clock pulse at the exact same time:

The 4-bit Synchronous Counter is represented in the following figure:

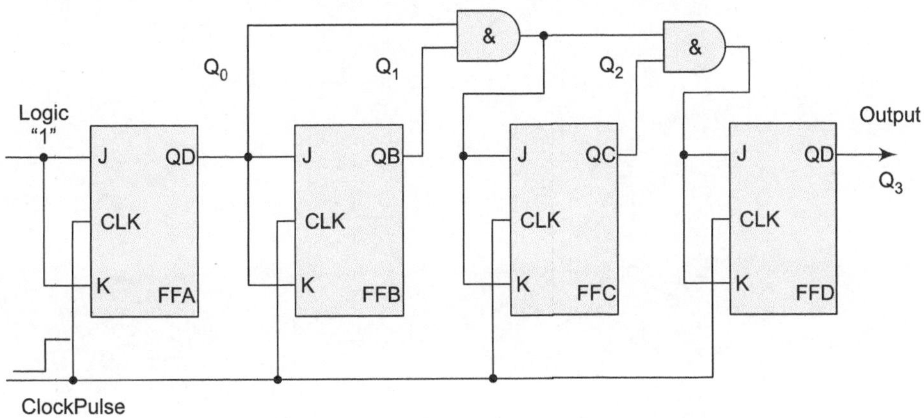

It can be seen that the external clock pulses (pulses to be counted) are fed directly to each *J-K flip-flop* in the counter chain and that both the J and K inputs are all tied together, but only in the first flip-flop, flip-flop A (LSB) are they connected HIGH, logic "1" allowing the flip-flop to toggle on every clock pulse.

The J and K inputs of flip-flop B are connected to the output "Q" of flip-flop A, but the J and K inputs of flip-flops C and D are driven from AND gates which are also supplied with signals from the input and output of the previous stage. If we enable each J-K flip-flop to toggle based on whether or not all preceding flip-flop outputs (Q) are "HIGH" we can obtain the same counting sequence as with the asynchronous circuit but without the ripple effect, since each flip-flop in this circuit will be clocked at exactly the same time. As there is no propagation delay in synchronous counters because all the counter stages are triggered in parallel, the maximum operating frequency of this type of counter is much higher than that of a similar asynchronous counter.

EXERCISE

1. Explain various types of number system. What is the need of number system conversion?
2. How hexadecimal numbers are represented in binary form? Explain with example.
3. How a number from binary can be converted into hexadecimal and octal?
4. Subtract 32H from 45H using 2'S Complement method.
5. Show the internal steps performed by microprocessor to subtract the following unsigned numbers:
   - FAH-62H
   - 62H-FAH
6. Find the 2'S complement of the hexadecimal number 1CH.
7. Explain various binary arithmetic operations with examples.
8. What do you mean by flip flop? Explain various types of flip flop.
9. Explain the drawback of SR flip flop. How this can be eliminated in JK flip flop?
10. What is a counter? Explain difference between synchronous and asynchronous counter.
11. Explain various types of binary codes.
12. What is a the difference between edge triggered and level triggered flip flop?
13. Explain various computer data formats.
14. What is the need of buffer?

2 MICROPROCESSOR EVOLUTION

It is interesting to note that the microprocessor had existed for only 10 years prior to the creation of the PC! Intel invented the microprocessor in 1971; the PC was created by IBM in 1981. Now more than 30 years later, we are still using systems based more or less on the design of that first PC. The processors powering our PCs today are still backward compatible in many ways with the 8088 that IBM selected for the first PC in 1981.

November 15, 2001 marked the 30th anniversary of the microprocessor, and in those 30 years processor speed has increased more than 18,500 times (from 0.108MHz to 2GHz). The 4004 was introduced on November 15, 1971 and originally ran at a clock speed of 108KHz (108,000 cycles per second, or just over one-tenth a megahertz). The 4004 contained 2,300 transistors and was built on a 10-micron process. This means that each line, trace, or transistor could be spaced about 10 microns (millionths of a meter) apart. Data was transferred 4 bits at a time, and the maximum addressable memory was only 640 bytes. The 4004 was designed for use in a calculator but proved to be useful for many other functions because of its inherent programmability. For example, the 4004 was used in traffic light controllers, blood analyzers, and even in the NASA Pioneer 10 deep space probe!

In April 1972, Intel released the 8008 processor, which originally ran at a clock speed of 200 KHz (0.2MHz). The 8008 processor contained 3,500 transistors and was built on the same 10-micron process as the previous processor. The big change in the 8008 was that it had an 8-bit data bus, which meant it could move data 8 bits at a time twice as much as the previous chip. It could also address more memory, up to 16KB. This chip was primarily used in dumb terminals and general-purpose calculators.

The next chip in the lineup was the 8080, introduced in April 1974, running at a clock rate of 2MHz. Due mostly to the faster clock rate, the 8080 processor had 10 times the performance of the 8008. The 8080 chip contained 6,000 transistors and was built on a 6-micron process. Similar to the previous chip, the 8080 had an 8-bit data bus, so it could transfer 8 bits of data at a time. The 8080 could address up to 64KB of memory, significantly more than the previous chip.

2.1 HUMAN ANALOGY OF COMPUTER

The human energy system is somewhat like a computer. This does not mean that people are mechanically logical. Rather, in the same way that human are made in the image of God, computers are made in the image of humans. Every work of art is a self-portrait; creation can only be formed from what is available within the artist. Any creation is always going to reflect what the artist is like, how s/he thinks, feels and sees the world. If humans are made in the image of God, it's a matter not of body parts but of characteristics: intelligence, creativity, glory. Being made in the image of its human designers, a computer will express something of the nature of human beings. Thus, we can learn quite a bit about the human energy system by using computers as an analogy.

Now, a computer is constructed of physical components that we can see and touch, including but not limited to:

- an input device (mouse or keyboard)
- a processing unit (CPU with calculation capability, usually RAM running at a certain speed)
- a monitoring device so that you can see what is going on (a screen, for example)
- permanent and temporary data banks (*e.g.*, hard, floppy or CD-ROM drives)
- connective wiring and a power source.

Humans have equivalent components. Our senses gather data, and our brains function as CPUs. Instead of screens, we have facial expressions, body language and verbal language to express what is happening inside. Our "hard drive" is the bio-magnetic field of charged particles surrounding the body. Connective wiring is found in our nervous system and meridians; the power source is always God, on the spiritual level, and the combination of oxygen, water, and food on the physical level. In both humans and computers, all these components have to be compatible and correctly wired up for a computer to work at all.

However, the best components in the world only create an environment for the work you wish to do. In order to create, you also require data and programs. There are structural and programming aspects to a computer that are equally vital to its function as a calculating machine.

2.2 TYPES OF COMPUTERS

Computers come in a variety of types designed for different purposes, with different capabilities and costs.

2.2.1 Microcomputer

A **microcomputer** is a computer that has a microprocessor chip as its CPU. They are often called **personal computers** because they are designed to be used by one person at a time. Personal computers are typically used at home, at school, or at a business. Popular uses for microcomputers include word processing, surfing the web, sending and receiving e-mail, spreadsheet calculations, database management, editing photographs, creating graphics, and playing music or games.

Personal computers come in two major varieties, desktop computers and laptop computers. **Desktop computers** are larger and not meant to be portable. They usually sit in one place on a desk or table and are plugged into a wall outlet for power. The cabinet of the computer holds the motherboard, drives, power supply, and expansion cards. This cabinet may lay flat on the desk, or it may be a **tower** that stands vertically(Figure). The computer has a separate monitor (either a CRT or LCD). A separate keyboard and mouse allow the user to input data and commands.

A Desktop Computer System

Laptop or **notebook** computers are small and lightweight enough to be carried around with the user. They run on battery power, but can also be plugged into a wall outlet. They typically have a built-in LCD display that folds down to protect the display when the computer is carried around. They also feature a built-in keyboard and some kind of built-in pointing device (such as a touch pad).

While some laptops are less powerful than typical desktop machines, this is not true in all cases. Laptops, however, cost more than desktop units of equivalent processing power because the smaller components needed to build laptops are more expensive.

A Laptop Computer

Workstations/Servers

A **workstation** is a powerful, high-end microcomputer. They contain one or more microprocessor CPUs. They may be used by a single-user for applications requiring more power than a typical PC (rendering complex graphics, or performing intensive scientific calculations).

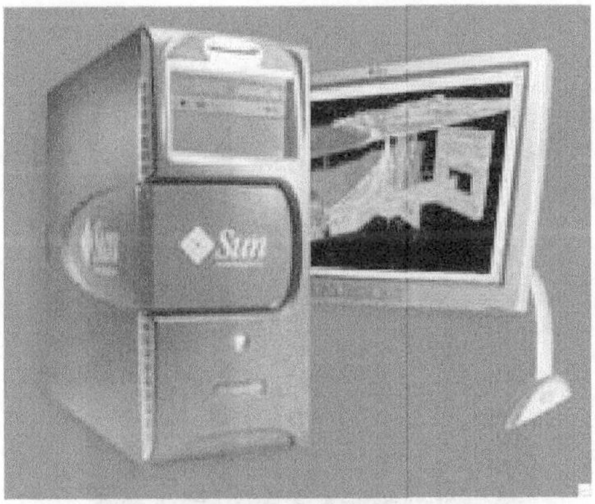

A Workstation

Alternately, workstation-class microcomputers may be used as **server** computers that supply files to **client** computers over a network. This class of powerful microcomputers can also be used to handle the processing for many users simultaneously who are connected via terminals; in this respect, high-end workstations have essentially supplanted the role of minicomputers (see below).

The term "workstation" also has an alternate meaning: In networking, any client computer connected to the network that accesses server resources may be called a **workstation**. Such a network client workstation could be a personal computer or even a "workstation" as defined here .

Note: Dumb terminals are not considered to be network workstations (client workstations on the network are capable of running programs independently of the server, but a terminal is not capable of independent processing).

2.2.2 Minicomputers

A **minicomputer** is a multi-user computer that is less powerful than a mainframe. This class of computers became available in the 1960's when large scale integrated circuits made it possible to build a computer much cheaper than the then existing mainframes. The niche previously filled by the minicomputer has been largely taken over by high-end microcomputer workstations serving multiple users.

2.2.3 Mainframe Computer

A **mainframe** computer is a large, powerful computer that handles the processing for many users simultaneously (up to several hundred users). The name mainframe originated after minicomputers appeared in the 1960's to distinguish the larger systems from the smaller minicomputers.

Users connect to the mainframe using terminals and submit their tasks for processing by the mainframe. A **terminal** is a device that has a screen and keyboard for input and output, but it does not do its own processing (they are also called **dumb terminals** since they cannot process data on their own). The processing power of the mainframe is time-shared between all of the users. (Note that a personal computer may be used to "emulate" a dumb terminal to connect to a mainframe or minicomputer; you run a program on the PC that pretends to be a dumb terminal).

Mainframes typically cost several hundred thousand dollars. They are used in situations where a company wants the processing power and information storage in a centralized location. Mainframes are also now being used as high-capacity server computers for networks with many client workstations.

A Mainframe Computer

2.2.4 Supercomputer

A **supercomputer** is a mainframe computer that has been optimized for speed and processing power. The most famous series of supercomputers were designed by the company founded and named after Seymour Cray. The **Cray-1** was built in the 1976 and installed at Los Alamos National Laboratory. Supercomputers are used for extremely calculation-intensive tasks such as simulating nuclear bomb detonations, aerodynamic flows, and global weather patterns. A supercomputer typically costs several million dollars.

Recently, some supercomputers have been constructed by connecting together large numbers of individual processing units (in some cases, these processing units are standard microcomputer hardware).

A Supercomputer

2.3 EVOLUTION OF A MICROPROCESSOR

A **microprocessor** incorporates most or all of the functions of a computer's central processing unit (CPU) on a single integrated circuit (IC, or microchip). The first microprocessors emerged in the early 1970s and were used for electronic calculators, using binary-coded decimal (BCD) arithmetic on 4-bit words. Other embedded uses of 4- and 8-bit microprocessors, such as terminals, printers, various kinds of automation etc, followed rather quickly. Affordable 8-bit microprocessors with 16-bit addressing also led to the first general purpose microcomputers in the mid-1970s.

Computer processors were for a long period constructed out of small and medium-scale ICs containing the equivalent of a few to a few hundred transistors. The integration of the whole CPU onto a single chip therefore greatly reduced the cost of processing capacity. From their humble beginnings, continued increases in microprocessor capacity have rendered other forms of computers almost completely obsolete, with one or more microprocessor as processing element in everything from the smallest embedded systems and handheld devices to the largest mainframes and supercomputers.

Since the early 1970s, the increase in capacity of microprocessors has been known to generally follow Moore's Law, which suggests that the complexity of an integrated circuit, with respect to minimum component cost, doubles every two years.

In the late 1990s, and in the high-performance microprocessor segment, heat generation (TDP), due to switching losses, static current leakage, and other factors, emerged as a leading developmental constraint.

2.3.1 Discrete Implementation

A discrete OTS (off the shelf) microprocessor solution is the "old" or normal way to go. It is the traditional approach to electronic "microprocessor powered" designs. There are many vendors of traditional processors at the moment (like there are Atmel, microchip, Arm,...) all those vendors have added a wide range of features to their specific microprocessor families. A discrete microprocessor is implemented as an ASIC with a specific peripheral set along with the processor core.

Selecting this kind of processor is a time consuming process, you have to meet the application's cost and the functional requirements to the design. Sometimes the processor will even not meet those functional requirements. For instance when an application needs custom logic or a significant amount of peripheral functionality that is not available in a discrete solution.

In this case, the logical place to look is at a processor and peripheral set that can be tailored to the application and included with the custom logic that is needed for the application.

2.3.2 SSI Implementation

The first integrated circuits contained only a few transistors. Called "**Small-Scale Integration**" (**SSI**), digital circuits containing transistors numbering in the tens provided a few logic gates. For example, early linear ICs such as the Plessey SL201 or the Philips TAA320 had as few as two transistors. The term Large Scale Integration was first used by IBM scientist Rolf Landauer when describing the theoretical concept, from there came the terms for SSI, MSI, VLSI, and ULSI.

SSI circuits were crucial to early aerospace projects, and vice-versa. Both the Minuteman missile and Apollo program needed lightweight digital computers for their inertial guidance systems; the Apollo guidance computer led and motivated the integrated-circuit technology, while the Minuteman missile forced it into mass-production.

These programs purchased almost all of the available integrated circuits from 1960 through 1963, and almost alone provided the demand that funded the production improvements to get the production costs from $1000/circuit (in 1960 dollars) to merely $25/circuit (in 1963 dollars). They began to appear in consumer products at the turn of the decade, a typical application being FM inter-carrier sound processing in television receivers.

2.3.3 MSI Implementation

The next step in the development of integrated circuits, taken in the late 1960s, introduced devices which contained hundreds of transistors on each chip, called "**Medium-Scale Integration**" (**MSI**).

They were attractive economically because while they cost little more to produce than SSI devices, they allowed more complex systems to be produced using smaller circuit boards, less assembly work (because of fewer separate components), and a number of other advantages.

2.3.4 LSI Implementation

Further development, driven by the same economic factors, led to "**Large-Scale Integration**" (**LSI**) in the mid 1970s, with tens of thousands of transistors per chip.

Integrated circuits such as 1K-bit RAMs, calculator chips, and the first microprocessors, that began to be manufactured in moderate quantities in the early 1970s, had under 4000 transistors. True LSI circuits, approaching 10000 transistors, began to be produced around 1974, for computer main memories and second-generation microprocessors.

2.4 GENERATIONS OF MICROPROCESSOR

In this section we will discuss the various generations of Microprocessors:

2.4.1 First Generation of Processors

The first **generation** of processors represents the series of chips from Intel that were found in the first PCs. IBM, as the architect of the PC at the time, chose Intel processors and support chips to build

the PC motherboard, setting a standard that would hold for many subsequent processor generations to come.

(a) 8088 and 8086 Processors

Intel introduced a revolutionary new processor called the 8086 back in June of 1978. The 8086 was one of the first 16-bit processor chips on the market, at the time virtually all other processors were 8-bit designs. The 8086 had 16-bit internal registers and could run a new class of software using 16-bit instructions. It also had a 16-bit external data path, which meant it could transfer data to memory 16 bits at a time.

The address bus was 20 bits wide, meaning that the 8086 could address a full 1MB (2^{20}) of memory. This was in stark contrast to most other chips of that time that had 8-bit internal registers, an 8-bit external data bus, and a 16-bit address bus allowing a maximum of only 64KB of RAM (2^{16}).

Unfortunately, most of the personal computer world at the time was using 8-bit processors, which ran 8-bit CP/M (Control Program for Microprocessors) operating systems and software. The board and circuit designs at the time were largely 8-bit as well. Building a full 16-bit motherboard and memory system would be costly, pricing such a computer out of the market.

The cost was high because the 8086 needed a 16-bit data bus rather than a less expensive 8-bit bus. Systems available at that time were 8-bit, and slow sales of the 8086 indicated to Intel that people weren't willing to pay for the extra performance of the full 16-bit design. In response, Intel introduced a kind of crippled version of the 8086, called the 8088. The 8088 essentially deleted 8 of the 16 bits on the data bus, making the 8088 an 8-bit chip as far as data input and output were concerned. However, because it retained the full 16-bit internal registers and the 20-bit address bus, the 8088 ran 16-bit software and was capable of addressing a full 1MB of RAM.

For these reasons, IBM selected the 8-bit 8088 chip for the original IBM PC. Years later, IBM was criticized for using the 8-bit 8088 instead of the 16-bit 8086. In retrospect, it was a very wise decision. IBM even covered up the physical design in its ads, which at the time indicated its new PC had a "high-speed 16-bit **microprocessor.**" IBM could say that because the 8088 still ran the same powerful 16-bit software the 8086 ran, just a little more slowly. In fact, programmers universally thought of the 8088 as a 16-bit chip because there was virtually no way a program could distinguish an 8088 from an 8086. This allowed IBM to deliver a PC capable of running a new **generation** of 16-bit software, while retaining a much less expensive 8-bit design for the hardware. Because of this, the IBM PC was actually priced less at its introduction than the most popular PC of the time, the Apple II. For the trivia buffs out there, the IBM PC listed for $1,265 and included only 16KB of RAM, while a similarly configured Apple II cost $1,355.

The original IBM PC used the Intel 8088. The 8088 was introduced in June 1979, but the IBM PC did not appear until August 1981. Back then, there was often a significant lag time between the introduction of a new processor and systems that incorporated it. That is unlike today, when new processors and systems using them are often released on the same day.

The 8088 in the IBM PC ran at 4.77MHz, or 4,770,000 cycles (essentially computer heartbeats) per second. Each cycle represents a unit of time to the system, with different instructions or operations requiring one or more cycles to complete. The average instruction on the 8088 took 12 cycles to complete.

Computer users sometimes wonder why a 640KB conventional-memory barrier exists if the 8088 chip can address 1MB of memory. The conventional-memory barrier exists because IBM reserved 384KB of the upper portion of the 1,024KB (1MB) address space of the 8088 for use by adapter cards and system BIOS. The lower 640KB is the conventional memory in which DOS and software applications execute.

(b) 80186 and 80188 Processors

After Intel produced the 8086 and 8088 chips, it turned its sights toward producing a more powerful chip with an increased instruction set. The company's first efforts along this line—the 80186 and 80188—were unsuccessful. But incorporating system components into the CPU chip was an important idea for Intel because it led to faster, better chips, such as the 286.

The relationship between the 80186 and 80188 is the same as that of the 8086 and 8088; one is a slightly more advanced version of the other. Compared CPU to CPU, the 80186 is almost the same as the 8088 and has a full 16-bit design. The 80188 (like the 8088) is a hybrid chip that compromises the 16-bit design with an 8-bit external communications interface. The advantage of the 80186 and 80188 is that they combine on a single chip 15 to 20 of the 8086–8088 series system components—a fact that can greatly reduce the number of components in a computer design. The 80186 and 80188 chips were used for highly intelligent peripheral adapter cards of that age, such as network adapters.

(c) 8087 Coprocessor

Intel introduced the 8086 processor in 1976. The math coprocessor that was paired with the chip—the 8087—often was called the numeric data processor (NDP), the math coprocessor, or simply the math chip. The 8087 is designed to perform high-level math operations at many times the speed of the main processor. The primary advantage of using this chip is the increased execution speed in number-crunching programs, such as spreadsheet applications.

2.4.2 Second Generation Processors

The only chip generally considered second generation is the 80286. Intel also made a chip between the 8088 and the 80286, called the 80186. Largely due to a lack of performance enhancements compared to the 286, in addition to compatibility problems with support chips, the 80186 was never really used in PCs. It has however found a strong niche role in the design of embedded microcontrollers, even to this day.

Intel 80286

The 80286 was the chip used in IBM's AT (advanced technology) system. The 286 was the first major step up in PC processors, providing significant performance increases over the 8088 and 8086—double or more performance at the same clock speed. The 286 also widened the address bus to allow access to 16 MB of memory, and introduced protected mode operation. It was originally available in 6 MHz and 8 MHz versions, but was later expanded to faster versions, all the way up to 20 MHz.

The 286 opened up the PC world to many users, but still was used mainly as the equivalent of a "turbo-charged 8088". At this time DOS was still the virtually exclusive operating system, and the protected mode the 286 offered was largely ignored.

The 80286 is of course considered obsolete today.

2.4.3 Third Generation Processors

The third generation processors (the 386 family) represent another step forward from the second generation class. Intel started with this generation of chips to create "subfamilies" of related chips with different capabilities, using the "DX" and "SX" designations.

Intel 80386DX

The Intel 80386DX was the first true 32-bit processor used on the PC platform. Its internal register size was increased to 32 bits, and its data and address buses were as well, doubling data path width to the processor and increasing addressable memory to 4 GB theoretical. The 80386 family of chips offered more performance than the 80286s they replaced, largely through processor speed increases. The 386 did not offer the very large improvements over the 286 that the 286 did over the 8088.

The 386DX's increased power and the improved processor modes it offered (including full protected mode and virtual real modes) spurred the introduction of GUI-based operating systems on the PC, such as Microsoft Windows (although they are still quite slow on 386 chips). The instruction set of the 386 has set the standard for what is now called "x86" and hasn't changed very much since it was introduced. The 80386DX was the first to use pipelining to allow much improved processor performance through the use of much higher clock frequencies.

Invented by Intel, the 386 was also "cloned" by AMD and Cyrix. These are very good copies with no real compatibility problems. Intel only produced the 386DX up to 33 MHz (presumably to prevent overlapping into the 486's performance range) while AMD and Cyrix produced a 40 MHz version as well. This latter chip produced very good performance (for the time), comparable to many lower-end 486s. They are still however obsolete by today's standards, of course.

Intel 80386SX

The 80386SX is a "lite" version of the 80386DX chip. It uses only a 16-bit data bus, the same external bus width as the 80286. It also only can address 16 MB of memory, like the 80286. The SX version of the 386 chip was in fact released well after the DX, and in some ways was intended to move the market away from the 286 since it had roughly the same interfaces but better performance. It was actually introduced several *years* after the DX.

Note: It is a common myth that the 80386SX can be substituted in place of an 80286 in a 286 motherboard. While the chips are compatible in terms of their external interfaces they use different packaging. A 386SX could be put into a 286 motherboard if a proper adapter were used.

The narrower data bus width of the 386SX creates a reduction in performance of about 20-25% compared to an equivalent-speed 386DX; a significant difference but not excessive. The 80386SX chip was a popular choice for the first small notebook computers, especially the 386SL variant that introduced the SMM power management features to the Intel line. The 386SX is still a 32-bit processor internally and will run 32-bit software (really slowly). It is available in speeds from 16 to 33 MHz; a 40 MHz version was not produced for the SX, and is also available in cloned versions from AMD and Cyrix.

The 386SX chip is today considered obsolete.

2.4.4 Fourth Generation Processors

The fourth generation of processors saw rapid growth in the CPUs' power and capabilities, and the introduction of several new technologies as well. It was here that AMD and Cyrix had their first real early successes in not just emulating Intel's designs but coming up with real value leaders of their own, at the top end of the fourth generation scale. Also, it was at this time that the new phenomenon of upgradable processors and standardized motherboard sockets began to become prevalent.

Intel 80486DX

The Intel 80486DX was the first member of the 486 family (which has many more members than the 386 family did). It provided a very significant increase in power over the 80386DX processor, in fact, far more proportionately than the 386 did over the 286. A 486DX processor provides approximately 100 to 150% more performance than a 386DX of the same clock speed. The 80486 brought GUIs to the mainstream on PCs; it is the minimum processor that most people consider "usable" for running an operating system like Microsoft Windows.

Interestingly, the 486 does not provide its performance improvements by widening any of the buses, as had been the case in the previous two generations: it is still a 32-bit processor with 32-bit data and address buses, just like the 386DX. However, internally, the 486 incorporates several significant improvements over the 386:

- **Faster Overall Instruction Execution Time:** The core of the chip can execute instructions in less time than earlier processors.
- **Deeper Pipeline:** The execution pipeline was increased by one step.
- **Primary Cache:** The 486 processor was the first to incorporate level 1 cache on the chip, to reduce the number of required accesses to memory.
- **Integrated Floating Point Unit:** The chip includes an integrated math coprocessor (not on the SX version however). In addition, the coprocessor provides much more performance than the optional 80387 used with 386 chips, in part because it is integrated into the chip.
- **Burst Mode:** The 486 introduced the use of burst mode to reduce wait time on memory accesses.
- **Power Management:** SL power management enhancements as an option (instead of a specialized SL chip as was the case for the 386SL).
- **Improved Support Architecture:** In general, 486 motherboards were more efficient than 386 ones, and began to use secondary cache as well. This improves processor performance significantly.

Interestingly, the 80487SX coprocessor that is intended for use with the SX version of the 80486, is in fact a full-blown 80486DX processor. This is part of how Intel structured its fourth generation family's upgrade path. See the description of the 80486SX for more details on this.

AMD and Cyrix did produce clones of the original Intel 80486 processors, but they were not a big player in the 486 clone market until the higher-speed DX2 and DX4 processors.

The 486DX is considered obsolete, although the chip still has fairly good power for performing a wide variety of light tasks, such as word processing and some older games, and light Internet access. The 486DX-50, which runs on a 50 MHz system bus, provides performance comparable to the 486DX2-66 in many ways, because the latter uses only a 33 MHz system bus. The 486DX-50 was not used in nearly as many systems as the other processor speeds were. It should not be confused with the 486DX2-50, which runs at the same processor clock speed but is clock-doubled relative to the system bus (which runs at 25 MHz).

Note: The 486DX processor normally was purchased as part of a new system only, not as part of an upgrade. Most early 486 systems used a 168-pin socket for the chip, which predates the numbered standardized socket system that Intel created. The 486DX will fit into a Socket 1, Socket 2 or Socket 3 however.

Intel 80486SX

The Intel 80486SX is the same chip as the 80486DX with one exception: the lack of an integrated math coprocessor (floating point unit). Note that this is a different kind of difference between the SX and DX versions than is the case with the 386; the 386SX and 386DX both had no coprocessor and the 386SX had narrower data and address buses than the 386DX. The 486DX and 486SX have the same bus widths. Since it is the same chip except for the floating point processor, the 486SX has the same advantages over the 386 that the 486DX does. Note that the 486SX was made available in slower clock speeds than the 486DX; the SX comes in 16, 20, 25 and 33 MHz versions, while the DX is 25, 33 and 50 MHz.

Intel made some rather interesting (some would say *bizarre*) decisions about how to market the 486 line. The 486SX was in many ways a marketing gimmick only; it was in fact a 486DX with the math coprocessor disabled! Why did Intel bother to sell a full 486DX and a "crippled" one (the 486SX)? Mainly for market targeting reasons: they wanted to be able to fill the niche of low-end buyers without dropping the price on their "top of the line" 486DX. The 486SX was a popular choice for laptops due to its lower cost and lower power consumption.

The 80486SX uses a "math coprocessor" called the 80487SX (there is no 80487DX at all so don't be confused by that). The 80487SX is, in fact, a fully functional 80486*DX* chip! What Intel wanted people to think was that (like with its earlier coprocessors) you would put the 80487SX in and it would handle the math functions. In fact, when inserted, the 80487SX *shuts down* the 80486SX and handles both integer and floating point operations (since it is internally a 80486DX, which does both). This makes no difference from a performance standpoint but is kind of a technical curiosity.

With the large number of very cheap 486DX and faster chips and systems around, the 486SX is really quite obsolete. In addition, it does not come in a 50 MHz version. If you are not doing any floating point operations at all it is comparable to a 486DX of the same speed, but there is usually no need to have to worry about this given that even the 486DX is quite obsolete at this time anyway.

Note: The 486SX processor normally was purchased as part of a new system only, not as part of an upgrade. Most early 486 systems used a 168-pin socket for the chip, which predates the numbered standardized socket system that Intel created. The 486SX will fit into a Socket 1, Socket 2 or Socket 3 however.

Intel 80486DX2 and 80486DX2 Overdrive

The 80486DX2 was the first chip to use "clock doubling" technology, where the processor runs at a faster speed than the memory bus it talks to. This was done to allow the processor speed to be increased without having to deal with the much more difficult task of increasing motherboard speed. Chips that run at faster than memory bus speed improve performance but at a diminishing rate as the multiplier increases, due to the processor waiting for data from memory.

Intel produced 50 and 66 MHz DX2 chips, intended for use in 25 and 33 MHz system bus systems. These chips have been sold as regular chips intended for use in new systems, which generally come in 168 pin packages to go in the original 168 pin socket used in 486 systems. They have also been made in 169 pin Overdrive versions to go in Socket 1 (the original Overdrive socket). These can be used to upgrade older 486DX or 486SX systems.

AMD and Cyrix not only cloned the 66 MHz DX2 processor, they took Intel one step further with the 80486DX2-80, running at 80 MHz. This uses a 40 MHz system bus, which isn't a speed that is normally used by Intel systems but that became more popular late in the 486 life cycle due to the performance increase it gives over 33 MHz bus systems. In addition, the AMD (enhanced version) and Cyrix chips have several advantages over the Intel chips (they had the benefit of developing them well after Intel):

- **Power Management:** They have SMM (power management) built in.
- **Write-Back Cache:** They have write-back capability for the primary cache which provides a small boost in performance.
- **Lower Power:** They use less power because they run at 3 volts, but will still work in standard 5 volt motherboard sockets without a converter or regulator (they are said to be "5 volt tolerant").

Other than clock speed, the 80486DX2 is virtually identical to the 80486DX. They are obsolete due to the availability of faster, very inexpensive processors such as the 5x86-133 that go in the same sockets, but 486DX2 systems are perfectly viable for many uses, including routine office word processing and spreadsheet work under DOS and Windows 3x. The 80486DX2-66 is by far the most common version of this chip; a great number of these systems were produced and many are still in use today, especially in small businesses.

Note: The 486DX2 was the first processor to really *require* a heat sink in order to operate reliably. The increased speed of this chip means that it runs very hot (at least the Intel 5 volt versions).

Intel 80486DX4 and 80486DX4 Overdrive

The 80486DX4 continued the trend started by the 80486DX2 toward faster clock speed processors on slower motherboards. The DX4 uses "clock tripling", where the processor runs at three times the speed of the memory bus (motherboard). Note that despite the name the DX4 does not run at four times the memory bus speed ("DX3" refers to 2.5 times the memory bus, but never became a shipping product for the 486.)

Intel's DX4 runs at two speeds: 75 MHz (for the 25 MHz bus) and 100 MHz (for the 33 MHz), with the 100 being by far the most popular. In order to keep power and heat to a manageable level the voltage of these chips is reduced to 3.3 volts. The AMD and Cyrix versions are very similar to their DX2 cousins, in that they are 5 volt tolerant, meaning they can handle being put in a 5 volt motherboard; the Intel chip is not 5 volt tolerant and cannot be put into a 5 volt board (well, you can physically put them into the board, but you shouldn't do it! :)). AMD and Cyrix also make a 120 MHz version that runs on a 40 MHz bus; this is relatively rare. The Intel, AMD and Cyrix versions of the DX4 chip are quite similar. One difference is that the AMD and Cyrix support write-back cache while the Intel does not; however the Intel has its level 1 cache doubled to 16 KB. All three chips support power management.

The 486DX4 was most commonly put into systems as Overdrive processors for older, slower systems. Intel did not originally make the DX4 available as a stand-alone chip, presumably since it shipped after the Pentium had already been introduced. However, they did make the chip available stand-alone later on. New 486DX4-100 systems were often made with the AMD version of the chip. The Overdrive version is generally required for older motherboards that don't support 3.3 volt power.

Like the 80486DX2, the 80486DX4 is technically obsolete, but the 100 MHz version especially has good processing power for routine applications. It is also acceptable for some other uses, but is not found in modern systems.

AMD 5x86 (80486DX5)

The 486DX4-100 was the fastest 486 processor made by Intel before they decided to leave the fourth generation and concentrate on the Pentium. AMD took clock multiplying one increment further with what it calls the "5x86" chip. Despite the implication that it is a fifth-generation chip, it is not—it is a high speed, clock-quadrupled 486 processor, that runs in 486 motherboards. It does use a rather advanced 0.35 micron CMOS process (advanced for a fourth-generation chip that is).

The AMD 5x86 was made available in one speed only, 133 MHz, for use in 33 MHz motherboards. The processor runs at four times the system clock and fits into a Socket 3. Since having a clock multiplier of four was not part of the original Socket 3 design, AMD made the 5x86 look for a *two times* setting from the motherboard and interpret that as four times instead. In other words, to use the 5x86 you want to set the motherboard to the 2x setting. This will actually cause the 5x86 to run at 4x. The chip will actually physically fit into an older 486 socket such as a socket 1 or 2 or the original 168-pin 486 socket, but doing this requires a voltage regulator since the AMD chip runs at 3.3 volts.

AMD calls this chip the "5x86-P75" because it offers performance comparable to low-end fifth-generation chips. In fact, it is comparable in integer performance to a 75 MHz Pentium. It is also called the 80486DX5-133, which is a more accurate reflection of what the chip really is. Architecturally, the chip is virtually identical to the 80486DX4, except that it matches Intel's 16 KB level 1 cache (the AMD 80486DX4 has only 8 KB). Note that this is not the case with Cyrix's 5x86 chip, which is quite different.

The 5x86-133 is the most powerful 486-class chip available (the Cyrix 5x86-120 is actually faster but was discontinued by Cyrix many months before AMD stopped mass-producing the 5x86-133). The 5x86-133 has traditionally been an excellent choice for an economy PC, especially for home-builders, because not only is the chip itself very inexpensive, so are the motherboards that use it. With 200+ MHz Pentium and Pentium-compatible chips coming well down in price, the 5x86-133 is following the rest of the fourth-generation chips to obsolescence, quickly. However, since it is the "king of the hill" for 486 motherboards, it remains a good choice for those who want to upgrade their 486 systems without replacing the motherboard. The chips are still out there, though they are getting much more difficult to find now.

Cyrix 5x86 ("M1sc")

Despite having the same name as AMD's 5x86 processor, the Cyrix 5x86 is a totally different animal. While AMD designed its 5x86 by further increasing the clock on the 486DX4, Cyrix took the opposite approach by modifying its M1 processor core (used for the 6x86 processor) to make a "lite" version to work on 486 motherboards. As such, the Cyrix 5x86 in some ways resembles a Pentium OverDrive (which is a Pentium core modified to work in a 486 motherboard) internally more than it resembles the AMD 5x86. This chip is probably the hardest to classify as either fourth or fifth generation.

The 5x86 employs several architectural features that are normally found only in fifth-generation designs. The pipeline is extended to six stages, and the internal architecture is 64 bits wide. It has a larger (16 KB) primary cache than the 486DX4 chip. It uses branch prediction to improve performance.

The 5x86 was available in two speeds, 100 and 120 MHz. The 5x86-120 is the most powerful chip that will run in a 486 motherboard—it offers performance comparable to a Pentium 90 or 100. The 5x86 is still a clock-tripled design, so it runs in 33 and 40 MHz motherboards. (The 100 MHz version will actually run at 50x2 as well, but normally was run at 33 MHz.) It is a 3 volt design and is intended for a Socket 3 motherboard. It will run in an earlier 486 socket if a voltage regulator is used. I have heard that some motherboards will not run this chip properly so you may need to check with Cyrix if trying to use this chip in an older board. These chips have been discontinued by Cyrix but are still good performers, and for those with a compatible motherboard, as good as you can get. Unfortunately, they are extremely difficult to find now.

2.4.5 Fifth Generation Processors

The fifth generation of processors saw several changes from earlier CPU families, and several trends continue as well. Chips continued to get faster and faster, and architectural changes were made to increase overall system speed as well. AMD and Cyrix developed their own compatible processors instead of just trying to clone Intel's, leading to more variety and choice in the marketplace. The Pentium and the compatibles that followed it opened up the world of computers for millions of users and propelled computing to the next level.

Intel Pentium ("P5"/"P54C")

Intel's new fifth-generation chip was expected to be called the 586, following their earlier naming conventions. However, with the rise of AMD and Cyrix, Intel wanted to be able to register as a trademark the name of their new CPU, and numbers can't be trademarked. Thus, the Pentium was born. It is now one of the most recognized trademarks in the computer world, one reason why Intel doesn't seem to ever want to make another processor whose name doesn't have "Pentium" in it somewhere.

The Pentium is the defining processor of the fifth generation. It has in fact had several generations itself; the first Pentiums are different in many ways from the latest ones. It has been the target for compatibility for AMD's K5 and Cyrix's 6x86 chips, as well as generations that have followed. The chip itself is instruction set compatible with earlier x86 CPUs, although it does include a few new (rarely used) instructions.

The Pentium provides greatly increased performance over the 486 chips that precede it, due to several architectural changes. Roughly speaking, a Pentium chip is double the speed of a 486 chip of the same clock speed. In addition, the Pentium goes to much higher clock speeds than the 486 ever did. The following are the key architectural enhancements made in the Pentium over the 486-class chips (note that some of these are present in Cyrix's 5x86 processor, but that chip was developed *after* the Pentium):

- **Superscalar Architecture:** The Pentium is the first superscalar processor; it uses two parallel execution units. Some people have likened the Pentium to being a pair of 486s in the same chip for this reason, though this really isn't totally accurate. It is really only partially superscalar because the second execution unit doesn't have all the capabilities of the first; some instructions won't run in the second pipeline. In order to take advantage of the dual pipelines, code must be optimized to arrange the instructions in a way that will let both pipelines run at the same time. This is why you sometimes see reference to "Pentium optimization". Regardless, the performance is much higher than the single pipeline of the 486.
- **Wider Data Bus:** The Pentium's data bus is doubled to 64 bits, providing double the bandwidth for transfers to and from memory.
- **Much Faster Memory Bus:** Most Pentiums run on 60 or 66 MHz system buses; most 486s run on 33 MHz system buses. This greatly improves performance. Pentium motherboards also incorporate other performance-enhancing features, such as pipelined burst cache. The Pentium processor was also the first specifically designed to work with the (then new) PCI bus.
- **Branch Prediction:** The Pentium uses branch prediction to prevent pipeline stalls when branches are encountered.
- **Integrated Power Management:** All Pentiums have built in SMM power management (optional on most of the 486s).
- **Split Level 1 Cache:** The Pentium uses a split level 1 cache, 8 KB each for data and instructions. The cache was split so that the data and instruction caches could be individually tuned for their specific use.
- **Improved Floating Point Unit:** The floating point unit of the Pentium is significantly faster than that of the 486.

The Pentium is available in a wide variety of speeds, and in regular and OverDrive versions. It is also available in several packaging styles, although the pin grid array (PGA) is still the most prevalent. The original Pentiums, the 60 and 66 MHz versions, were very different than the later versions that are used in most PCs; they used older, 5 volt technology and had significant problems with heat. Intel solved this with later (75-200 MHz) versions by going to a smaller circuit size and 3.3 volt power.

Pentiums use three different sockets. The original Pentium 60 and 66 use Socket 4. Pentiums from 75 to 133 will fit in either socket 5 or socket 7; Pentium 150s, 166s and 200s require Socket 7. Intel makes Pentium OverDrives that allow the use of faster Pentiums in older Pentium sockets (in addition to OverDrives that go in 486 motherboards).

The Pentium processor achieved a certain level of "fame" as a result of the bug that was discovered in its floating point unit not long after it was released. This is commonly known as the "FDIV" bug after the instruction (floating point divide) that it most commonly turns up in. While bugs in processors are relatively common, they usually are minor and don't have a direct impact on computation results. This one did, and achieved great notoriety in part because Intel didn't own up to the problem and offer to correct it immediately. Intel does offer a replacement on affected processors, which were only found in early versions (60 to 100) sold in 1994 and earlier.

If you suspect your Pentium of having the FDIV bug, try this computation test using a spreadsheet or calculator program: take the number 4,195,835 and divide it by 3,145,727. Then take the result and multiply it by the same number again (3,145,727). You should of course get the same 4,195,835 back that you started with. On a PC with the FDIV bug you will get 4,195,579 (an error of 256), but beware that some operating systems and applications have been *patched* to compensate for this bug, so a simple math test isn't necessarily conclusive.

For many years, the Pentium processor was the mainstream processor of choice, but finally the Pentium with MMX has driven it to the economy market. With the regular Pentium maxing out at 200 MHz and the Pentium with MMX 166 dropping well below $200, the "Pentium Classic" doesn't make nearly as much sense as it used to for new PCs. The 60 and 66 are obsolete due to their slow speed and older technology, and the 75 to 150 are obsolete because their performance is much lower than the 166 and 200, for almost the same amount of money.

The entire classic Pentium line is now technically obsolete, due to the availability of inexpensive, faster Pentium with MMX chips (as well as comparable offerings from AMD and Cyrix). The non-MMX Pentium is no longer generally used in new systems. However, since the Pentium with MMX requires split rail voltage, the classic Pentium 200 remains a great chip for those who have socket 7 motherboards and want to upgrade, but who do not have split rail voltage support.

Cyrix 6x86 ("M1")

Cyrix entered the fifth generation processor market in a big way with its (IMO poorly-named) 6x86 processor, formerly called the M1. This is an advanced processor that is an alternative to the Intel Pentium, and is pin- and voltage-compatible with it. Cyrix apparently gave it the "6x86" moniker in reference to some of its more advanced features, which it calls "sixth generation". In reality, the processor is comparable in power (and in many ways, architecturally as well) to the fifth-generation Pentium. It is, like the Pentium, a superscalar, native x86 execution CPU.

The 6x86, like the AMD K5, is *not* a Pentium "clone", in the traditional sense of the word. Clones are exact or near-exact copies, usually being reversed engineered or based on licensed code. The 6x86 is based on an original design of Cyrix's and is not 100% Pentium compatible. (It is 100% x86 compatible, but the Pentium uses some extensions to the base x86 instruction set that the 6x86 doesn't support).

The 6x86 incorporates several advanced architectural features that allow it to outperform a Pentium of equal clock speed. For example, a Cyrix running at 133 MHz will typically equal or outperform a Pentium 166 (not Pentium with MMX 166). For this reason, and because the public seems overly obsessed with thinking that clock speed is all that matters, Cyrix helped invent the "P rating" system. This was a good idea, to help people make more valid comparisons. Unfortunately however, it has caused a great deal of confusion of its own, because Cyrix named the chips with their P rating and not their clock speed. Many users (and even vendors!) confuse the P rating with the clock speed and try to set their motherboards to run, say, a 6x86-PR166+ to run at 166 MHz. This is overclocking the chip by 25% and the usual result is that the chip fails. It is important to make sure that the chip is set to its correct clock speed.

Compared to the Pentium (classic), the Cyrix 6x86 has the following enhancements or improvements that contribute to its greater performance to clock speed ratio (it also has all the advantages over the 486 that are mentioned in the section discussing the original Pentium):

- **Deepening of Internal Pipelines:** Both of the internal integer executions units are increased from 5 to 7 stages.
- **Out of Order Completion:** Instructions flowing down the two integer pipelines can complete out of order.
- **Superior Branch Prediction Unit:** The branch prediction unit of the 6x86 can handle multiple branches and has better performance than the Pentium's.
- **Improved Cache Mapping:** The primary cache in the 6x86 is 4-way set associative instead of the Pentium's 2-way. (It is also unified instead of being split like the Pentium's, though I don't think this is really an enhancement).
- **Register Renaming:** This feature improves parallel performance of the pipelines.
- **Speculative Execution:** The 6x86 uses speculative execution to reduce pipeline stall time.

Of course the 6x86 is also cheaper than a Pentium of the same performance rating. Often, it is less than half the cost. This is in fact the main reason that the 6x86 is purchased instead of the Pentium. Compared to the Pentium (classic), the 6x86 has the following disadvantages:

- **Software Compatibility Problems:** This is probably the biggest problem with the 6x86 and one reason that many people won't use it, even though it provides excellent value for many applications. Some rather parochial software has been written by programmers who think Intel is the only company making CPUs, and therefore they have used Pentium-specific instructions (recall that the Pentium has extensions to the standard x86 instruction set). This causes problems with the 6x86 because the 6x86 is not 100% Pentium compatible. Some companies have made "patches" available to allow their software to run on the 6x86. This problem is most pronounced with games and high-end graphical utilities that have been heavily optimized for the Pentium. One of the most publicized problems with the 6x86 is that its early versions had a bug that causes Windows NT to run more slowly than it should have. Cyrix has addressed this issue and is offering chip replacements to affected users. New systems shouldn't have this problem.
- **Processor Identification Problems:** In addition to programs that won't work with the 6x86, there are those that would work with the 6x86 if only they knew what it was. There is a special instruction called "CPUID" that is used to identify the make and model of the processor. Some programs use this to detect if the computer is running on a Pentium, and will refuse to run (or run in a reduced mode) if they don't find one. These tests were originally used by software requiring a lot of power, to distinguish between Pentiums and 486s (which the software would be too slow to run on). The 6x86 by default does not respond to the "CPUID" instruction so some programs think it is a 486. Again, this problem is most pronounced with games.

Also, early versions of Windows 95 will not identify the 6x86 properly since they predate it—they think it is a 486. This doesn't cause any performance problems however, other than programs (games) running under Windows 95 looking at the information that Windows 95 detects to decide if the machine is a Pentium or not . This problem is also fixed with the OEM SR2 release, which detects the 6x86 properly. Various utilities are available to "trick" the system into identifying the chip as a Pentium, if needed. This can solve many of these problems, although not all of them.

- **Motherboard Compatibility Problems:** Some motherboards (especially cheaper ones) have quirks that cause them not to work correctly with the 6x86. *Cyrix* maintains an approved motherboard list to help 6x86 buyers choose boards that are compatible with the chip. There are some boards that will work with it despite not being on the list, but the ones on it are known to work.
- **Heat and Power Problems:** The 6x86 uses significantly more power and generates more heat than the Pentium. This causes problems in some boxes and with some motherboards. Later versions address this issue and the low-power 6x86L eliminates it almost completely (but requires a motherboard with 2.8 volt core power). The best way to avoid this problem is to use an approved motherboard (as mentioned just above) since often the heat problems are caused by motherboards with insufficient voltage regulators. You should also use an active heat sink specifically meant for use on the 6x86; special cooling units are designed for the 6x86 and cool better than generic Pentium fans.
- **No Multi-Processor Support:** The implementation of SMP (symmetric multiprocessing) supported by today's motherboards is customized to Intel processors and will not work with the 6x86.
- **Lower-Power Floating Point Unit:** While the 6x86 equals or exceeds a Pentium of its "P rating" at integer operations, it lags far behind in floating-point performance. Cyrix intentionally decided not to spend time optimizing the FPU due to its lack of importance for most applications (which is true). A 6x86 generally has floating point performance slightly below a Pentium of its *clock speed* (as opposed to P rating). So a 6x86-PR166+ has integer performance equal to a Pentium 166 or higher, but floating point performance is more comparable to a Pentium 120 or even 100. The most famous problem this causes is with Quake, the popular game that uses floating point instructions. Most games actually do not use the FPU much at all.

Many of the compatibility problems with the 6x86 are a result of the fact that most software was written when Intel was the "only game in town". With AMD and Cyrix establishing themselves as serious players, it is likely that in the future programmers will be more comprehensive in their choice of coding methods. However, the 6x86 really isn't totally Pentium compatible, which the buyer needs to take into consideration.

The 6x86 chip is designed for Socket 7, and is available in several different clock speeds, some of them rather unusual. All of the chips use a 2x multiplier, and they support bus speeds of 50, 55, 60, 66 and 75 MHz. Of course only 3 of those will be readily recognizable to those familiar with Pentium processors. The 55 MHz bus speed of the PR133+ is in particular an oddball; many (if not most) motherboards will not support this speed.

The 75 MHz bus speed is an interesting advance in many ways, as it pushes the traditional 66 MHz memory speed limit forward. The system bus speed is for most systems a bottleneck that severely limits overall performance; increasing it can increase performance greatly. This in fact is one major reason that the 150 MHz 6x86-PR200+ performs at the level of a Pentium 200. However, there are several additional caveats associated with the 75 MHz bus speed used by the 6x86-PR200+:

- **Chipset Support:** Intel is the major producer of Pentium-class chipsets, which are rated to 66 MHz, the fastest memory bus speed that Intel processors use. Not surprisingly, Intel is not in a big hurry to support a higher bus speed used only by a Cyrix chip. This leaves two choices for motherboard manufacturers: overclock an Intel chipset (which some do) or use an alternative.
- **Motherboard Support:** Largely because of the chipset problem, motherboards that support the 75 MHz bus speed of the PR200+ are harder to find than regular motherboards.
- **PCI Device Overclocking:** Since the PCI bus runs at half of the memory bus speed, when you increase the memory bus to 75 MHz you are increasing the PCI bus to 37.5 MHz. When you do this, you are effectively overclocking all of your PCI cards by 12.5%. This is not usually a problem, but you should bear in mind that *many* peripheral makers do not do extensive testing at the higher bus speed because of the very small market of the Cyrix chips. This will hopefully change in the future as higher bus speeds become more prevalent.

The Cyrix 6x86 is manufactured by IBM, which in fact also sells the 6x86 under its own name. (Cyrix doesn't have its own manufacturing facilities). These chips are identical; they are manufactured in the same IBM fabrication plant, and independently packaged and tested. The rumour is that the IBM version is superior due to better testing, but I've found nothing reliable to prove this.

Responding to complaints about the heat and power problems of the 6x86, Cyrix came out with a lower powered version called the 6x86L. This chip reduces power consumption by 25% or more compared to the 6x86. It uses a smaller, 0.35 micron circuit size, and split rail voltage much like the Pentium with MMX (3.3V external, 2.8V internal). This version is preferred for those seeking to reduce the power problems of earlier models.

Overall, the 6x86 is a decent processor that provides a lot of value, but at the cost of some headaches that users will have to deal with. When it first came out, it was an excellent alternative to the Intel Pentium CPU, because it was selling for several hundred dollars less. As Pentium prices dropped, so did the spread in cost between it and the 6x86, and there was increasingly little reason to use the Cyrix chip. Today the 6x86 is essentially obsolete; the main reason is the inexpensive availability of Cyrix's own 6x86MX MMX chip.

AMD K5 ("K5"/"5k86")

AMD's entry in the fifth generation processor sweepstakes is called the K5. This processor was eagerly awaited and it was hoped that it would provide a viable alternative to the Pentium early in the Pentium's life cycle. Unfortunately, AMD delivered the processor over a year late and at much lower clock speeds than had been originally anticipated. As a result instead of being the "Pentium killer" AMD had hoped for, the K5 was positioned as a low-cost Pentium alternative, much like the Cyrix 6x86.

The K5 is, internally, a very advanced processor, the most advanced of the fifth-generation chips. Internally it is more comparable to the Pentium Pro. It is an x86 translation/emulation processor, decoding x86 instructions into RISC-like microinstructions and executing them on a 6-pipeline internal core. This allows the K5 to achieve higher performance than a Pentium of the same speed. In many ways, the K5 is sixth-generation, but its performance level is held back to fifth-generation levels due its low clock speeds.

The K5 is not a Pentium "clone" in the traditional sense of the word, because it is a totally different internal design. It is also not 100% Pentium compatible because it is not an exact clone. However, it is a very close approximation, and suffers from fewer compatibility problems than the 6x86.

Due to its advanced internal architecture, the K5 (like the 6x86) is labeled according to its "P rating", which attempts to label the processor according to what speed Pentium the chip's performance approximates. This is a good way for AMD to position its products in a marketplace that overvalues the clock speed number. The K5 is available in P rating speeds from 75 to 166 and goes in a Socket 5 or Socket 7 motherboard. It is produced on a 0.35 micron CMOS process, until recently state of the art. AMD manufactures its own chips.

Compared to the Pentium (classic), the K5 has the following architectural improvements that contribute to its greater performance to clock speed ratio (it also has all the advantages over the 486 that are mentioned in the section discussing the Pentium):

- **RISC-Based Internal Architecture:** The K5 is internally a highly-parallel RISC processor using an x86 decoding front-end.
- **Six Execution Units:** The K5 has five integer units and one floating point unit for processing micro-instructions, while the Pentium has only two integer units.
- **Out of Order Completion:** Instructions flowing down the two integer pipelines can complete out of order.
- **Superior Branch Prediction Unit:** The branch target buffer is four times the size of the Pentium's (although K5 claims a lower accuracy with it than Intel does with the Pentium. Seems odd to me.)
- **Larger Primary Cache:** The K5 has a 16 KB instruction cache instead of the Pentium's 8 KB. The data cache is unchanged at 8 KB.
- **Improved Cache Mapping:** The primary cache in the K5 is 4-way set associative instead of the Pentium's 2-way.
- **Register Renaming:** This feature improves parallel performance of the pipelines.
- **Speculative Execution:** The K5 uses speculative execution to reduce pipeline stall time in its RISC core.

While these improvements are impressive, they are the reason that the K5 has P ratings well above its internal clock speeds; if it didn't have the higher P ratings the K5 could not compete with the Pentium at all. Like the Cyrix 6x86, the K5's main's attraction is significantly lower price than the Pentium. Unlike the 6x86, the K5 has a reputation for being reasonably compatible and has far fewer problems than the Cyrix chip does. There aren't the heat and motherboard compatibility problems of the 6x86 associated with the K5. However, it (arguably) does not offer as high a level of performance as the 6x86, and there is no K5 chip near the power level of the 6x86-PR200+.

The K5 can have some compatibility problems due to the fact that it is not a Pentium. Some Pentium-specific software can cause problems with the K5, but these programs are relatively rare. The "CPUID" problem associated with the 6x86 under Windows 95 is not an issue for the K5.

While the K5 exceeds or beats a Pentium of its "P rating" at integer operations, it lags behind in floating-point performance. The K5 falls somewhere between the 6x86 and the Pentium in terms of its FPU, with performance roughly equal to a Pentium 25% slower than the K5's P rating. This really isn't much of an issue for most people since few applications make extensive use of the FPU (although Quake is one high-profile game that does.)

The 6x86 suffers from the problem where it is occasionally overclocked by accident because users set the motherboard to the clock speed of the P rating instead of the actual clock speed of the chip itself. The K5 avoids this by making two design decisions. First, all of the K5 chips run on the same memory bus speed as the equivalent Pentium of its P rating. So a K5-PR133 runs on a 66 MHz bus speed, like the Pentium 133, and unlike the 6x86-PR133+. Second, all K5 chips are designed so that if you configure them with the multiplier that an equivalent-rated Pentium would use, the chip will run with the correct internal multiplier. The combination of these two characteristics means that to set up a K5 you jumper your motherboard as if it were a Pentium and it will work. This avoids a lot of confusion.

2.4.6 Sixth Generation Processors

New and innovative architectural designs are employed in these CPUs, as the easy performance improvements such as widening buses and increasing clock speeds get harder to continue. Clock speeds do increase, but new ways of getting more done in each clock cycle are also the focus.

The line between fifth and sixth generation is much blurrier here than between previous generations, due to sixth generation chips running in fifth generation motherboards, and the mixing of technologies between different chip families. I consider the AMD K6 and the Cyrix 6x86MX to be sixth generation chips despite the fact that they run in fifth generation motherboards. This is based on assessing them technologically and in terms of their performance, along with when they were released to market.

Intel Pentium II ("Klamath")

Intel's first new chip since the Pentium Pro took almost a year and a half to produce, and speculation ran rampant as to what exactly it was going to be like. The rumours regarding the difficulties that Intel was having producing the Pentium Pro were running wild, fueled by a shortage of Pentium Pros and corresponding price run-up in the fall of 1996. Information began to leak out that Intel was "abandoning" the Pentium Pro design in favour of something that did not have an integrated secondary cache, and from there the guessing games began as to what Intel would come up with.

Codenamed for the Oregon river "Klamath", Intel officially released the Pentium II in May of 1997. Taking a page from the big movie houses, Intel leveraged the many millions it had spent on making "Pentium" a household word, and gave its new chip the name of a sequel. (I personally think "Return of the Pentium" would have been better, but hey, I'm no marketing guy). This moniker creates a bit of confusion, because it makes it seem as if the Pentium II is the next chip after the Pentium, when of course it isn't—the Pentium Pro has been around for a couple of years.

In fact, the Pentium II is very much an evolutionary step from the Pentium Pro, which only served to increase the speculation that one of Intel's primary goals in making the PII was to get away from the expensive integrated level 2 cache that was so hard to manufacture on the PPro. Architecturally, the Pentium II is not very different from the Pentium Pro, with a similar x86 emulation core and most of the same features. (You may want to reference the section on the Pentium Pro for a list of some of the improvements that that chip made over the Pentium). The Pentium II improves over the Pentium Pro architecturally in the following areas (many of these having been done to compensate in part for the reduced speed of the Pentium II level 2 cache compared to the Pentium Pro):

- **Doubled Level 1 Cache:** The level 1 cache is increased in size from 16 KB to 32 KB total (16 KB + 16 KB).
- **Segment Register Caches:** These are special caches that are used to help the Pentium II process 16-bit code somewhat more efficiently. The Pentium Pro was optimized for 32-bit processing, meaning that it did not deal with 16-bit code quite as well. Probably in response to the dominance of the Windows 95 operating system (which still uses a lot of 16-bit code) there was a desire to improve performance in this respect.
- **Deeper Write Buffers:** The write buffers were increased in size, producing a small performance improvement.

Architecturally at least, that's basically for improvements on the Pentium II. The real improvements were in clock speed, and the addition of the MMX instruction set extensions. The speed that the Pentium II was going to be released at was a matter of some speculation. At first Intel was expected to release only a 233 MHz version, and a rumour of the 266 version being moved up became a hot topic of conversation. Intel surprised the industry a bit in releasing not only these two but also a 300 MHz Pentium II in May 1997. However, the 300 MHz chip was originally more of a marketing routine than anything else; availability was non-existent and the pricing insane (when the announced initial price of $1,970 was first released, a hot argument ensued about whether or not this was obviously a typo—it wasn't!). In later months Intel improved its production process further, and the 300 MHz chip dropped rapidly in price and became much more "mainstream".

Without a doubt, the most talked about aspect of the new Pentium II wasn't even the chip itself—it was how it was packaged. The integrated Pentium Pro secondary cache, running at full processor speed, was one of its great performance-enhancing features—a "fan favorite" if you will. Intel replaced this on the Pentium II with a special small circuit board containing the processor and 512 KB of secondary cache, running at half the processor's speed. This assembly, termed a single-edge cartridge (SEC), was designed to fit into a 242-pin slot on the Pentium II motherboard. This change of design led to quite a bit of controversy in several regards!

The decision to "downgrade" the secondary cache from processor-speed to half-processor-speed led to some criticism of Intel, as this was viewed as a step backwards by many. Some people said "well, the Pentium II at 266 may be faster than the Pentium Pro at 200, but a Pentium Pro at 266 would have been even better". This is true to some extent, but the matters of clock speed and architecture are not independent. It is quite possible that the change in the cache is what allowed the higher speeds to be achieved, and the Pentium II *is* still faster than the Pentium Pro. The secondary cache on the Pentium II is still transactional, and maintains the independent backside bus of the Pentium Pro. At higher speeds, Intel plans to add error detection and correction (ECC) to the cache bus, to ensure data integrity.

But the most controversial debate was over Intel's apparent decision to lock out its rivals AMD and Cyrix from following in its footsteps, as AMD and Cyrix had done in creating the K5 and 6x86 chips to compete with the Pentium. The slot 1 interface designed for the Pentium II was patented by Intel, all but assuring that the other two companies would not use it for their new CPUs. This has led to a firestorm of criticism from PC users concerned about competition, upgradability and other issues. With AMD and Cyrix deciding to stick with Socket 7, the market has basically "split" here, and we will have to see what happens in the future. There is already talk of a "slot 2" and this is making many people decide to hold off before investing in the Pentium II, due to concerns about being stuck with another dead-end motherboard design. This is something you should seriously consider, keeping in mind what happened to the buyers of the initial Pentium 60/66 (*i.e.*, they were stuck with dead-end technology after less than a year).

The initial incarnations of the Pentium II run on motherboards using the Intel 440FX "Natoma" chipset, originally designed for the Pentium Pro. This shows you how similar the processors really are internally. In fact, some motherboard manufacturers have made boards that will run either the Pentium II or Pentium Pro, by making the board with slot 1 and providing an "SEC" for the Pentium Pro which is just an empty circuit board with a Pentium Pro socket on it.

If there is one weakness to the Pentium II, it is probably cost. Even discounting the laughable initial price on the Pentium II 300, even the 233 and 266 MHz versions are extremely pricey considering how close the AMD K6 and Cyrix 6x86MX are in terms of performance. The price of the

Pentium II continues to drop, but it's still a rather expensive chip. When you add in the more expensive slot 1 motherboard (the non-Intel chips run in abundant and inexpensive socket 7 Pentium motherboards), you have the Pentium II not very competitive from a price/performance standpoint, where the K6, 6x86MX, or even Intel's Pentium with MMX all come out ahead. Looking at the upgrade market, it's not much of a contest, since the other chips work in many of the zillions of socket 7 motherboards out there and the Pentium II does not. Still though, despite the best efforts from AMD and Cyrix, Intel remains on the top of the heap in the race for pure performance.

In early 1998 Intel released the 333 MHz Pentium II, codenamed "Deschutes". This is the first Pentium II using the new 0.25 micron fabrication process, which dramatically reduces the die size and power consumption of the chip. The 333 MHz offering does not really offer much of a performance improvement over the 300 MHz Pentium II. The 333 will likely be the last Intel chip to use the standard 66 MHz system bus speed; the next speeds for the Pentium II will be 350 MHz and 400 MHz, running on the new 100 MHz system bus.

AMD K6

After AMD's plans to conquer Intel with the K5 turned into a debacle, there was a great deal of skepticism when AMD again said that they had the solution to make them more than just "discount alternatives" to Intel. Still, the promise was enough to make most PC users salivate: a chip with the power of a Pentium Pro, with MMX instruction support and intended to fit into a standard Pentium socket 7! AMD delivered its much anticipated K6 chip in April of 1997, beating Intel's Pentium II to the market by a month and assuring themselves of at least a shot at the brass ring.

AMD showed their commitment to establishing themselves in the market when they purchased NexGen in 1996 and with it, the design for the NX686 processor. With AMD's K5 over a year late to market and basically a disappointment, AMD needed a solution *now* and got it with the NX686. NexGen had been intending to market this chip in its own socket, but AMD changed the design to fit the standard socket 7, added MMX support, and renamed it the K6. Despite the name implying a design evolving from the K5, it is in fact a totally different design that was created by the NexGen team and adapted after the AMD purchase.

The K6 is a very advanced processor internally. It uses x86 translation/emulation, decoding x86 instructions into what it calls RISC86 operations. It compares internally very similarly to the Pentium Pro and the Pentium II processors, which work in a similar way. The K6 has several key improvements and a few minor ones. The most important performance-enhancing architectural features are:

- **Large Level 1 Cache:** The K6 uses a full 64 KB of level 1 cache, a number that a few years ago was large for *level 2* cache. AMD pretty much had to do this in order to ensure that the K6 would be able to compete with the Pentium Pro and Pentium II, because the K6 still uses the conventional, 66 MHz, motherboard-based second-level cache.
- **Extra Decoders:** The K6 has four x86 instructions decoders, compared with only three for the Pentium Pro and Pentium II.
- **Large Branch Prediction Table:** The K6's branch history table contains 8,192 entries, while most other processors top out at 1/8th this size. Due to this large size the branch prediction accuracy is estimated at 95% by AMD.
- **Six Integer Execution Units:** The K6 has more internal execution units than any other x86 processor right now, allowing for more parallelism and efficiency.

One trend that AMD has continued quite well with the K6 is its reputation for compatibility. The K6 is not a direct clone of the Pentium family at all, because it is its own design. However, K6 chips work with what are at least to my knowledge very few if any compatibility issues. Since the K6 does run on socket 7 motherboards, it is an excellent choice for upgrading provided that the voltage issues are dealt with. A BIOS upgrade is often also required to enable the K6 to be recognized by the BIOS.

A nice thing about the K6 is that unlike the K5, there are no silly "P" ratings to be concerned with. The K6 competes close enough to the Pentium Pro and Pentium II that AMD decided clock-for-clock was close enough. So a K6-200 really does run at 200 MHz, unlike most of the K5 series. (Not to mention the Cyrix chips.)

The voltages on the K6 can be a bit confusing. The K6 is a split-rail voltage chip, much like the Pentium with MMX. It runs at 3.3 volts externally; internally, the 166 and 200 run at 2.9V and the 233 at 3.2V. The newest K6, the 266, drops the internal voltage all the way down to 2.2V, as a result of its using the state of the art 0.25 micron process. Like the Pentium with MMX 233, the K6-233 runs at a 3.5x clock multiplier, which is selected when you set the motherboard to a 1.5x clock multiplier (since 3.5x isn't a clock option for socket 7 motherboards). As an aside, when you consider that the core voltage of the K6-233 is only 0.1V away from the external voltage, the K6-233 might in theory work without split rail, since 0.1V is within the margin of tolerance for the input voltage, according to AMD's own specifications (although it is *right* on the edge). Combined with the 1.5x setting, this may mean that the K6-233 could function in a regular Pentium 100 motherboard? Might be interesting to try that some day if I am feeling particularly adventurous...)

In terms of performance, the K6 is quite similar to the Pentium Pro at the same clock speed, and tends to be slightly faster than the Pentium with MMX. Despite the fact that AMD would like to think the K6 competes directly with the Pentium II, in fact it really does not. If you consider relative performance, and the fact that the K6 and the Pentium with MMX both go into socket 7 motherboards with split rail voltage, it's obvious what people are trying to decide between. The Pentium II is still a step above the K6, but in terms of price/performance the K6 is pretty hard to beat. When looking at upgrading, the K6 has the Pentium II beat hands down, competing only with the Pentium with MMX (unless money really is no object for you, and you don't mind the idea of a motherboard upgrade).

2.5 SYSTEM ARCHITECTURE OF ADVANCED MICROPROCESSOR

From their humble beginning 25 years ago, microprocessors have proliferated into an astounding range of chips, powering devices ranging from telephones to supercomputers. Today, microprocessors for personal computers get widespread attention—and have enabled Intel to become the world's largest semiconductor maker. In addition, embedded microprocessors are at the heart of a diverse range of devices that have become staples of affluent consumers worldwide.

The past decade has seen the evolution of microprocessor packaging from a simple protective scheme to a complex combination of different elements that enable microprocessor performance while still providing the basic function of protection. Packaging today's microprocessor on the one hand entails tailoring the package to enable microprocessor performance, a complex task considering the rapid rate of microprocessor performance growth. This challenge is in terms of schedule and technical complexity. On the other hand, the package forms the interface between the microprocessor and the external world of the motherboard and the computing system. In this capacity, package design must allow for an easy interface and must meet a diverse set of form factor requirements.

The package provides a conduit for the microprocessor through a space transformation allowing small-scale features on the silicon to be electrically connected to the external environment. This is a challenging geometrical problem and requires that packaging interconnection densities must closely track the evolution of microprocessor interconnection densities. In connecting the die to the motherboard, the package must also ensure that the connections do not unduly inhibit the microprocessor performance by introducing unnecessary electrical impediments usually referred to as package "parasitics." As microprocessors have evolved, they have increased in speed, which in turn needs increasingly sophisticated power delivery schemes. Another consequence of microprocessor evolution has been increasing power dissipation. Package design must now provide a path for thermal dissipation, requiring a better understanding of the thermal characteristics of packaging materials and design. Package design also requires a good understanding of the structural characteristics of the package to ensure it is designed for reliability and robustness. Attention is increasingly focussed today on understanding the electrical, thermal and mechanical characteristics of packaging to optimize all these aspects.

The package is also the interface that connects the microprocessor to the motherboard. In this capacity it must have a compatible interface to allow for easy acceptance on the motherboard as well as the system design. The form factor of the package is a critical element for easy interface to the motherboard. The requirements are usually different in different market segments and often drive the need for form factors that are tailored to these different segments. For instance, the height of the package is critical to enable a microprocessor in a mobile market where a slim and low weight package is critical to success. On the other hand, the ability to dissipate high power, and hence features that enables this, are critical in a server or desktop market segment. Cost, compatibility and fit within the computer system are key parameters that must be designed for in making a microprocessor successful. This challenges us into concurrently developing multiple solutions and technologies geared towards specific market segments.

Aside from the challenges of package design, there is a need to develop efficient and cost-effective manufacturing processes that allow us to meet the schedule and volume demands of today's market places. These have presented us with interesting challenges in understanding the manufacturability, testability and reliability of packaging. Some of these issues are discussed in greater detail in this issue.

MAJC is an example of the design architecture of today's microprocessor. We can easily relate to the current trend by analyzing the architecture of MAJC.

MAJC (pronounced "magic") is an acronym for "Microprocessor Architecture for Java Computing." MAJC is a microprocessor architecture designed to meet the broadband demands of the 21st century. Addressing the challenge of high bandwidth and the need for state-of-the-art computational performance, MAJC architecture is characterized by:

- Scalability to take full advantage of advances in semiconductor technology.
- Broad scalability to systems with large numbers of processors.
- A new standard of performance for applications with DSP or New Media computational needs.
- Focus on bandwidth throughput.

2.5.1 Processor Needs into the 21st Century

- Several microprocessor trends were identified and accommodated in the design of the MAJC Architecture:
- Convergence of communication media and computers (audio, video, and data) require processors to compute information at wire speed.

- Advancements in semiconductor technology will provide rapidly-increasing resources on each microprocessor chip.
- As microprocessors are used in increasingly disparate applications from smart cards to supercomputers there is great value in the ability to create a wide span of implementations from a given processor architecture.
- Software, over time, will become independent of specific instruction sets; Just-In-Time (JIT) compilation techniques are expected to predominate for general-purpose processors and eliminate binary compatibility issues.
- Bandwidth between processors, memory, and I/O devices needs to be available to move information in real-time.
- The content processed by computers is becoming increasingly media-rich; DSP-like functions are required to process this media content.

2.5.2 Features of Today's Microprocessors

The following are the features of Today's Microprocessor:

(*a*) **Modular Architecture:** To support the creation of a wide range of implementations the architecture supports modular implementations. A basic implementation might comprise a single processor unit with four functional units. By replicating those design elements, an implementation can be built that includes a few or even hundreds of processors, each with four functional units, each of which can operate on many data items simultaneously with parallel-operation (SIMD) instructions. Conversely, a tiny application-specific implementation can be derived from the basic one by trimming the complement of functional units down to one or two and/or removing hardware support for any instructions not needed in its target application.

(*b*) **Software Portability:** The architecture was designed to efficiently execute code generated by installation-time or just-in-time (JIT) compilation techniques. It may be the first commercial architecture designed without a requirement for binary compatibility between generations. This allows implementations to evolve over time without accumulating the baggage required to support old binaries, as traditional architectures have always done. Instead, software portability across implementations is obtained through use of architecture-neutral means of software distribution.

(*c*) **Multiple Levels of Parallelism:** The architecture provides the ability to exploit parallelism at many levels—at the data word level through SIMD instructions, at the instruction level through multiple functional units per processor, at the thread-of-execution level through support for multithreaded software, and at the system level through its intrinsic support for "MPs-on-a-chip" (multiple processor units per implementation). A implementation with more than one functional unit per processor unit provides MSIMD: multiple single-instruction multiple-data parallelism.

(*d*) **Multiple Processor Units per Cluster:** Although a MAJC implementation can be a single processor unit, the architecture explicitly incorporates the concept of multiple processors per implementation. Given 21st century semiconductor density, each such array of processor units or "processor cluster" can be implemented on a single chip. As semiconductor technology advances, clusters with more processors per chip can be be implemented.

(*e*) **Multiple Functional Units per Processor Unit:** Every MAJC processor unit can issue multiple instructions simultaneously, one to each of its functional units. Most implementations are expected to provide two to four functional units per processor unit.

(*f*) **Multithreaded Software:** Execution of multithreaded software comes naturally given the architecture's ability to execute multiple threads simultaneously on multiple processor units. MAJC implementations with hardware support for vertical micro threading can efficiently execute multiple threads on each processor unit.

(*g*) **SIMD Instructions:** At the lowest level of parallelism, MAJC architecture provides SIMD (Single Instruction/ Multiple Data) or "vector" instructions. A SIMD instruction executing in a single functional unit could perform the same operation on multiple data items simultaneously.

(*h*) **Integral Support for Media-Rich Data:** The MAJC architecture is particularly well-suited for processing media-rich content because it directly supports common media data types and can process multiple simultaneous operations on that data. Processing power is multiplied on three levels: Single Instruction/Multiple Data (SIMD) DSP-like instructions in each functional unit, multiple functional units per processor unit, and multiple processor units per processor cluster.

(*i*) **Balanced Performance: Processor versus Memory and I/O:** A MAJC implementation is designed to utilize several techniques to balance processor speed with access to external memory and I/O devices:

- 100's of general-purpose registers per processor unit, which reduce the frequency of memory accesses
- Load-Group instructions, which increase bandwidth into the processor by simultaneously loading multiple registers from memory or an I/O device
- Store buffering, which increases bandwidth out of the processor by optimizing Store operations initiated by software

(*j*) **Data Type-Independent Registers:** The general-purpose register file in a MAJC implementation is data type-agnostic: any register can hold information of any data type and be accessed by any instruction. In particular, there is no distinction between integer and floating-point registers. This allows registers to be allocated as needed by each application, without restrictions imposed by hardware partitioning of the register set.

(*k*) **Instruction Grouping:** Grouping instructions across multiple functional units can be performed dynamically in hardware (as in a superscalar processor), statically by a compiler, or by some combination of the two. Rather than devoting valuable chip area to hardware grouping logic, MAJC relies primarily on software compilers to group instructions across functional units.

(*l*) **Data and Address Size:** A MAJC implementation may implement either 32- or 64-bit addressing and data operations, as dictated by the needs of its target applications.

(*m*) **Context Switch Optimization:** Process (task) context switch time can be reduced by using the architecture's "register dirty bits", which allow an operating system to minimize the number of registers saved and restored during a context switch.

(*n*) **Memory Byte Order:** The MAJC architecture's native byte order is "big-endian"; that is, multisystem values are stored in memory with the most significant byte at the lowest address and the least significant byte at the highest address. However, a MAJC implementation can manipulate data stored in any memory byte-order (notably "little-endian"). The BYTESHUFFLE instruction can reorder bytes efficiently in an arbitrary manner. Also, an implementation may define an Alternate Space Identifier (ASI) dedicated to performing automatic byte reordering whenever corresponding Load and Store from Alternate Address Space instructions are executed.

2.6 MICROPROCESSOR LOGIC

To understand how a microprocessor works, it is helpful to look inside and learn about the logic used to create one. In the process you can also learn about **assembly language** – the native language of a microprocessor – and many of the things that engineers can do to boost the speed of a processor.

A microprocessor executes a collection of machine instructions that tell the processor what to do. Based on the instructions, a microprocessor does three basic things:

- Using its ALU (Arithmetic/Logic Unit), a microprocessor can perform mathematical operations like addition, subtraction, multiplication and division. Modern microprocessors contain complete floating point processors that can perform extremely sophisticated operations on large floating point numbers.
- A microprocessor can move data from one memory location to another.
- A microprocessor can make decisions and jump to a new set of instructions based on those decisions.
- There may be very sophisticated things that a microprocessor does, but those are its three basic activities. The following diagram shows an extremely simple microprocessor capable of doing those three things:

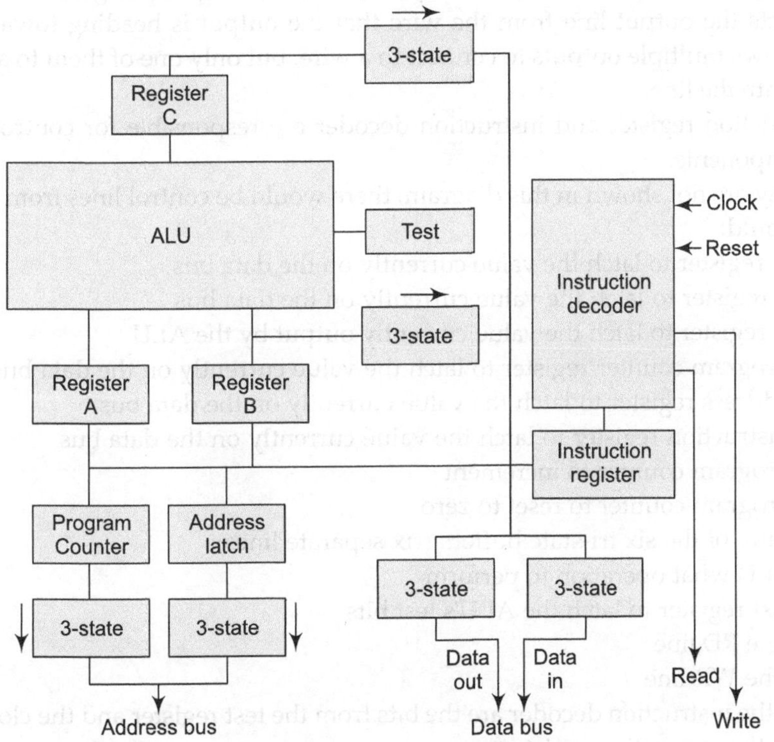

This is about as simple as a microprocessor gets. This microprocessor has:

- An **address bus** (that may be 8, 16 or 32 bits wide) that sends an address to memory
- A **data bus** (that may be 8, 16 or 32 bits wide) that can send data to memory or receive data from memory
- An **RD** (read) and **WR** (write) line to tell the memory whether it wants to set or get the addressed location

- A **clock line** that lets a clock pulse sequence the processor
- A **reset line** that resets the program counter to zero (or whatever) and restarts execution

Let's assume that both the address and data buses are 8 bits wide in this example.

Here are the components of this simple microprocessor:

- Registers A, B and C are simply latches made out of flip-flops.
- The address latch is just like registers A, B and C.
- The program counter is a latch with the extra ability to increment by 1 when told to do so, and also to reset to zero when told to do so.
- The ALU could be as simple as an 8-bit adder, or it might be able to add, subtract, multiply and divide 8-bit values. Let's assume the latter here.
- The test register is a special latch that can hold values from comparisons performed in the ALU. An ALU can normally compare two numbers and determine if they are equal, if one is greater than the other, etc. The test register can also normally hold a carry bit from the last stage of the adder. It stores these values in flip-flops and then the instruction decoder can use the values to make decisions.
- There are six boxes marked "3-State" in the diagram. These are **tri-state buffers**. A tri-state buffer can pass a 1, a 0 or it can essentially disconnect its output (imagine a switch that totally disconnects the output line from the wire that the output is heading toward). A tri-state buffer allows multiple outputs to connect to a wire, but only one of them to actually drive a 1 or a 0 onto the line.
- The instruction register and instruction decoder are responsible for controlling all of the other components.

Although they are not shown in this diagram, there would be control lines from the instruction decoder that would:

- Tell the A register to latch the value currently on the data bus
- Tell the B register to latch the value currently on the data bus
- Tell the C register to latch the value currently output by the ALU
- Tell the program counter register to latch the value currently on the data bus
- Tell the address register to latch the value currently on the data bus
- Tell the instruction register to latch the value currently on the data bus
- Tell the program counter to increment
- Tell the program counter to reset to zero
- Activate any of the six tri-state buffers (six separate lines)
- Tell the ALU what operation to perform
- Tell the test register to latch the ALU's test bits
- Activate the RD line
- Activate the WR line

Coming into the instruction decoder are the bits from the test register and the clock line, as well as the bits from the instruction register.

2.7 MICROPROCESSOR MEMORY

The previous section talked about the address and data buses, as well as the RD and WR lines. These buses and lines connect either to RAM or ROM — generally both. In our sample microprocessor, we have an address bus 8 bits wide and a data bus 8 bits wide. That means that the microprocessor can address (2^8) 256 bytes of memory, and it can read or write 8 bits of the memory at a time. Let's assume that this simple microprocessor has 128 bytes of ROM starting at address 0 and 128 bytes of RAM starting at address 128.

ROM Chip

ROM stands for read-only memory. A ROM chip is programmed with a permanent collection of pre-set bytes. The address bus tells the ROM chip which byte to get and place on the data bus. When the RD line changes state, the ROM chip presents the selected byte onto the data bus.

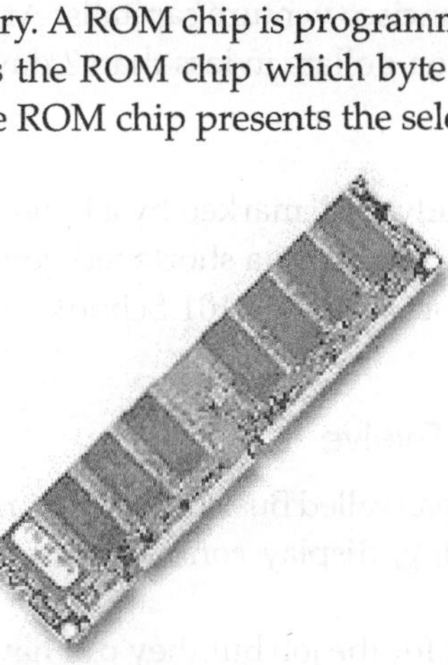

RAM stands for random-access memory. RAM contains bytes of information, and the microprocessor can read or write to those bytes depending on whether the RD or WR line is signaled. One problem with today's RAM chips is that they forget everything once the power goes off. That is why the computer needs ROM.

By the way, nearly all computers contain some amount of ROM (it is possible to create a simple computer that contains no RAM — many microcontrollers do this by placing a handful of RAM bytes on the processor chip itself — but generally impossible to create one that contains no ROM). On a PC, the ROM is called the BIOS (Basic Input/output System). When the microprocessor starts, it begins executing instructions it finds in the BIOS. The BIOS instructions do things like test the hardware in the machine, and then it goes to the hard disk to fetch the **boot sector**. This boot sector is another small program, and the BIOS stores it in RAM after reading it off the disk. The microprocessor then begins executing the boot sector's instructions from RAM. The boot sector program will tell the microprocessor to fetch something else from the hard disk into RAM, which the microprocessor then executes, and so on. This is how the microprocessor loads and executes the entire operating system.

2.8 EVOLUTION OF INTEL PROCESSOR

In November, 1971, a company called Intel publicly introduced the world's first single chip microprocessor, the Intel 4004 (U.S. Patent #3,821,715), invented by Intel engineers Federico Faggin, Ted Hoff, and Stan Mazor. After the invention of integrated circuits revolutionized computer design, the only place to go was down in size that is. The Intel 4004 chip took the integrated circuit down one step further by placing all the parts that made a computer think (*i.e.*, central processing unit, memory, input and output controls) on one small chip. Programming intelligence into inanimate objects had now become possible.

The History of Intel

In 1968, Bob Noyce and Gordon Moore were two unhappy engineers working for the Fairchild Semiconductor Company who decided to quit and create their own company at a time when many Fairchild employees were leaving to create start-ups. People like Noyce and Moore were nicknamed the "Fairchildren".

Bob Noyce typed himself a one page idea of what he wanted to do with his new company, and that was enough to convince San Francisco venture capitalist Art Rock to back Noyce's and Moore's new venture. Rock raised $2.5 million dollars in less than 2 days.

Intel Trademark

The name "Moore Noyce" was already trademarked by a hotel chain, so the two founders decided upon the name "Intel" for their new company, a shortened version of "Integrated Electronics".

Intel's first money making product was the 3101 Schottky bipolar 64-bit static random access memory (SRAM) chip.

One Chip Does the Work of Twelve

In late 1969, a potential client from Japan called Busicom, asked to have twelve custom chips designed. Separate chips for keyboard scanning, display control, printer control and other functions for a Busicom-manufactured calculator.

Intel did not have the manpower for the job but they did have the brainpower to come up with a solution. Intel engineer, Ted Hoff decided that Intel could build one chip to do the work of twelve. Intel and Busicom agreed and funded the new programmable, general-purpose logic chip.

Federico Faggin headed the design team along with Ted Hoff and Stan Mazor, who wrote the software for the new chip. Nine months later, a revolution was born. At 1/8th inch wide by 1/6th inch long and consisting of 2,300 MOS (metal oxide semiconductor) transistors, the baby chip had as much power as the ENIAC, which had filled 3,000 cubic feet with 18,000 vacuum tubes.

Cleverly, Intel decided to buy back the design and marketing rights to the 4004 from Busicom for $60,000. The next year Busicom went bankrupt, they never produced a product using the 4004. Intel followed a clever marketing plan to encourage the development of applications for the 4004 chip, leading to its widespread use within months.

The Intel 4004 Microprocessor

The 4004 was the world's first universal microprocessor. In the late 1960s, many scientists had discussed the possibility of a computer on a chip, but nearly everyone felt that integrated circuit technology was not yet ready to support such a chip. Intel's Ted Hoff felt differently; he was the first person to recognize that the new silicon-gated MOS technology might make a single-chip CPU (central processing unit) possible.

Hoff and the Intel team developed such architecture with just over 2,300 transistors in an area of only 3 by 4 millimeters. With its 4-bit CPU, command register, decoder, decoding control, control monitoring of machine commands and interim register, the 4004 was one heck of a little invention. Today's 64-bit microprocessors are still based on similar designs, and the microprocessor is still the most complex mass-produced product ever with more than 5.5 million transistors performing hundreds of millions of calculations each second—numbers that are sure to be outdated fast.

EXERCISE

1. Explain the various generations of microprocessor.
2. Explain the various types of computers.
3. What do you mean by client server architecture? Explain in detail.
4. What do you mean by scale of integration? Explain various types of integration.
5. What are the features of Pentium processor?
6. What are the current trends in microprocessor?
7. What do you mean by modular architecture and software portability?
8. Explain the microprocessor memory.
9. Explain the purpose of address bus and data bus.
10. What do you mean by register?

3

Introduction to Microprocessor

A microprocessor or processor is the heart of the computer and it performs all the computational tasks, calculations and data processing etc. inside the computer. Microprocessor is the brain of the computer. In the computers, the most popular type of the processor is the Intel Pentium chip and the Pentium IV is the latest chip by Intel Corporation. The microprocessors can be classified based on the following features:

(*a*) **Instruction Set:** It is the set of the instructions that the Microprocessor can execute.

(*b*) **Bandwidth:** The number of bits processed by the processor in a single instruction.

(*c*) **Clock Speed:** Clock speed is measured in the MHz and it determines that how many instructions a processor can process.

The speed of the microprocessor is measured in the MHz or GHz. The processor is also known as the CPU (Central Processing Unit). It contains the control unit and the arithmetic unit and both work together to process the commands. CPU is used in every computer whether it is a workstation, server or a laptop. CPU is a complete computational engine that is designed as a chip. It starts the work when you turn on your computer.

CPU is designed to perform the arithmetic and logical operations inside the computer. Common operations inside the computer include adding, subtracting, multiplying, comparing the values and fetching the different numbers to process them. The higher the CPU clocks' speed the more efficient will be the performance of the computer.

Computer's performance is also influenced by the system bus architecture, memory used, type of the processor and software program being running. Pentium 4 is the fastest type of the Intel's processor that contains 125,000,000 transistors and operates at the speed of 3.6 GHz.

3.1 MICROPROCESSOR CONCEPTS

The following are the basic concepts related to Microprocessor:

3.1.1 Instruction Execution

A microprocessor typically breaks the execution of an instruction into smaller operations. Each of these operations completes a portion of the entire instruction's execution. Typically the instruction is broken down into the following operations:

D-Decode

Looks at the opcode in the instruction to determine the type of instruction to be executed.

A-Address

Generates the address for any memory based operands.

T-Translate

Modifies the address based on table values setup in memory. These tables typically are used by an operating system to change where data is stored. This allows multiple programs that do stores to the same address to really write to different locations of physical memory.

B-Buffer

Fetches operands needed to execute an instruction. These typically come from the general registers or for memory based operands, the cache (a.k.a. buffer).

E-Execute

Performs the ALU operation. These are simple mathematical operations such as addition, and subtraction. For execution of many instructions, the ALU performs an operation that has no effect, such as adding zero. This is used for instructions that have a result the same as the source operand. For example, a store instruction takes the value of a register and stores it, unmodified, at a specified memory location. For this instruction the ALU needs to perform an operation that produces a result equal to the source operand.

W-Write

Writes the result.

Example 1. For this example, we will demonstrate the execution of the following instruction: 58204060

With the following assumptions:

- The CPU implements the S/370 architecture.
- Memory Locations 105060 — 105063 have the values 89, AB, BA, 98.
- General Purpose Register (GPR) 4 is equal to 00005000.
- The Operating System has "setup tables" so that the number 100000 (HEX) is added to all of the memory addresses specified by the program to determine the location in physical memory.

Solution:

In the S/370 architecture, the first 8 bits (sometimes the first 16) of an instruction specify the instruction type. In this case, the first 8 bits are equal to 58 Hex (0101 1000 Binary). During the design of the S/370, this value was assigned as the indication of a Load (L) instruction. The purpose of this instruction is to load a 4 byte value from memory into a specified GPR. A Load instruction is defined to use the RX instruction format. Therefore, the instruction is composed of the following fields:

Opcode – 58 R1 – 2 X 2 – 0 B2 – 4 D2 – 060

As specified above, opcode specifies the type of the instruction, in this case, it's a Load instruction. For this format (RX), the R1 field specifies the GPR that the value loaded is to be stored into. The last three fields (X2, B2, D2) are used to generate (produce) the address of where the first byte of the value to be loaded is at. The X2 and B2 fields specify registers that contain the index and base parts of the address unless the field is equal to 0. When the X2 or B2 fields are equal to 0 then it is implied that the corresponding index or base is equal to 0, instead of the contents of GPR0. The last field for address generation is D2. It specifies a displacement for the address generated from the index and base values to the start of the actual operand. The address is calculated by adding the index, base and displacement values together. For this example there is an implied index of 0 since the X2 field is 0, a base of 5000 (the contents of GPR4) and a displacement of 060. With this the address that will be generated for this instruction is 5060. Given the I-Unit pipeline described in this section, this instruction would be executed like this:

D-Stage

Determines the instruction to be executed. This is done by looking at the first 8 (sometimes 16) bits of the instruction. In this case, the first 8 bits are 58, which indicates it's a Load instruction in the S/370 architecture.

A-Stage

Performs Effective Address (EA) generation. This will determine the location of the value to be loaded from memory. For this example, the index and base fields are equal to 0 and 4, respectively so the index is an implied value of 0, while the base comes from GPR4. Based on the assumptions stated above, the contents of GPR4 are equal to 5000. The implied index and base values are added with the displacement value (60) from the instruction to produce an EA of 5060.

T-Stage

Translates the EA to its equivalent address in physical memory. This is performed based on the values of tables located in memory. The contents of these tables are normally controlled by the operating system. For this example, we assume that these tables have been setup so that the data in physical memory is always located in an address 100000 Hex greater than the EA. In this case, the EA of 5060 is translated to a physical address of 105060.

B-Stage

Obtains the source operand(s). For a Load instruction, there is a single source operand, which is the value to be loaded from memory. For this example, the physical address (determined during the T-Stage) was 105060, so during this stage, the CPU needs to obtain the 4 bytes starting at that address. Based on the assumptions above, the 4 byte value starting at that address is 89ABBA98.

E-Stage

Calculates the result of the instruction. For a Load instruction, the result is the same as the input. This is done by the ALU performing an operation that produces a result as equal to the source operand (*e.g.*, Add 0). During the B-Stage, an OP1 value of 89ABBA98 was obtained for this instruction. An operation that has no effect on the source will be used to produce the result of 89ABBA98.

W-Stage

Writes the result from the E-Stage (89ABBA98). For a Load instruction, the result is written to the GPR specified by the instruction's R1 field. In this case, it's GPR2. So the value of 89ABBA98 is written into GPR2.

Example 2 – Store : For this example, we will demonstrate the execution of the following instruction: 506040C0

With the following assumptions:

- The CPU implements the S/370 architecture.
- GPRs 4 and 6 are equal to 00005000 and 12345678 respectively.

- The Operating System has "setup tables" so that the number 100000 (HEX) is added to all of the memory addresses specified by the program to determine the location in physical memory.

As with the previous example, the first 8 bits of the instruction specify the instruction to be executed. In this case, the first 8 bits are equal to 50 Hex (0101 0000 Binary). For the most part, this instruction performs the opposite operation of the Load instruction shown in the previous example. Namely, instead of loading a value from memory into a GPR, it will store the contents of a GPR into memory. As with the previous example, this instruction also uses the RX format, so it is composed of the following fields:

Opcode – 50 R1 – 6 X2 – 0 B2 – 4 D2 – 0C0

This instruction would be executed using the sample pipeline as follows:

D-Stage

Examine the first 8 (sometimes 16) bits of the instruction to determine the instruction to be executed. In this case, they are equal to 50, which in the S/370 architecture indicate it's a Store instruction.

A-Stage

Calculates the EA. Performs Effective Address (EA) generation. For a Store instruction, this specifies the location of the value to be stored. For this example, the index and base fields are equal to 0 and 4, respectively so the index is an implied value of 0, while the base comes from GPR4. Based on the assumptions stated above, the contents of GPR4 are equal to 5000. The implied index and base values are added with the displacement value (60) from the instruction to produce an EA of 5060.

T-Stage

As with the previous example, the EA of 005060 is translated to it's equivalent physical address of 105060.

B-Stage

Obtains the source operand(s). For a Store instruction, there is a single source operand, which is the value of the GPR specified by the R1 field of the instruction. In this example, the R1 field is equal to 6, so the single source operand is the contents of GPR6. The single source operand will be equal to 12345678.

E-Stage

Calculates the result of the instruction. Just as with a Load instruction, a Store instruction has a result the same as the input. During this stage, the ALU performs an operation that has no effect on the source operand. For this example, the result will be equal to 12345678.

W-Stage

Writes the result from the E-Stage (12345678). For a Store instruction, the result is written into the physical location determined from the result of the T-Stage. In this case, the 4 byte value of 12345678 will be stored at 105060.

3.2 MICROPROCESSOR ARCHITECTURE AND ITS COMPONENTS

A **microprocessor** incorporates most or all of the functions of a computer's central processing unit (CPU) on a single integrated circuit (IC, or microchip). The first microprocessors emerged in the early 1970s and were used for electronic calculators, using binary-coded decimal (BCD) arithmetic on 4-bit words. Other embedded uses of 4- and 8-bit microprocessors, such as terminals, printers, various kinds of automation etc, followed rather quickly. Affordable 8-bit microprocessors with 16-bit addressing also led to the first general purpose microcomputers in the mid-1970s.

Computer processors were for a long period constructed out of small and medium-scale ICs containing the equivalent of a few to a few hundred transistors. The integration of the whole CPU onto a single chip therefore greatly reduced the cost of processing capacity. From their humble beginnings, continued increases in microprocessor capacity have rendered other forms of computers almost completely obsolete , with one or more microprocessor as processing element in everything from the smallest embedded systems and handheld devices to the largest mainframes and supercomputers.

Since the early 1970s, the increase in capacity of microprocessors has been known to generally follow Moore's Law, which suggests that the complexity of an integrated circuit, with respect to minimum component cost, doubles every two years.

In the late 1990s, and in the high-performance microprocessor segment, heat generation (TDP), due to switching losses, static current leakage, and other factors, emerged as a leading developmental constraint.

A microprocessor is a computer processor on a microchip. It's sometimes called a logic chip. It is a complete computation engine that is fabricated on a single chip. A microprocessor is designed to perform arithmetic and logic operations that make use of small number-holding areas called registers. This sits on the motherboard of each Personal computer. The motherboard stands for the card in the PC that controls almost all the other peripherals.

The following figure represents the general architecture of 8085 Microprocessor:

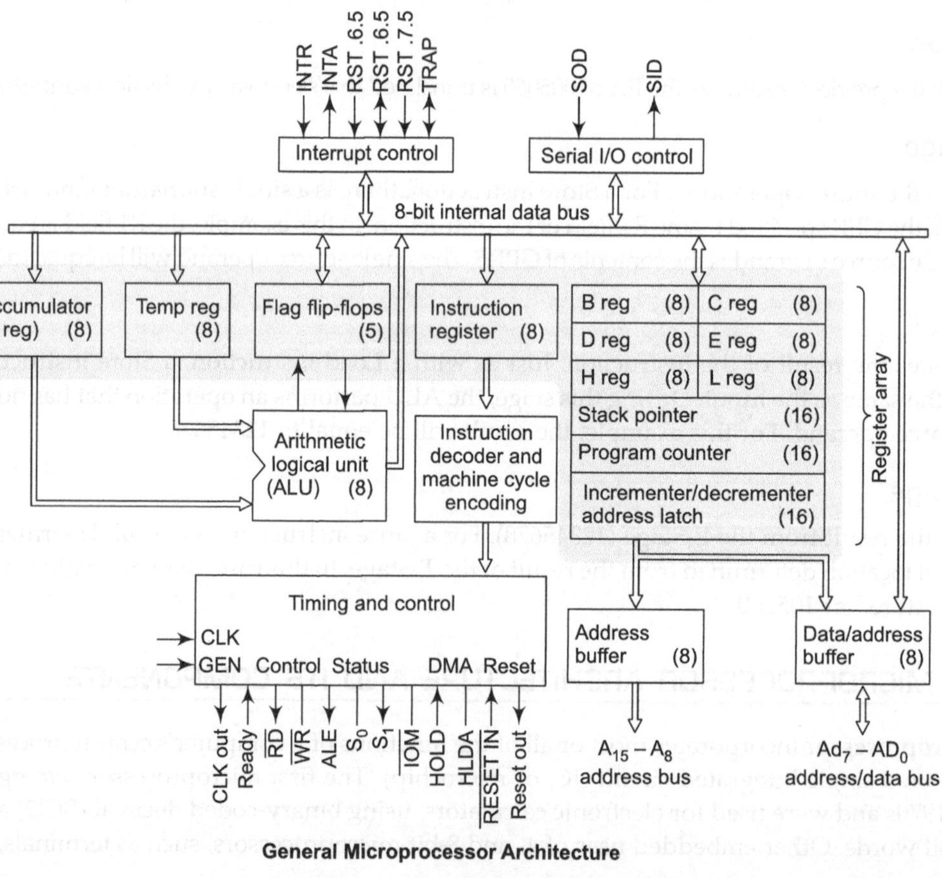

General Microprocessor Architecture

3.2.1 Microprocessor Components

Some of the common components of a microprocessor are:

- Control Unit
- I/O Units
- Arithmetic Logic Unit (ALU)
- Registers
- Cache

We will give a brief introduction to these components below.

(a) Control Unit

The control unit is the circuitry that controls the flow of information through the processor, and coordinates the activities of the other units within it. In a way, it is the "brain within the brain", as it controls what happens inside the processor, which in turn controls the rest of the PC.

The functions performed by the control unit vary greatly by the internal architecture of the CPU, since the control unit really implements this architecture. On a regular processor that executes x86 instructions natively, the control unit performs the tasks of fetching, decoding, managing execution and then storing results. On a processor with a RISC core the control unit has significantly more work to do. It manages the translation of x86 instructions to RISC micro-instructions, manages scheduling the micro-instructions between the various execution units, and juggles the output from these units to make sure they end up where they are supposed to go. On one of these processors the control unit may be broken into other units (such as a *scheduling unit* to handle scheduling and a *retirement unit* to deal with results coming from the pipeline) due to the complexity of the job it must perform.

(b) Input/output Units

In computing, **input/output**, or **I/O**, refers to the communication between an information processing system (such as a computer), and the outside world possibly a human, or another information processing system. Inputs are the signals or data received by the system, and outputs are the signals or data sent from it. The term can also be used as part of an action; to "perform I/O" is to perform an input or output operation. I/O devices are used by a person (or other system) to communicate with a computer. For instance, a keyboard or a mouse may be an input device for a computer, while monitors and printers are considered output devices for a computer. Devices for communication between computers, such as modems and network cards, typically serve for both input and output.

Note that the designation of a device as either input or output depends on the perspective. Mice and keyboards take as input physical movement that the human user outputs and converts it into signals that a computer can understand. The output from these devices is input for the computer. Similarly, printers and monitors take as input signals that a computer outputs. They then convert these signals into representations that human users can see or read. For a human user the process of reading or seeing these representations is receiving input. These interactions between computers and humans is studied in a field called human–computer interaction.

In computer architecture, the combination of the CPU and main memory (*i.e.*, memory that the CPU can read and write to directly, with individual instructions) is considered the brain of a computer, and from that point of view any transfer of information from or to that combination, for example to or from a disk drive, is considered I/O. The CPU and its supporting circuitry provide memory-mapped I/O that is used in low-level computer programming in the implementation of device drivers.

An I/O algorithm is one designed to exploit locality and perform efficiently when data reside on secondary storage, such as a disk drive.

Interface

I/O Interface is required whenever the I/O device is driven by the processor. The interface must have necessary logic to interpret the device address generated by the processor. Handshaking should be implemented by the interface using appropriate commands like (BUSY, READY, WAIT), and the processor can communicate with I/O device through the interface. If different data formats are being exchanged, the interface must be able to convert serial data to parallel form and vice-versa. There must be provision for generating interrupts and the corresponding type numbers for further processing by the processor if required

A computer that uses memory-mapped I/O accesses hardware by reading and writing to specific memory locations, using the same assembler language instructions that computer would normally use to access memory.

(c) Arithmetic Logic Unit (ALU)

The arithmetic-logic unit (ALU) performs all arithmetic operations (addition, subtraction, multiplication, and division) and logic operations. Logic operations test various conditions encountered during processing and allow for different actions to be taken based on the results. The data required to perform the arithmetic and logical functions are inputs from the designated CPU registers and operands. The ALU relies on basic items to perform its operations. These include number systems, data routing circuits (adders/subtracters), timing, instructions, operands, and registers. The following Figure shows a representative block diagram of an ALU of a microcomputer.

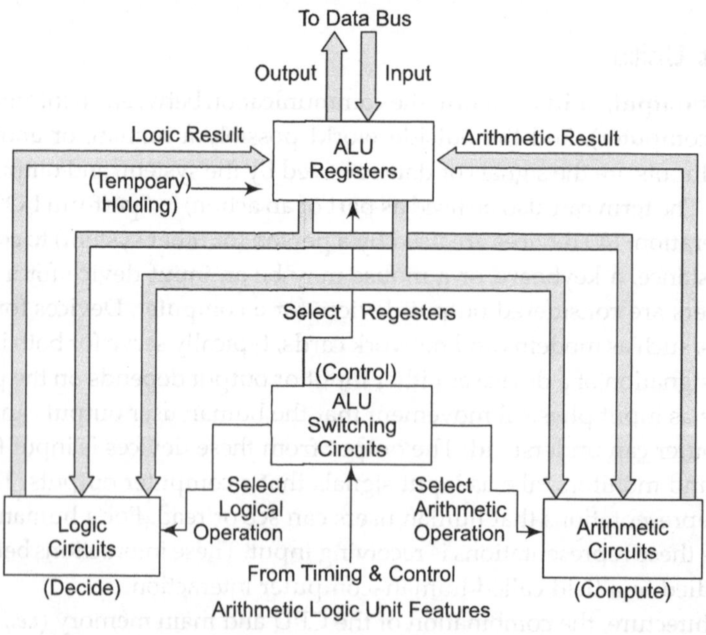

Arithmetic Logic Unit Features

(d) Register

Register is a, special, high-speed storage area within the CPU. All data must be represented in a register before it can be processed. For example, if two numbers are to be multiplied, both numbers must be in registers, and the result is also placed in a register. (The register can contain the address of a memory location where data is stored rather than the actual data itself.)

The number of registers that a CPU has and the size of each (number of bits) help determine the power and speed of a CPU. For example a 32-bit CPU is one in which each register is 32 bits wide. Therefore, each CPU instruction can manipulate 32 bits of data.

Usually, the movement of data in and out of registers is completely transparent to users, and even to programmers. Only assembly language programs can manipulate registers. In high-level languages, the compiler is responsible for translating high-level operations into low-level operations that access registers.

(e) Cache

In computer science, a **cache** is a component that improves performance by transparently storing data such that future requests for that data can be served faster. The data that is stored within a cache might be values that have been computed earlier or duplicates of original values that are stored elsewhere. If requested data is contained in the cache (cache hit), this request can be served by simply reading the cache, which is comparably faster. Otherwise (cache misses), the data has to be recomputed or fetched from its original storage location, which is comparably slower. Hence, the more requests can be served from the cache the better the overall system performance is.

As opposed to a buffer, which is managed explicitly by a client, a cache stores data transparently: This means that a client who is requesting data from a system is not aware that the cache exists! Therefore the name cache (from French "cacher", to conceal).

To be cost efficient and to enable an efficient lookup of data, caches are comparably small. Nevertheless, caches have proven extremely effective in many areas of computing because access patterns in typical computer applications have locality of reference. References exhibit temporal locality if data is requested again that has been recently requested already. References exhibit spatial locality if data is requested that is physically stored close to data that has been requested already.

3.3 COMPUTER BUS

A **bus**, in computing, is a set of physical connections (cables, printed circuits, etc.) which can be shared by multiple hardware components in order to communicate with one another.

The purpose of buses is to reduce the number of "pathways" needed for communication between the components, by carrying out all communications over a single data channel. This is why the metaphor of a "data highway" is sometimes used.

Graphically it can be represented as follows:

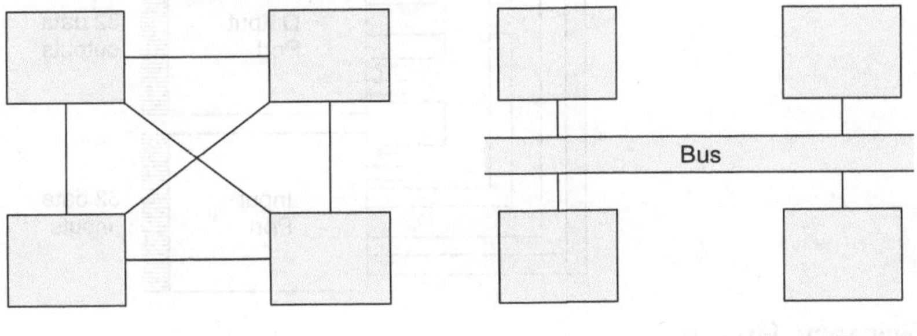

If only two hardware components communicate over the line, it is called a **hardware port** (such as a serial port or parallel port).

3.3.1 Characteristics of a Bus

A bus is characterized by the amount of information that can be transmitted at once. This amount, expressed in bits, corresponds to the number of physical lines over which data is sent simultaneously. A 32-wire ribbon cable can transmit 32 bits in parallel. The term "**width**" is used to refer to the number of bits that a bus can transmit at once.

Additionally, the bus speed is also defined by its **frequency** (expressed in Hertz), the number of data packets sent or received per second. Each time that data is sent or received is called a **cycle**.

This way, it is possible to find the maximum **transfer speed** of the bus, the amount of data which it can transport per unit of time, by multiplying its width by its frequency. A bus with a width of 16 bits and a frequency of 133 MHz, therefore, has a transfer speed equal to:

$16 * 133.10^6 = 2128*10^6$ bit/s,

Or $2128*10^6/8 = 266*10^6$ bytes/s

Or $266*10^6/1000 = 266*10^3$ KB/s

Or $259.7*10^3/1000 = 266$ MB/s

3.3.2 Bus Subassembly

In reality, each bus is generally constituted of 50 to 100 distinct physical lines, divided into three subassemblies:

- The **address bus** (sometimes called the *memory bus*) transports memory addresses which the processor wants to access in order to read or write data. It is a unidirectional bus.
- The **data bus** transfers instructions coming from or going to the processor. It is a bidirectional bus.
- The **control bus** (or *command bus*) transports orders and synchronization signals coming from the control unit and travelling to all other hardware components. It is a bidirectional bus, as it also transmits response signals from the hardware.

Graphically it can be represented as follows:

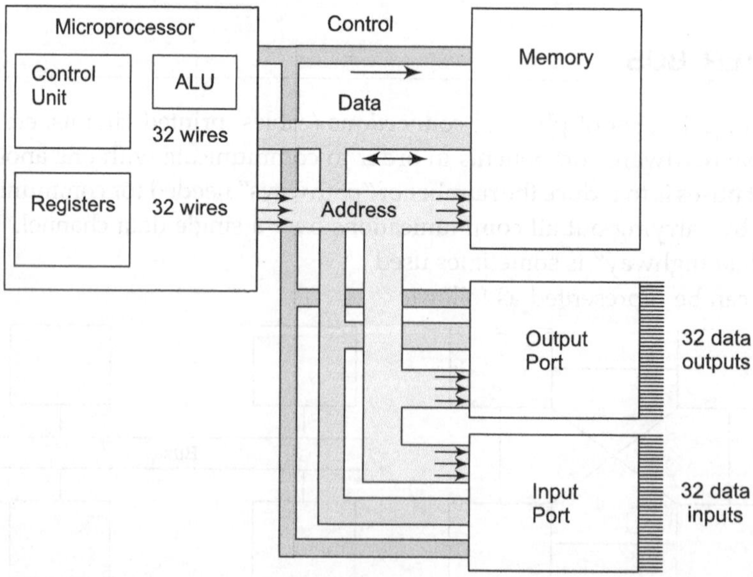

3.3.3 The Primary Buses

There are generally two buses within a computer:

- The **internal bus** (sometimes called the *front-side bus*, or *FSB* for short). The internal bus allows the processor to communicate with the system's central memory (the RAM).
- **The expansion bus** (sometimes called the *input/output bus*) allows various motherboard components (USB, serial, and parallel ports, cards inserted in PCI connectors, hard drives, CD-ROM and CD-RW drives, etc.) to communicate with one another. However, it is mainly used to add new devices using what are called **expansion slots** connected to the input/output bus.

3.3.4 The Chipset

A **chipset** is the component which routes data between the computer's buses, so that all the components which make up the computer can communicate with each other. The **chipset** originally was made up of a large number of electronic chips, hence the name. It generally has two components:

- The **NorthBridge** (also called the *memory controller*) is in charge of controlling transfers between the processor and the RAM, which is the way it is located physically near the processor. It is sometimes called the **GMCH**, for *Graphic and Memory Controller Hub*.
- The **SouthBridge** (also called the *input/output controller* or *expansion controller*) handles communications between peripheral devices. It is also called the **ICH** (*I/O Controller Hub*). The term Bridge is generally used to designate a component which connects two buses.

Graphically the process is represented as follows:

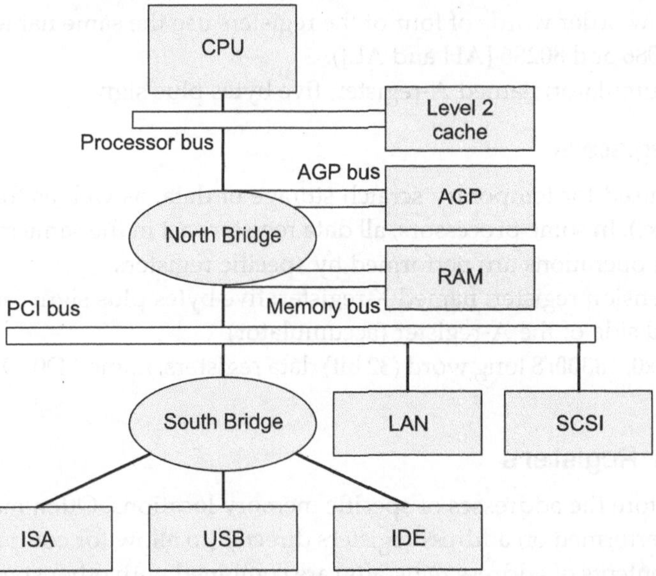

It is interesting to note that, in order to communicate, two buses must have the same width. The explains why RAM modules sometimes have to be installed in pairs (for example, early Pentium chips, whose processor buses were 64-bit, required two memory modules each 32 bits wide).

3.4 REGISTER SET

Registers are fast memory, almost always connected to circuitry that allows various arithmetic, logical, control, and other manipulations, as well as possibly setting internal flags.

Most early computers had only one data register that could be used for arithmetic and logic instructions. Often there would be additional special purpose registers set aside either for temporary fast internal storage or assigned to logic circuits to implement certain instructions. Some early computers had one or two address registers that pointed to a memory location for memory accesses (a pair of address registers typically would act as source and destination pointers for memory operations). Computers soon had multiple data registers, address registers, and sometimes other special purpose registers. Some computers have general purpose registers that can be used for both data and address operations. Every digital computer using von Neumann architecture has a register (called the program counter) that points to the next executable instruction. Many computers have additional control registers for implementing various control capabilities. Often some or all of the internal flags are combined into a flag or status register.

3.4.1 Accumulators

Accumulators are registers that can be used for arithmetic, logical, shift, rotate, or other similar operations. The first computers typically only had one accumulator. Many times there were related special purpose registers that contained the source data for an accumulator. Accumulators were replaced with data registers and general purpose registers. Accumulators reappeared in the first microprocessors.

- **Intel 8086/80286:** one word (16 bit) accumulator; named AX (high order byte of the AX register is named AH and low order byte of the AX register is named AL).
- **Intel 80386:** one double word (32 bit) accumulator; named EAX (low order word uses the same names as the accumulator on the Intel 8086 and 80286 [AX] and low order and high order bytes of the low order words of four of the registers use the same names as the accumulator on the Intel 8086 and 80286 [AH and AL]).
- **MIX:** one accumulator; named A-register; five bytes plus sign.

3.4.2 Data Registers

Data registers are used for temporary scratch storage of data, as well as for data manipulations (arithmetic, logic, etc.). In some processors, all data registers act in the same manner, while in other processors different operations are performed by specific registers.

- **MIX:** one extension register; named X-register; five bytes plus sign; can be concatenated on the right hand side of the A-register (accumulator)
- **Motorola 680x0, 68300:** 8 long word (32 bit) data registers; named D0, D1, D2, D3, D4, D5, D6, and D7

3.4.3 Address Registers

Address registers store the addresses of specific memory locations. Often many integer and logic operations can be performed on address registers directly (to allow for computation of addresses).

Sometimes the contents of address register(s) are combined with other special purpose registers to compute the actual physical address. This allows for the hardware implementation of dynamic memory pages, virtual memory, and protected memory.

The number of bits of an address register (possibly combined with information from other registers) limits the maximum amount of addressable memory. A 16-bit address register can address 64K of physical memory. A 24-bit address register can address address 16 MB of physical memory. A 32-bit address register can address 4 GB of physical memory. A 64-bit address register can address 1.8446744×10^{19} of physical memory. Addresses are always unsigned binary numbers.

- **MIX:** one jump registers; named J-register; two bytes and sign is always positive
- **Motorola 680x0, 68300:** 8 long word (32 bit) address registers; named A0, A1, A2, A3, A4, A5, A6, and A7 (also called the stack pointer)

3.4.4 General Purpose Registers

General purpose registers can be used as either data or address registers.

- **DEC VAX:** 16 word (32 bit) general purpose registers; named R0 through R15.
- **IBM 360/370:** 16 full word (32 bit) general purpose registers; named 0, 1, 2, 3, 4, 5, 6, 7, 8, 9, A (or 10), B (or 11), C (or 12), D (or 13), E (or 14), and F (or 15).
- **Intel 8086/80286:** 8 word (16 bit) general purpose registers; named AX, BX, CX, DX, BP, SP, SI, and DI (high order bytes of the AX, BX, CX, and DX registers have the names AH, BH, CH, and DH and low order bytes of the AX, BX, CX, and DX registers have the names AL, BL, CL, and DL).
- **Intel 80386:** 8 double word (32 bit) general purpose registers; named EAX, EBX, ECX, EDX, EBP, ESP, ESI, and EDI (low order words use the same names as the general purpose registers on the Intel 8086 and 80286 and low order and high order bytes of the low order words of four of the registers use the same names as the general purpose registers on the Intel 8086 and 80286).
- **Motorola 88100:** 32 word (32 bit) general purpose registers; named r0 through r31.

3.4.5 Constant Registers

Constant registers are special read-only registers that store a constant. Attempts to write to a constant register are illegal or ignored. In some RISC processors, constant registers are used to store commonly used values (such as zero, one, or negative one) — for example, a constant register containing zero can be used in register to register data moves, providing the equivalent of a clear instruction without adding one to the instruction set. Constant registers are also often used in floating point units to provide such value as pi or e with additional hidden bits for greater accuracy in computations.

- **Motorola 88100:** r0 (general purpose register 0) contains the constant 32 bit integer zero.

3.4.6 Floating Point Registers

Floating point registers are special registers set aside for floating point math.

3.4.7 Index Registers

Index registers are used to provide more flexibility in addressing modes, allowing the programmer to create a memory address by combining the contents of an address register with the contents of an index register (with displacements, increments, decrements, and other options). In some processors, there are specific index registers (or just one index register) that can only be used only for that purpose. In some processors, any data register, address register, or general register (or some combination of the three) can be used as an index register.

- **IBM 360/370:** any of the 16 general purpose registers may be used as an index register.
- **Intel 80x86:** 7 of the 8 general purpose registers may be used as an index register (the ESP is the exception).
- **MIX:** five index registers; named I-registers I1, I2, I3, I4, and I5; five bytes plus sign.
- **Motorola 680x0, 68300:** any of the 8 data registers or the 8 address registers may be used as an index register.

3.4.8 Base Registers

Base registers or **segment registers** are used to segment memory. Effective addresses are computed by adding the contents of the base or segment register to the rest of the effective address computation.

In some processors, any register can serve as a base register. In some processors, there are specific bases or segment registers (one or more) that can only be used for that purpose. In some processors with multiple base or segment registers, each base or segment register is used for different kinds of memory accesses (such as a segment register for data accesses and a different segment register for program accesses).

- **IBM 360/370:** any of the 16 general purpose registers may be used as a base register.
- **Intel 80x86:** 6 dedicated segment registers: CS (code segment), SS (stack segment), DS (data segment), ES (extra segment, a second data segment register), FS (third data segment register), and GS (fourth data segment register).
- **Motorola 680x0, 68300:** any of the 8 address registers may be used as a base register.

3.4.9 Control Registers

Control registers control some aspect of processor operation. The most universal control register is the program counter.

3.4.10 Program Counter

Almost every digital computer ever made uses a **program counter**. The program counter points to the memory location that stores the next executable instruction. Branching is implemented by making changes to the program counter. Some processor designs allow software to directly change the program counter, but usually software only indirectly changes the program counter (for example, a JUMP instruction will insert the operand into the program counter). An assembler has a **location counter**, which is an internal pointer to the address (first byte) of the next location in storage (for instructions, data areas, constants, etc.) while the source code is being converted into object code.

The VAX uses the 16th of 16 general purpose registers as the program counter (PC). Almost the entire instruction set can directly manipulate the program counter, allowing a very rich set of possible kinds of branching.

The program counter in System/360 and 370 machines is contained in bits 40-63 of the program status word (PSW), which is directly accessible by some instructions.

- **IBM 360/370:** program counter is bits 40-63 of the program status word (PSW)
- **Intel 8086/80286:** 16-bit instruction pointer (IP)
- **Intel 80386:** 32-bit instruction pointer (EIP)
- **Motorola 680x0, 68300:** 32-bit program counter (PC).

3.4.11 Stack Pointer

Stack pointers are used to implement a processor stack in memory. In many processors, address registers can be used as generic data stack pointers and queue pointers. A specific stack pointer or address register may be hardwired for certain instructions. The most common use is to store return addresses, processor state information, and temporary variables for subroutines.

- **IBM 360/370:** any of the 16 general purpose registers may be used as a stack pointer.
- **Intel 8086/80286:** dedicated stack pointer (SP) combined with stack segment pointer (SS) to create address of stack.
- **Intel 80386:** dedicated stack pointer (ESP) combined with stack segment pointer (SS) and the stack-frame base pointer (EBP) to create address of stack.

- **Motorola 680x0, 68300:** dedicated user stack pointer (USP, A7) and system stack pointer (SSP, A7) for implicit stack pointer operations, as well as allowing any of the 8 address registers to be as explicit stack pointers.

3.5 ADDRESSING MODES

Addressing modes are an aspect of the instruction set architecture in most central processing unit (CPU) designs. The various addressing modes that are defined in a given instruction set architecture define how machine language instructions in that architecture identify the operand (or operands) of each instruction. An addressing mode specifies how to calculate the effective memory address of an operand by using information held in registers and/or constants contained within a machine instruction or elsewhere.

In computer programming, addressing modes are primarily of interest to compiler writers and to those who write code directly in assembly language.

3.5.1 Types of Addressing Modes

Each instruction of a computer specifies an operation on certain data. There are various ways of specifying address of the data to be operated on. These different ways of specifying data are called the addressing modes. The most common addressing modes are:

- Immediate addressing mode
- Direct addressing mode
- Indirect addressing mode
- Register addressing mode
- Register indirect addressing mode
- Displacement addressing mode
- Stack addressing mode

To specify the addressing mode of an instruction several methods are used. Most often used are :

(*a*) Different operands will use different addressing modes.

(*b*) One or more bits in the instruction format can be used as mode field. The value of the mode field determines which addressing mode is to be used.

The effective address will be either main memory address of a register.

(*i*) **Immediate Addressing:** This is the simplest form of addressing. Here, the operand is given in the instruction itself. This mode is used to define a constant or set initial values of variables. The advantage of this mode is that no memory reference other than instruction fetch is required to obtain operand. The disadvantage is that the size of the number is limited to the size of the address field, which most instruction sets is small compared to word length.

The operand is specified in the instruction. Can be used e.g. to initialize register to constant value (immediate operand).

(*ii*) **Direct Addressing:** In direct addressing mode, effective address of the operand is given in the address field of the instruction. It requires one memory reference to read the operand from the given location and provides only a limited address space. Length of the address field is usually less than the word length.

In this addressing mode the effective address is equal to the address part of the instruction. The operand resides in this address.

Ex : Move P, Ro, Add Q, Ro P and Q are the address of operand.

(*iii*) **Indirect Addressing:** Indirect addressing mode, the address field of the instruction refers to the address of a word in memory, which in turn contains the full length address of the operand. The advantage of this mode is that for the word length of N, an address space of 2N can be addressed. The disadvantage is that instruction execution requires two memory references to fetch the operand. Multilevel or cascaded indirect addressing can also be used. In the indirect addressing the address field gives the address where the effective address is stored in memory.

(*iv*) **Register Addressing:** Register addressing mode is similar to direct addressing. The only difference is that the address field of the instruction refers to a register rather than a memory location 3 or 4 bits are used as address field to reference 8 to 16 generate purpose registers. The advantages of register addressing are Small address field is needed in the instruction. In the Register Addressing mode Operands are in registers that reside within the CPU. The particular register is selected with the register field of the instruction.

(*v*) **Register Indirect Addressing:** This mode is similar to indirect addressing. The address field of the instruction refers to a register. The register contains the effective address of the operand. This mode uses one memory reference to obtain the operand. The address space is limited to the width of the registers available to store the effective address.
In the Register Indirect Addressing Mode the content of a register specifies the address of the operand in memory.

(*vi*) **Displacement Addressing:** In displacement addressing mode there are 3 types of addressing mode. They are:

1. Relative addressing
2. Base register addressing
3. Indexing addressing.

This is a combination of direct addressing and register indirect addressing. The value contained in one address field A is used directly and the other address refers to a register whose contents are added to A to produce the effective address.

(*a*) **Relative Addressing:** In the relative addressing the content of the program counter (PC) is added to the address part of the instruction in order to obtain the effective address *i.e.*, the effective address is relative to the address of the next instruction. This address mode can be used in branchtype instructions when the branch address is in the area surrounding the instruction word.

(*b*) **Base Register Addressing:** In the base register addressing the content of a base register is added to the address part of the instruction to obtain the effective address. The address part of the instruction gives the displacement relative to the base address.

(*c*) **Indexing Addressing:** In the indexing addressing the content of an index register (special CPU register) is added to the address part of the instruction to obtain the effective address. The index register can be incremented for accessing consecutive operands.

(*vii*) **Stack Addressing:** Stack is a linear array of locations referred to as last-in first out queue. The stack is a reserved block of location, appended or deleted only at the top of the stack. Stack pointer is a register which stores the address of top of stack location. This mode of addressing is also known as implicit addressing.

Example 1. When running a program, the CPU first loads an instruction.
It then processes data using this instruction.
Addressing modes tell the CPU where to find the data.

The 6502 has several addressing modes.
A sample is given below.
IMMEDIATE addressing.
In the program, the data follows the instruction.
ADD 99
Add 99 to the contents of the accumulator.
ABSOLUTE addressing.
In the program, the 16 bit address of the data follows the instruction.
STA 1243
Store the contents of the accumulator at address 1243.
ZERO PAGE addressing.
In the program, the instruction is followed by an 8 bit address.
This means that only addresses 0-255 (in page zero) can be used.
CMP 230
Compare the contents of the accumulator with the contents of address 0230.
Other addressing modes are:
IMPLIED
RELATIVE
ACCUMULATOR
INDEXED
INDIRECT

Example 2. For each possible mode, calculate the effective address and the operand that must be loaded into AC. Use the instruction defined in the following Figure. Following Table shows the tabular list of the numerical example.

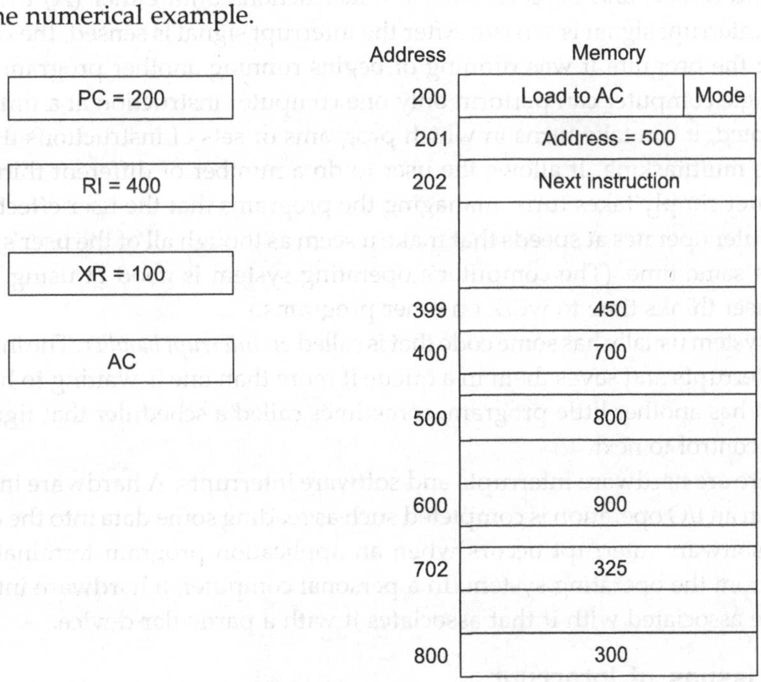

Numerical example for addressing modes

Tabular List of Numerical Example

Addressing Mode	Effective Address	Content of AC
Direct address	500	800
Immediate operand	201	500
Indirect address	800	300
Relative address	702	325
Indexed address	600	900
Register	—	400
Register indirect	400	700
Autoincrement	400	700
Autodecrement	399	450

decrements R1 prior the execution → Autodecrement

Tabular list of the numerical example

3.6 INTERRUPTS

An interrupt is a signal from a device attached to a computer or from a program within the computer that causes the main program that operates the computer (the operating system) to stop and figure out what to do next. Almost all personal (or larger) computers today are *interrupt-driven* — that is, they start down the list of computer instruction s in one program (perhaps an application such as a word processor) and keep running the instructions until either (A) they can't go any further or (B) an interrupt signal is sensed. After the interrupt signal is sensed, the computer either resumes running the program it was running or begins running another program.

Basically, a single computer can perform only one computer instruction at a time. But, because it can be interrupted, it can take turns in which programs or sets of instructions that it performs. This is known as multitasking. It allows the user to do a number of different things at the same time. The computer simply takes turns managing the programs that the user effectively starts. Of course, the computer operates at speeds that make it seem as though all of the user's tasks are being performed at the same time. (The computer's operating system is good at using little pauses in operations and user thinks time to work on other programs.)

An operating system usually has some code that is called an *interrupt handler*. The interrupt handler prioritizes the interrupts and saves them in a queue if more than one is waiting to be handled. The operating system has another little program, sometimes called a scheduler that figures out which program to give control to next.

In general, there are hardware interrupts and software interrupts. A hardware interrupt occurs, for example, when an I/O operation is completed such as reading some data into the computer from a tape drive. A software interrupt occurs when an application program terminates or requests certain services from the operating system. In a personal computer, a hardware interrupt request (IRQ) has a value associated with it that associates it with a particular device.

3.6.1 Basic Issues of Interrupt

Some of the basic issues of interrupt are as follows:

- What are the different kinds of interrupts?
- What are the advantages of having an interruption mechanism?
- How is the CPU informed about the occurrence of an interrupt?
- What does the CPU do on occurrence of an interrupt?

The following table gives the list of some common interrupts and events that cause the occurrence of those interrupts.

Interrupt Condition	Occurrence of Event
Interrupt are generated by executing program itself (also called traps)	• Division by Zero • The number exceeds the maximum allowed • Attempt of executing an illegal/privileged instruction • Trying to reference memory location other than allowed for that program
Interrupt generated by clock in the processor	Generally used on expiry of time allocated for a program in multiprogramming operating systems
Interrupts generated by I/O devices and their interfaces	• Request of starting an Input/Output operation • Normal completion of an Input/Output operation • Occurrence of an error in Input/Output operation
Interrupts on Hardware failure	• Power failure • Memory parity error

3.6.2 Interrupts Hardware

Hardware interrupts are not very different in behaviour. But still, when considering the path traveled by hardware interrupt from the instant it leaves the hardware device until it reaches the processor, you must take into account the actions taken by the interrupt controller (or the 8259A if you prefer).

When the processor executes an INT instruction, it retrieves the interrupt code from the opcode itself. This code is later used by the processor to index the **IVT** (interrupt vector table) and find the address of the interrupt handler. On the other hand, when a hardware interrupt is detected, the interrupt controller sends the interrupt code to the processor via the data bus. When the code is finally acquired by the processor (either from the INT opcode or from the interrupt controller) the steps taken are equivalent to those discussed earlier.

3.6.3 Types of Interrupts

The following are the types of Interrupts:

(*a*) External Interrupts
(*b*) Internal Interrupts
(*c*) Software Interrupts

(*a*) **External Interrupts:** An interrupt caused by an external source such as the computer operator, external sensor or monitoring device, or another computer is called External Interrupt.

(*b*) **Internal Interrupts:** An interrupt that is caused by processing, for example, a request for input or output or an arithmetic overflow error is called Internal Interrupt.

(*c*) **Software Interrupts:** A *software interrupt*, also called an *exception*, is an *interrupt* that is caused by software, usually by a program in *user mode.*

An interrupt is a signal to the *kernel* (*i.e.*, the core of the operating system) that an *event* has occurred, and this results in changes in the sequence of instructions that is executed by the CPU (central processing unit). One of the two main types of interrupts, a *hardware interrupt*, is a signal to the system from an event that has originated in hardware, such as the pressing of a key on the keyboard, a movement of the mouse or a progression in the system clock.

3.6.4 Interrupt Handler or Interrupt Service Routine

An **interrupt handler**, also known as an **interrupt service routine** (**ISR**), is a callback subroutine in an operating system or device driver whose execution is triggered by the reception of an interrupt. Interrupt handlers have a multitude of functions, which vary based on the reason the interrupt was generated and the speed at which the Interrupt Handler completes its task.

An interrupt handler is a low-level counterpart of event handlers. These handlers are initiated by either hardware interrupts or interrupt instructions in software, and are used for servicing hardware devices and transitions between protected modes of operation such as system calls.

Overview

In several operating systems—Linux, Microsoft Windows, and some other operating systems in the past, interrupt handlers are divided into two parts: the **First-Level Interrupt Handler (FLIH)** and the **Second-Level Interrupt Handlers (SLIH)**. FLIHs are also known as *hard interrupt handlers* or *fast interrupt handlers*, and SLIHs are also known as *slow/soft interrupt handlers*, Deferred Procedure Call.

A FLIH implements at minimum platform-specific interrupt handling similarly to *interrupt routines*. In response to an interrupt, there is a context switch, and the code for the interrupt is loaded and executed. The job of a FLIH is to quickly service the interrupt, or to record platform-specific critical information which is only available at the time of the interrupt, and schedule the execution of a SLIH for further long-lived interrupt handling.

FLIHs cause jitter in process execution. FLIHs also mask interrupts. Reducing the jitter is most important for real-time operating systems, since they must maintain a guarantee that execution of specific code will complete within an agreed amount of time. To reduce jitter and to reduce the potential for losing data from masked interrupts, programmers attempt to minimize the execution time of a FLIH, moving as much as possible to the SLIH. With the speed of modern computers, FLIHs may implement all device and platform-dependent handling, and use a SLIH for further platform-independent long-lived handling.

FLIHs which service hardware typically mask their associated interrupt (or keep it masked as the case may be) until they complete their execution. An (unusual) FLIH which unmasks its associated interrupt before it completes is called a reentrant interrupt handler. Reentrant interrupt handlers might cause a stack overflow from multiple preemptions by the same interrupt vector, and so they are usually avoided. In a priority interrupt system, the FLIH also (briefly) masks other interrupts of equal or lesser priority.

A SLIH completes long interrupt processing tasks similarly to a process. SLIHs either have a dedicated kernel thread for each handler, or are executed by a pool of kernel worker threads. These threads sit on a run queue in the operating system until processor time is available for them to perform processing for the interrupt. SLIHs may have a long-lived execution time, and thus are typically scheduled similarly to threads and processes.

In Linux, FLIHs are called *upper half*, and SLIHs—*lower half or bottom half*. This is different from naming used in other Unix-like systems, where both are a part of *bottom half*.

3.7 DATA TRANSFER SCHEMES

After going through the details of the device interfaces, the next point to be discussed is how the interface may be used to support input/output from devices. Binary information received from an external device is usually stored in memory for later processing. Information transferred from the central computer into an external device originates in the memory unit. Data transfer between the central computer and I/O devices may be handled in a variety of modes. Three techniques are possible for I/O operation. These are as follows:

- Programmed input/output
- Interrupt driven input/output
- Direct memory access

The following figure gives an overview of these three techniques:

	Interrupt Required	I/O interface to/from memory transfer
Programmed I/O	No	Through CPU
Interrupt-driven I/O	Yes	Through CPU
DMA	Yes	Direct to Memory

Overview of the three Input/Output Techniques

3.7.1 Programmed I/O

In programmed I/O, the I/O operations are completely controlled by the processor. The processor executes a program that initiates, directs and terminates an I/O operation. It requires a little special I/O hardware, but is quite time consuming for the processor since the processor has to wait for slower I/O operations to complete.

Programmed input/output is a useful I/O method for computers where hardware costs need to be minimized. The input or output operation in such cases may involve:

(*a*) Transfer of data from I/O device to the processor registers.

(*b*) Transfer of data from processor registers to memory.

With the programmed I/O method, the responsibility of the processor is to constantly check the status of the I/O device to check whether it is free or it has finished inputting the data. Thus, this method is very time consuming where the processor wastes a lot of time in checking and verifying the status of an I/O device. The following Figure gives an example of the use of programmed I/O to read in a block of data from a peripheral device into memory.

3.7.1.1 I/O Commands

There are four types of I/O commands that an I/O interface may receive when it is addressed by a processor:

(*a*) **Control:** These commands are device specific and are used to provide specific instructions to the device, e.g. a magnetic tape requiring rewinding and moving forward by a block.

(*b*) **Test:** This command checks the status such as if a device is ready or not or is in error condition.

(*c*) **Read:** This command is useful for input of data from input device.
(*d*) **Write:** This command is used for output of data to output device.

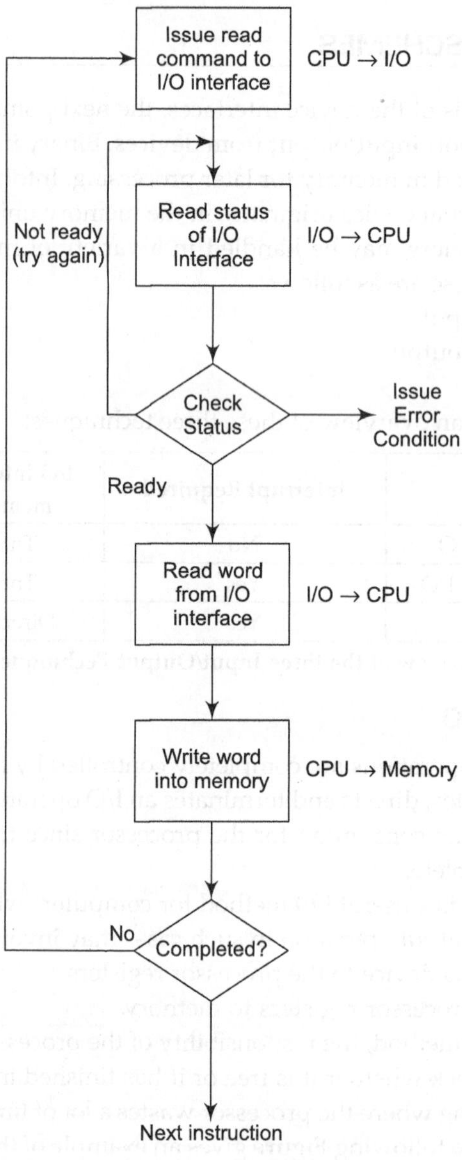

3.7.1.2 I/O Instructions

An I/O instruction is stored in the memory of the computer and is fetched and executed by the processor producing an I/O-related command for the I/O interface. With programmed I/O, there is a close correspondence between the I/O-related instructions and the I/O commands that the processor issues to an I/O interface to execute the instructions. In systems with programmed I/O, the I/O interface, the main memory and the processors normally share the system bus. Thus, each I/O interface should interpret the address lines to determine if the command is for itself. There are two methods for doing so. These are called memory-mapped I/O and isolated I/O. With memory-mapped I/O, there is a single address space for memory locations and I/O devices. The processor treats the status and data registers of I/O interface as memory locations and uses the same machine instructions to access both memory and I/O devices. For a memory-mapped I/O only a single read and a single write line are needed for memory or I/O interface read or write operations. These lines are activated by the processor for either memory access or I/O device access. The following figure shows the memory-mapped I/O system structure.

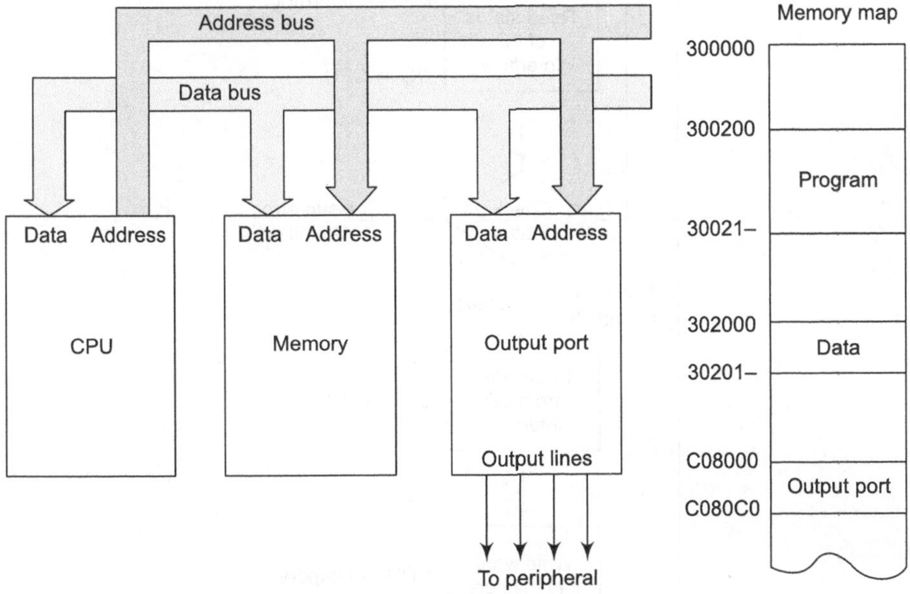

With isolated I/O, there are separate control lines for both memory and I/O device read or write operations. Thus a memory reference instruction does not affect an I/O device. In isolated I/O, the I/O devices and memory are addressed separately; hence separate input/output instructions are needed which cause data transfer between addressed I/O interface and processor.

3.7.2 Interrupt Initiated I/O

The problem with programmed I/O is that the processor has to wait a long time for the I/O interface to see whether a device is free or wait till the completion of I/O. The result is that the performance of the processor goes down tremendously. What is the solution? What about the processor going back to do other useful work without waiting for the I/O device to complete or get freed up? But how will the processor be intimated about the completion of I/O or a device is ready for I/O? A well-designed mechanism was conceived for this, which is referred to as interrupt-driven I/O. In this mechanism, provision of interruption of processor work, once the device has finished the I/O or when it is ready for the I/O, has been provided. The interrupt-driven I/O mechanism for transferring a block of data is shown in the following Figure. Please note that after issuing a read command (for input) the CPU goes off to do other useful work while I/O interface proceeds to read data from the associated device. On the completion of an instruction cycle, the CPU checks for interrupts (which will occur when data is in data register of I/O interface and it now needs CPU's attention). Now CPU saves the important register and processor status of the executing program in a stack and requests the I/O device to provide its data, which is placed on the data bus by the I/O device. After taking the required action with the data, the CPU can go back to the program it was executing before the interrupt.

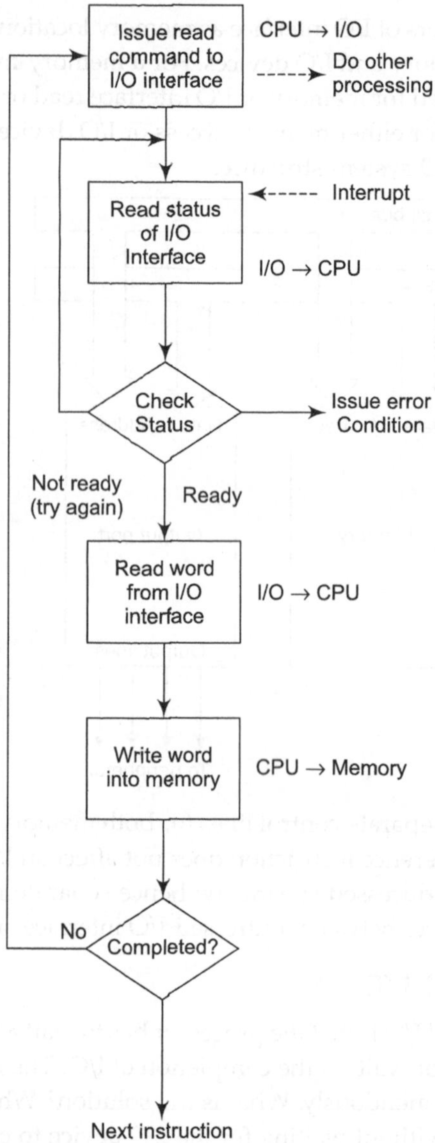

3.7.3 Interrupt-Processing

The occurrence of an interrupt fires a numbers of events, both in the processor hardware and software. The following figure shows a sequence.

When an I/O device completes an I/O operation, the following sequence of hardware events occurs:

1. The device issues an interrupt signal to the processor.
2. The processor finishes execution of the current instruction *before responding* to the interrupt.
3. The processor tests for the interrupts and sends an acknowledgement signal to the device that issued the interrupt.
4. The minimum information required to be stored for the task being currently executed, before the CPU starts executing the interrupt routine (using its registers) are:

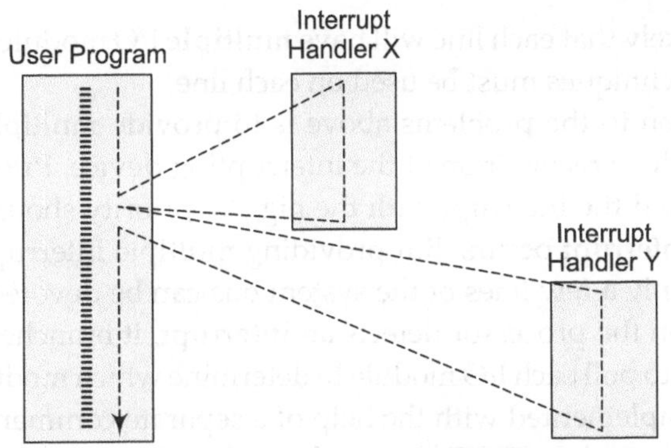

(*a*) The status of the processor, which is contained in the register called program status word (PSW), and
(*b*) The location of the next instruction to be executed, of the currently executing program, which is contained in the program counter (PC).

5. The processor now loads the PC with the entry location of the interrupt-handling program that will respond to this interrupting condition. Once the PC has been loaded, the processor proceeds to execute the next instruction that is the next instruction cycle, which begins with an instruction fetch. Because the instruction fetch is determined by the contents of the PC, the result is that control is transferred to the interrupt-handler program. The execution results in the following operations.
6. The PC & PSW relating to the interrupted program have already been saved on the system stack. In addition, the contents of the processor registers are also needed to be saved on the stacks that are used by the called Interrupt Servicing Routine because these registers may be modified by the interrupt-handler. Here a user program is interrupted after the instruction at location N. The contents of all of the registers plus the address of the next instruction (N+1) are pushed on to the stack.
7. The interrupt handler next processes the interrupt. This includes determining of the event that caused the interrupt and also the status information relating to the I/O operation.
8. When interrupt processing is complete, the saved register values are retrieved from the stack and restored to the registers.
9. The final step is to restore the values of PSW and PC from the stack. As a result, the instruction to be executed will be from the previously interrupted program.

Thus, interrupt handling involves interruption of the currently executing program, execution of interrupt servicing program and restart of interrupted program from the point of interruption.

3.7.3.1 Design Issues

Two design issues arise in implementing interrupt-driven I/O:

1. How does the processor determine which device issued the interrupt?
2. If multiple interrupts have occurred, how does the processor decide which one to be processed first?

To solve these problems, four general categories of techniques are in common use:

(*a*) **Multiple Interrupt Lines:** The most straight forward approach is to provide **multiple interrupt lines** between the processor and the I/O modules. It is impractical to dedicate more than a few bus **lines** or processor pins to **interrupt lines**. Thus, though **multiple interrupt lines** are needed, it is most likely that each line will have **multiple** I/O modules attached to it. Thus one of the other three techniques must be used on each line.

The simplest solution to the problems above is to provide multiple interrupt lines, which will result in immediate recognition of the interrupting device. Priorities can be assigned to various interrupts and the interrupt with the highest priority should be selected for service in case a multiple interrupt occurs. But providing multiple interrupt lines is an impractical approach because only a few lines of the system bus can be devoted for the interrupt.

(*b*) **Software Poll:** When the processor detects an **interrupt**, it branches to an **interrupt** service routine whose job is to poll each I/O module to determine which module caused the **interrupt**. The poll could be implemented with the help of a separate command line (e.g. TESTI/O). In this case, the processor raises TESTI/O and places the address of a particular I/O module on the address **lines**. The I/O module responds positively if it sets the **interrupt**.

Alternatively, each I/O module could contain an addressable status register. The processor branches to a device service routine specific to that device.

Once the correct module is identified, the processor branches to a device service routine specific to that device.

The main disadvantage of software poll is that it is time consuming. Processor has to check it is equal to the number of I/O modules.

In this scheme, on the occurrence of an interrupt, the processor jumps to an interrupt service program or routine whose job it is to poll (roll call) each I/O interface to determine which I/O interface has caused the interrupt. This may be achieved by reading the status register of the I/O interface. Once the correct interface is identified, the processor branches to a device-service routine specific to that device. The disadvantage of the software poll is that it is time consuming.

(*c*) **Daisy Chain:** In this method for interrupts all I/O modules share a common **interrupt** request **lines**. However the **interrupt** acknowledge line is connects in daisy sends out an **interrupt** acknowledge. The **interrupt** acknowledge signal propagates—through a series of I/O module until it gets to a requesting module.

The requesting module typically responds by placing a word on the data **lines**. This word is referred to as a vector and is either the address of the I/O module or some other unique identification. In either case, the processor uses the vector a pointer to the appropriate device service routine. This avoids the need to execute a general **interrupt** service routine first. This technique is referred to as a vectored **interrupt**.

This scheme provides a hardware poll. With this technique, an interrupt acknowledge line is chained through various interrupt devices. All I/O interfaces share a common interrupt request line. When the processor senses an interrupt, it sends out an interrupt acknowledgement. This signal passes through all the I/O devices until it gets to the requesting device. The first device which has made the interrupt request thus senses the signal and responds by putting in a word which is normally an address of interrupt servicing program or a unique identifier on the data lines. This word is also referred to as interrupt vector. This address or identifier in turn is used for selecting an appropriate interrupt servicing program. The daisy chaining has an in-built priority scheme, which is determined by the sequence of devices on interrupt acknowledge line.

(*d*) **Bus Arbitration:** In bus arbitration method, an I/O module must first gain control of the bus before it can raise the **interrupt** request line. Thus, only one module can raise the **interrupt** line at a time. When the processor detects the **interrupt**, it responds on the **interrupt** acknowledge line. The requesting module then places it vector on the data line.

In this scheme, the I/O interface first needs to control the bus and only after that it can request for an interrupt. In this scheme, since only one of the interfaces can control the bus, therefore only one request can be made at a time. The interrupt request is acknowledged by the CPU on response of which I/O interface places the interrupt vector on the data lines. An interrupt vector normally contains the address of the interrupt serving program.

An example of an interrupt vector can be a personal computer, where there are several IRQs (Interrupt request) for a specific type of interrupt.

3.7.4 Direct Memory Access

The transfer of data between a fast storage device such as magnetic disk and memory is often limited by the speed of the CPU. Removing the CPU from the path and letting the peripheral device manage the memory buses directly would improve the speed of transfer. This transfer technique is called direct memory access, (DMA). During DMA transfer; the CPU is idle and has no control of the memory buses. A DMA controller takes over the buses to manage the transfer directly between the I/O device and memory.

The CPU may be placed in an idle state in a variety of ways. One common method extensively used in microprocessors is to disable the buses through, special control signals. The following figure shows two control signals in the CPU that facilitate the DMA transfer. The bus request (BR) input is used by the DMA controller to request the CPU to relinquish control of the buses. When this input is active, the CPU terminates the execution of the current instruction and places the address bus, the data bus, and the read and write lines into a high-impedance state. The high-impedance state behaves like an open circuit, which means that the output is disconnected and does not have logic significance (see Sec. 4-3). The CPU activates the bus grant (BG) output to inform the external DMA that the buses are in the high-impedance state. The DMA that originated the bus request can now take control of the buses to conduct memory transfers without processor intervention. When the DMA terminates the transfer, it disables the bus request line. The CPU disables the bus grant, takes control of the buses, and returns to its normal operation.

When the DMA takes control of the bus system, it communicates directly with the memory. The transfer can be made in several ways. In DMA burst transfer, a block sequence consisting of a number of memory words is transferred in a continuous burst while the DMA controller is master of the memory buses. This mode of transfer is needed for fast devices such as magnetic disks, where data transmission cannot be stopped or slowed down until an entire block is transferred. An alternative technique called cycle stealing allows the DMA controller to transfer one data word at a time, after which it must return control of the buses to the CPU. The CPU merely delays its operation for one memory cycle to allow the direct memory I/O transfer to "steal" one memory cycle.

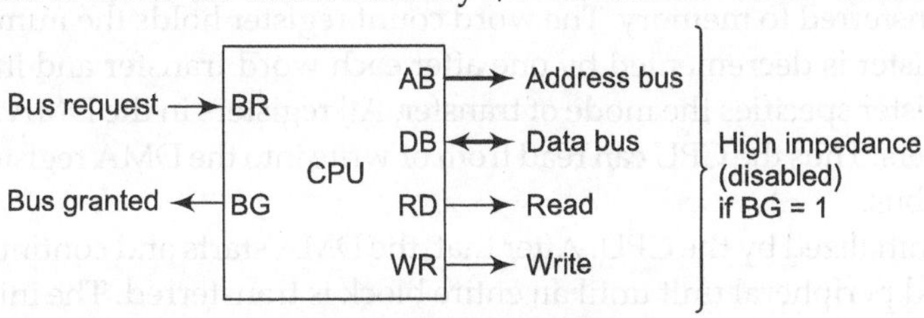

CPU Bus Control Signals

3.7.4.1 DMA Controller

The DMA controller needs the usual circuits of an interface to communicate with the CPU and I/O device. In addition, it needs an address register, a word count register, and a set of address lines. The address register and address lines are used for direct communication with the memory. The word count register specifies the number of words that must be transferred. The data transfer may be done directly between the device and memory under control of the DMA.

The following figure shows the block diagram of a typical DMA controller. The unit communicates with the CPU via the data bus and control lines. The registers in the DMA are selected by the CPU through the address bus by enabling the DS (DMA select) and RS (register select) inputs. The RD (read) and WR (write) inputs are bidirectional. When the BG (bus grant) input is 0, the CPU can communicate with the DMA registers through the data bus to read from or write to the DMA registers. When BG = 1, the CPU has relinquished the buses and the DMA can communicate directly with the memory by specifying an address in the address bus and activating the RD or WR control. The DMA communicates with the external peripheral through the request and acknowledge lines by using a prescribed handshaking procedure.

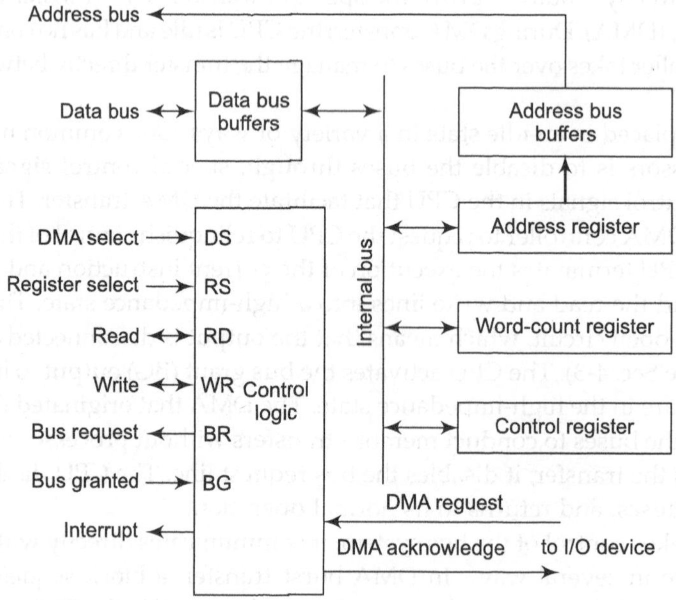

Block Diagram of a DMA Controller

The explanation of the above figure is as follows:

The DMA controller has three registers: an address register, a word count register, and a control register. The address register contains an address to specify the desired location in memory. The address bits go through bus buffers into the address bus. The address register is incremented after each word that is transferred to memory. The word count register holds the number of words to be transferred. This register is decremented by one after each word transfer and internally tested for zero. The control register specifies the mode of transfer. All registers in the DMA appear to the CPU as I/O interface registers. Thus the CPU can read from or write into the DMA registers under program control via the data bus.

The DMA is first initialized by the CPU. After that, the DMA starts and continues to transfer data between memory and peripheral unit until an entire block is transferred. The initialization process is essentially a program consisting of I/O instructions that include the address for selecting particular DMA registers. The CPU initializes the DMA by sending the following information through the data bus:

1. The starting address of the memory block where data are available (for read) or where data are to be stored (for write)

2. The word count, which is the number of words in the memory block
3. Control to specify the mode of transfer such as read or write
4. A control to start the DMA transfer

The starting address is stored in the address register. The word count is stored in the word count register, and the control information in the control register. Once the DMA is initialized, the CPU stops communicating with the DMA unless it receives an interrupt signal or if it wants to check how many words have been transferred.

3.7.4.2 DMA Transfer

The position of the DMA controller among the other components in a computer system is illustrated in the following figure. The CPU communicates with the DMA through the address and data buses as with any interface unit. The DMA has its own address, which activates the DS and RS lines. The CPU initializes the DMA through the data bus. Once the DMA receives the start control command, it can start the transfer between the peripheral device and the memory.

When the peripheral device sends a DMA request, the DMA controller activates the BR line, informing the CPU to relinquish the buses. The CPU responds with its BG line, informing the DMA that its buses are disabled. The DMA then puts the current value of its address register into the address bus, initiates the RD or WR signal, and sends a DMA acknowledge to the peripheral device. Note that the RD and WR lines in the DMA controller are bidirectional. The direction of transfer depends on the status of the BG line. When BG = 0, the RD and WR are input lines allowing the CPU to communicate with the internal DMA registers. When BG = 1 the RD and WR are output lines from the DMA controller to the random access memory to specify the read or write operation for the data.

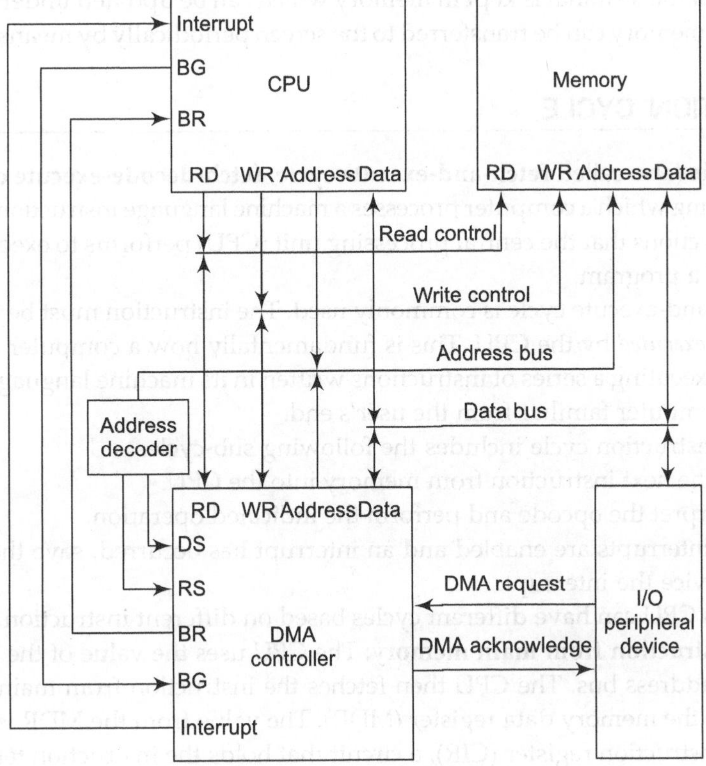

DMA Transfer in a Computer System

When the peripheral device receives a DMA acknowledge, it puts a word in the data bus (for write) or receives a word from the data bus (for read). Thus the DMA controls the read or write operations and supplies the address for the memory. The peripheral unit can then communicate with memory through the data bus for direct transfer between the two units while the CPU is momentarily disabled.

For each word that is transferred, the DMA increments its address register and decrements its word count register. If the word count does not reach zero, the DMA checks the request line coming from the peripheral. For a high-speed device, the line will be active as soon as the previous transfer is completed. A second transfer is then initiated, and the process continues until the entire block is transferred. If the peripheral speed is slower, the DMA request line may come somewhat later. In this case the DMA disables the bus request line so that the CPU can continue to execute its program. When the peripheral requests a transfer, the DMA requests the buses again.

If the word count register reaches zero, the DMA stops any further transfer and removes its bus request. It also informs the CPU of the termination by means of an interrupt. When the CPU responds to the interrupt, it reads the content of the word count register. The zero value of this register indicates that all the words were transferred successfully. The CPU can read this register at any time to check the number of words already transferred. A DMA controller may have more than one channel. In this case, each channel has a request and an acknowledge pair of control signals which are connected to separate peripheral devices. Each channel also has its own address register and word count register within the DMA controller. A priority among the channels may be established so that channels with high priority are serviced before channels with lower priority, DMA transfer is very useful in many applications. It is used for fast transfer of information between magnetic disks and memory. It is also useful for updating .the display in an interactive terminal. Typically, an image of the screen display of the terminal is kept in memory which can be updated under program control. The contents of the memory can be transferred to the screen periodically by means of DMA transfer.

3.8 INSTRUCTION CYCLE

An **instruction cycle** (also called **fetch-and-execute cycle, fetch-decode-execute cycle,** and **FDX**) is the time period during which a computer processes a machine language instruction from its memory or the sequence of actions that the central processing unit (CPU) performs to execute each machine code instruction in a program.

The name fetch-and-execute cycle is commonly used. The instruction must be *fetched* from main memory, and then *executed* by the CPU. This is fundamentally how a computer operates, with its CPU reading and executing a series of instructions written in its machine language. From this arise all functions of a computer familiar from the user's end.

Recall that the instruction cycle includes the following sub-cycles:

- Fetch: Read the next instruction from memory into the CPU.
- Execute: Interpret the opcode and perform the indicated operation.
- Interrupt: If interrupts are enabled and an interrupt has occurred, save the current process state and service the interrupt.

Each computer's CPU can have different cycles based on different instruction sets:

1. **Fetch the instruction from main memory:** The CPU uses the value of the program counter (PC) on the address bus. The CPU then fetches the instruction from main memory via the data bus into the memory data register (MDR). The value from the MDR is then placed into the current instruction register (CIR), a circuit that holds the instruction temporarily so that it can be decoded and executed.

2. **Decode the instruction:** The instruction decoder interprets and implements the instruction. The instruction register (IR) holds the current instruction, while the program counter (PC) holds the address in memory of the next instruction to be executed.
3. **Fetch data from main memory:** Read the effective address from main memory if the instruction has an indirect address. Fetch the required data from main memory to be processed and place it into data registers.
4. **Execute the instruction:** From the instruction register, the data forming the instruction is decoded by the control unit. It then passes the decoded information as a sequence of control signals to the relevant function units of the CPU to perform the actions required by the instruction

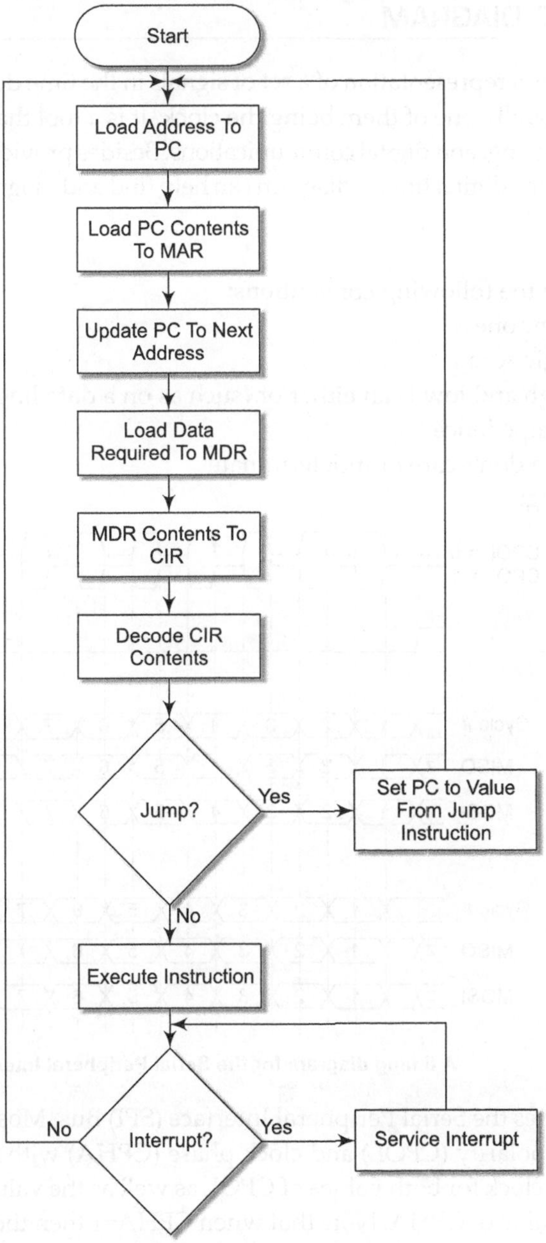

Fetch Execute Cycle Diagram

such as reading values from registers, passing them to the Arithmetic logic unit (ALU) to add them together and writing the result back to a register. A condition signal is sent back to the control unit by the ALU if it is involved.

5. **Store results:** Also called write back to memory. The result generated by the operation is stored in the main memory, or sent to an output device. Based on the condition feedback from the ALU, the PC is either incremented to address the next instruction or updated to a different address where the next instruction will be fetched. The cycle is then repeated.

A diagram of the Fetch Execute Cycle is represented on the previous page.

3.9 DIGITAL TIMING DIAGRAM

A **Digital timing diagram** is a representation of a set of signals in the time domain. A timing diagram can contain many rows, usually one of them being the clock. It is a tool that is ubiquitous in digital electronics, hardware debugging, and digital communications. Besides providing an overall description of the timing relationships, the digital timing diagram can help find and diagnose digital logic hazards.

Diagram Convention

Most timing diagrams use the following conventions:

- Higher value is a logic one
- Lower value is a logic zero
- A slot showing a high and low is an either or (such as on a data line)
- A Z indicates high impedance
- A greyed out slot is a don't-care or indeterminate

Example: SPI bus timing:

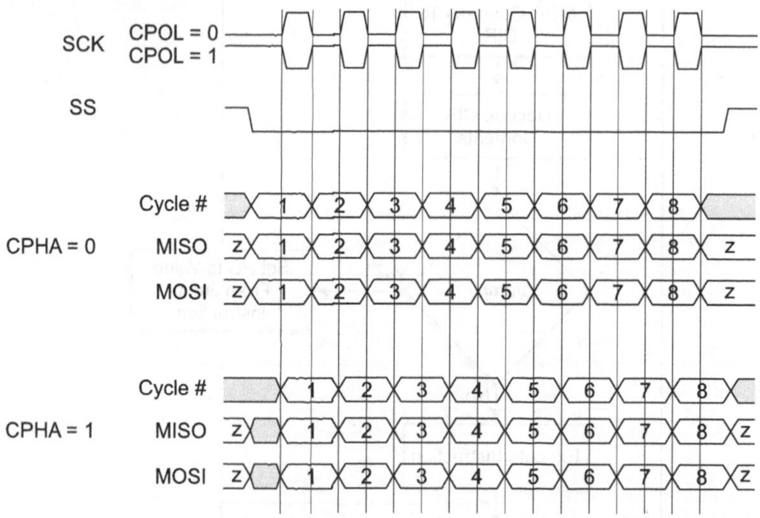

A timing diagram for the Serial Peripheral Interface Bus

Above examples describes the Serial Peripheral Interface (SPI) Bus. Most SPI master nodes have the ability to set the clock polarity (CPOL) and clock phase (CPHA) with respect to the data. This timing diagram shows the clock for both values of CPOL as well as the values for the two data lines (MISO & MOSI) for each value of CPHA. Note that when CPHA=1 then the data is delayed by one-half clock cycle.

SPI operates in the following way:

- The master determines an appropriate CPOL & CPHA value
- The master pulls down the slave select (SS) line for a specific slave chip
- The master clocks SCK at a specific frequency
- During each of the 8 clock cycles the transfer is full duplex:
- The master writes on the MOSI line and reads the MISO line
    - The slave writes on the MISO line and reads the MOSI line
- When finished the master can continue with another byte transfer or pull SS high to end the transfer.

When a slave's SS line is high then both of its MISO and MOSI line should be high impedance so as to avoid disrupting a transfer to a different slave. Prior to SS being pulled low, the MISO & MOSI lines are indicated with a "z" for high impedance. Also prior to the SS being pulled low the "cycle #" row is meaningless and is shown greyed-out.

3.10 INTERFACING

(a) Interfacing I/O Devices

I/O devices are accessed in much the same way as the memory, though they sometimes need addressing by an external addressing device or decoder. The most basic devices for it are simple latches or even buffers, enabled by a decoder or I/O line. In the case of the TIL311 display device shown on previous page, the package incorporates not only the LED segments, but also the binary to segment decoder AND a data latch! You can see the actual chip itself in the bottom half of the device. All in all, a very easy and convenient display to use, as all one does is connect it's four data lines straight up to the processors data bus, and the strobe to the I/O decoder. Voila! Note that it only has **four** data lines, so you have to decide whether to connect to the top half of the bus or the lower half. Of course if using two TIL311's, one *could* use the LOWER four bits for one display and the UPPER four bits for the other. By commoning the enable lines, one could then update them in the program simultaneously from the same byte! (Assuming an 8 bit processor).

Although PIO and PIA multi I/O LSI chips offer many different interfacing options, as a general rule, I tend to avoid them unless these facilities (such as interrupts) are really needed, as it is usually simpler to use either two or three tri-state latches or buffers. The ones I most commonly use are the 74xx373 or 374 octal latches, or the 74xx240 241 244 series octal buffers. Where *bi-directional* tri-state buffers are needed—such as that required with a multiplexed data / address bus, I generally opt for the 74xx245. Where space is at a premium and if only 4 bits are needed, the 74xx125 is a viable alternative tri—state device in a 14 pin pack. (The others all mentioned above are 20 pin.)

(b) Interfacing Memory Devices

Memory is something that all micro's need to interface to. It will be noticed that there is almost a 'standard' for the pin layout used by most of these—which can be very useful at times. Have a look at a few of the pinouts in the Data Sheets section to see for yourself.

Memory is connected by its ADDRESS and DATA lines to the processor and accessed via it's chip select and memory read / write lines. This is where the chips vary a little with capacities and type. Almost ALL will be accessed by taking their enable lines LOW. Read / Write lines invariably need the line taken low to WRITE.

A memory device may have more select lines than you need - *i.e.*, /CE1 (chip enable 1) /CE2 (chip enable 2) and /OE (output enable) etc. You only have to use those that you need. Simply hard-wire-enable the ones that you don't need, remembering that an output enable (for example) can't enable anything if the chip itself is not enabled with the /CE lines.

Graphically the process is represented as follows:

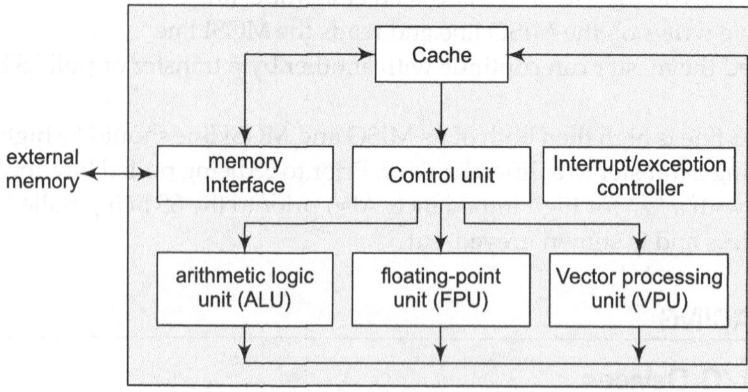

The ALU performs all basic computational operations: arithmetic, logical, and comparisons.

The control unit orchestrates the operation of the other units. It fetches instructions from the on-chip cache, decodes them, and then executes them. Each instruction has the control unit direct the other function units through a sequence of steps that carry out the instruction's intent. The execution path taken by the control unit can depend upon status bits produced by the arithmetic logic unit or the floating-point unit (FPU) after the instruction sequence completes. This capability implements conditional execution control flow, which is a critical element for general-purpose computation.

The memory interface enables the microprocessor to maintain two-way communication with off-chip semiconductor memory, which stores programs and data. This interface typically supports memory reads and writes in blocks of words (the number of bits that the processor operates on at one time). The block size facilitates burst data transfers to and from the chip's internal cache.

The interrupt or exception controller enables the microprocessor to respond to requests from the external environment or to error conditions by allowing interruptions of the ongoing operation. An interrupt might be an external peripheral requesting service, while an exception typically consists of a floating-point math error or an unrecognized instruction. The interrupt controller can prioritize and selectively handle these interrupts.

The internal cache is an on-chip memory storage area that holds recently used data values or instruction sequences that are likely to be used again in the near future. Since this information is already on-chip, it can be accessed rapidly, thereby accelerating the computation rate. Items not in the cache can take several or more extra operations to access, which significantly degrades the computation rate. Software writers often organize a program's code and data structures so that the most frequently used elements often occupy the cache, thus maintaining a high level of computational throughput.

The design of instruction sets (the commands that produce basic work when executed by the microprocessor) often influences the design of the microprocessor itself. Instruction sets—and as a consequence, the microprocessor architecture—are of two types: reduced instruction set computers (RISC) and complex instruction set computers (CISC). Because of the limits of early computer technology, most computers were by necessity RISC machines. Since most of the software was written in assembly language (that is, a programming language that represented the program's intent in actual machine instructions), there was a drive to build instruction sets of greater sophistication and complexity. These new CISC instruction sets made assembly language programming easier, but they also made it difficult to build high-speed computer hardware. First, CISC instructions were harder to decode. In addition, since CISC instructions involved long and complex operation sequences, they incurred a major cost by requiring more complicated logic to implement. Second, such instructions were also difficult to interrupt or abort if an exception occurred. Finally, such instructions usually carried many data dependencies that made it more difficult to support advanced architectural techniques. By returning to a RISC design, much faster computers can be built. In fact, an enhancement in performance by a factor of 2 to 3 has been attributed to this simple organizational change. To achieve these efficiencies, most of the RISC microprocessor's function units must be kept as busy as possible. This requires optimizing compilers that can translate a program's high-level source code and then reorder the resulting low-level instructions in such a way as to ensure the high throughput.

Microprocessors are found in virtually every consumer product that requires electric power, such as microwave ovens, automobiles, video recorders, cellular telephones, digital cameras, and hand-held computers. High-performance microprocessors implement the servers that store and distribute Web content, such as streaming audio and video, desktop computers, and the high-speed network switches that constitute the Web's infrastructure. More modest-powered microprocessors are at the heart of notebook computers and electronic games. Low-power microprocessors provide the control and flow logic of hand-held devices, digital cameras, cellular and cordless phones, pagers, and the diagnostic and pollution control of automobile engines.

EXERCISE

1. What is a bus? What are characteristics of bus?
2. Specify the function of the address bus and the direction of the information flow on the address bus.
3. Why is the data bus bidirectional?
4. What determines that Microprocessor is an 8, 16 or 32 bit?
5. What are the basic components of ALU? Describe the utility of accumulator and data register through examples.
6. What is the purpose of control unit?
7. What are the types of registers?
8. What is the purpose of Index registers?
9. What do you mean by cache memory?
10. Explain the different methods of data transfer possible between MPU and I/O.
11. What is the drawback of programmed input/output data transfer technique? How can it be removed?
12. What is interrupt? Give the difference between hardware and software interrupt.
13. Explain the basic blocks of microprocessor based system.
14. Explain the concept of Direct Memory Access (DMA).
15. What are instruction cycle, fetch cycle and execute cycle?
16. Explain the term timing diagram.
17. What do you mean by addressing modes? Explain various types of addressing modes.
18. What do you mean by interfacing I/O devices?
19. What is the difference between internal and external interrupt?
20. What are the various issues related to interrupt initiated I/O?

4 ARCHITECTURAL ADVANCEMENT OF MICROPROCESSOR

Everyone who works in the computer industry is well familiar with Moore's Law and the doubling of the number of transistors (an approximate measure of computer processing power) every 18 to 24 months. Until recently, overall microprocessor performance was often described in terms of processor clock speeds, expressed in megahertz (MHz) or gigahertz (GHz). Today there's far more than clock speed to consider when you're evaluating how a given processor will perform for a given application and where it fits on the performance scale. Microprocessor designers today are more focused on methods that leverage the latest silicon production processes and designs that minimize microprocessor footprint size, power consumption and heat generation. Designers are also concerned with micro architecture optimization, multiprocessing parallelism, reliability, designed-in security features, memory structure efficiency and better synergy between the hardware and accompanying software tools, such as compilers. The more attention that a designer devotes to refining the efficiency of the software code rather than making the hardware responsible for dynamic optimization, the higher the ultimate system performance will be.

As an example, the Intel® Itanium® processor family has been designed around small footprint cores that are remarkably compact in terms of transistor count, especially when one considers the amount of processing work that they accomplish. Itanium has taken instruction level parallelism to a new level, and this can be used in conjunction with thread level parallelism to leverage more processor cores and more threads per core to produce higher performance. Some microprocessor designs of the past have been overly complex and have relied on out-of-order logic to reshuffle and optimize software instructions. Going forward, microprocessor designers will continue to deliver better and better software tools, higher software optimization and better compilers. Because it is so efficient and so small and doesn't depend on out-of-order logic, the latest generation Itanium processor can deliver higher performance without creating thermal generation problems. This makes Itanium a very simple yet efficient and refined engine that enables more consistent long-term improvement in code execution via small improvements in software, thus reducing the need for significant advancements in hardware. These are becoming more and more difficult to accomplish as, even Gordon Moore believes, the exponential upward curve in microprocessor hardware advancements "can't continue forever."

4.1 RISC

RISC (reduced instruction set computer) is a microprocessor that is designed to perform a smaller number of types of computer instructions so that it can operate at a higher speed (perform more millions of instructions per second, or MIPS). Since each instruction type that a computer must perform requires additional transistors and circuitry, a larger list or set of computer instructions tends to make the microprocessor more complicated and slower in operation.

John Cocke of IBM Research in Yorktown, New York, originated the RISC concept in 1974 by proving that about 20% of the instructions in a computer did 80% of the work. The first computer to benefit from this discovery was IBM's PC/XT in 1980. Later, IBM's RISC System/6000, made use of the idea. The term itself (RISC) is credited to David Patterson, a teacher at the University of California in Berkeley. The concept was used in Sun Microsystems' SPARC microprocessors and led to the founding of what is now MIPS Technologies, part of Silicon Graphics. A number of current microchips now use the RISC concept.

The RISC concept has led to a more thoughtful design of the microprocessor. Among design considerations are how well an instruction can be mapped to the clock speed of the microprocessor (ideally, an instruction can be performed in one clock cycle); how "simple" an architecture is required; and how much work can be done by the microchip itself without resorting to software help.

4.1.1 Advantages

Besides performance improvement, some advantages of RISC and related design improvements are:

- A new microprocessor can be developed and tested more quickly if one of its aims is to be less complicated.
- Operating system and application programmers who use the microprocessor's instructions will find it easier to develop code with a smaller instruction set.
- The simplicity of RISC allows more freedom to choose how to use the space on a microprocessor.
- Higher-level language compilers produce more efficient code than formerly because they have always tended to use the smaller set of instructions to be found in a RISC computer.

4.1.2 The Microprocessor Revolution

The engine of the computer revolution is the microprocessor. It has led to new inventions, such as FAX machines and personal computers, as well as adding intelligence to existing devices, such as wristwatches and automobiles. Moreover, its performance has improved by a factor of roughly 10,000 in the 25 years since its birth in 1971! Has any other invention, so useful already at birth, undergone a similar improvement?

For the first 15 years of its life, the microprocessor improved its performance by an impressive 35% per year. As a result of research at universities and industrial laboratories, though, this rate increased in 1987 to about 55% per year — a doubling of performance every 18 months! Had the prior rate of 35% been maintained, computers today would be only one third the speed we actually enjoy. Stated differently, it's as if we had been granted a wish to use the computers of the year 2000 in 1995.

This increase coincided with the introduction of Reduced Instruction Set Computers (RISC). The *instruction set* is the hardware "language" in which the software tells the processor what to do. Surprisingly, reducing the size of the instruction set — eliminating certain instructions based upon a careful quantitative analysis, and requiring these seldom-used instructions to be emulated in software — can lead to higher performance, for several reasons:

- The vacated area of the chip can be used in ways that accelerate the performance of more commonly used instructions, more than compensating for the inevitably degraded performance of the seldom-used instructions.
- It becomes easier to optimize the design.

- It allows microprocessors to use techniques hitherto restricted to the largest computers.
- It simplifies translation from the high-level language in which people program into the instruction set that the hardware understands, resulting in a more efficient program.

RISC was also heralded a more quantitative approach to computer architecture, whereby careful experiments preceded the hardware design and sensible performance metrics were used to judge success. Previously, intuition guided many computer design projects, leading to disappointing results. As the quantitative approach replaced the less well-defined prior approach, people in all divisions of a computer company were trained in this new school of design; hence engineering, manufacturing, marketing, sales, and management acted in concert, further accelerating the rate of change.

4.1.3 The Genealogy of RISC

The roots of RISC lie in three research projects: the IBM 801, the Berkeley RISC processor, and the Stanford MIPS processor. These architectures attracted enormous interest because of claims of a performance advantage of anywhere from two to five times over contemporary machines using traditional architectures.

Begun in the late 1970s, the IBM project was the first to start but was the last to become public. The IBM machine was designed as a minicomputer made from hundreds of chips, while the university projects were both microprocessors. John Cocke is considered to be the father of the 801 design. In recognition of his contribution he received both the Turing award, the highest award in computer science and engineering, and the Presidential Medal of Technology.

In 1980, David A. Patterson and his colleagues at the University of California at Berkeley, sponsored by the Department of Defense Advanced Research Projects Agency, began the project that was to give this approach its name. They built two machines, called RISC-I and RISC-II. Because the IBM project was not widely known or discussed, the role played by the Berkeley group in promoting the RISC approach was critical to the acceptance of the technology.

In 1981, John L. Hennessy and his colleagues at Stanford published a description of the Stanford MIPS machine, also developed under DARPA sponsorship. Both university projects were interested in designing a simple machine that could be built as a microchip within the university environment. All three early RISC machines had similar "reduced" languages.

Importantly, the Berkeley and Stanford projects fit within the broad mosaic of the overall DARPA VLSI Program, a highly ambitious program which envisioned that integrated circuit technology could be made available to system designers — people with an overall view of the objectives and constraints of an entire hardware/software system — and that this would have tremendous impact. The DARPA VLSI Program developed the concept of the multichip wafer, which allowed multiple integrated circuit designs to share a single silicon fabrication run, dramatically reducing costs. It conceived of the MOSIS fabrication service, which created a multichip wafer from designs submitted electronically from multiple sites, allowing university system designers access to state-of-the-art silicon fabrication. It sponsored extensive developments in computer-aided design tools. It launched, in addition to the RISC revolution, a comparable revolution in cost-effective high-performance computer graphics through the Geometry Engine and Pixel Planes projects, now the basis of Silicon Graphics, Inc., and Ivex and Division. And it produced *people*.

4.1.4 Commercialization

In retrospect, it was important for the success of RISC to have three research projects. The RISC ideas were highly controversial at first — disbelievers were plentiful. In addition to exploring somewhat different avenues and making complementary discoveries, the three projects reinforced each others' results, and led to more voices advocating RISC in what was then a decidedly RISC-averse world.

In 1986 the computer industry began to announce commercial processors based on the technology explored by the three RISC research projects. Hennessy founded a company with the same name as his research project — MIPS. Hewlett-Packard converted their existing minicomputer line to RISC architectures. IBM never directly turned the 801 into a product. Instead, the ideas were adopted for a new, low-end architecture that was incorporated in the IBM RT-PC. Alas, this machine was a commercial failure, but subsequent RISC processors in which IBM has been involved (*e.g.*, the Apple/IBM/Motorola PowerPC) have been highly successful.

In 1987 Sun Microsystems began delivering machines based on the SPARC architecture, a derivative of the Berkeley RISC-II machine. In the view of many, it was Sun's success with RISC-based workstations that convinced the remaining skeptics that RISC was significant commercially. In particular, RISC advocates used Sun's success to get RISC restarted at IBM. IBM announced a new RISC architecture in 1990, as did DEC in 1993. Today, RISC is the foundation of a $15 billion industry.

Intel's microprocessors are used in the popular IBM PC, and hence are the most widely used microprocessors, but they predate RISC. RISC microprocessors have been the standard-bearers of performance, so Intel has embraced ideas from RISC and followed the quantitative approach. Thus both the ideas and the competition from RISC has benefited all computer users, since RISC has raised the performance target for the entire industry. With the announcement that Hewlett-Packard and Intel will move to a common instruction set in 1997, the end of the non-RISC architectures draws near.

4.1.5 The Future

Today's microprocessors are roughly 10,000 times faster than their ancestors. And microprocessor-based computer systems now cost only 1/40th as much as their ancestors, when inflation is considered. The result: an overall cost-performance improvement of roughly 1,000,000, in only 25 years! This extraordinary advance is why computing plays such a large role in today's world. Had the research at universities and industrial laboratories not occurred — had the complex interplay between government, industry, and academia not been so successful — a comparable advance would still be years away.

Microprocessor performance can continue to double every 18 months beyond the turn of the century. This rate can be sustained by continued research innovation. Significant new ideas will be needed in the next decade to continue the pace; such ideas are being developed by research groups today. The implications of such a breathtaking advance truly are limited only by our imaginations.

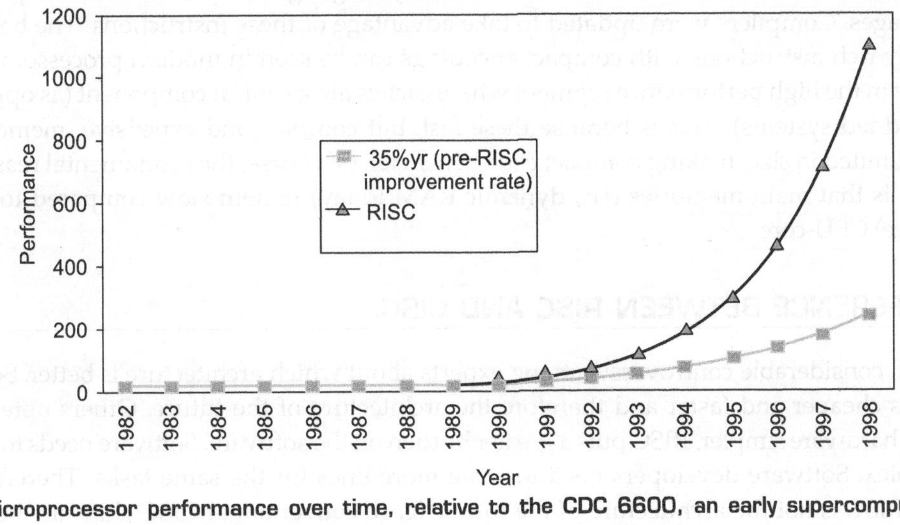

Microprocessor performance over time, relative to the CDC 6600, an early supercomputer

4.2 CISC

A **complex instruction set computer** is a computer instruction set architecture (ISA) in which each instruction can execute several low-level operations, such as a load from memory, an arithmetic operation, and a memory store, all in a single instruction. The term was retroactively coined in contrast to reduced instruction set computer (RISC).

Examples of CISC processor families are System/360, PDP-11, VAX, 68000, and x86.

Therefore fewer, simpler and faster instructions would be better, than the large, complex and slower CISC instructions. However, more instructions are needed to accomplish a task.

Another advantage of RISC is that - in theory - because of the more simple instructions, RISC chips require fewer transistors, which makes them easier to design and cheaper to produce.

Finally, it's easier to write powerful optimized compilers, since fewer instructions exist.

4.2.1 Incitements and Benefits

Before the RISC philosophy became prominent, many computer architects tried to bridge the so called semantic gap, *i.e.*, to design instruction sets that directly supported high-level programming constructs such as procedure calls, loop control, and complex addressing modes, allowing data structure and array accesses to be combined into single instructions. Instructions are also typically highly encoded in order to further enhance the code density. The compact nature of such instruction sets results in smaller program sizes and fewer (slow) main memory accesses, which at the time (early 1960s and onwards) resulted in a tremendous savings on the cost of computer memory and disc storage, as well as faster execution. It also meant good programming productivity even in assembly language, as high level languages such as Fortran or Algol were not always available or appropriate (microprocessors in this category are sometimes still programmed in assembly language for certain types of critical applications).

4.2.2 Performance

In the 70's analysis of high level languages indicated some complex machine language implementations and it was determined that new instructions could improve performance. Some instructions were added that were never intended to be used in assembly language but fit well with compiled high level languages. Compilers were updated to take advantage of these instructions. The benefits of semantically rich instructions with compact encodings can be seen in modern processors as well, particularly in the high performance segment where caches are a central component (as opposed to most embedded systems). This is because these fast, but complex and expensive, memories are inherently limited in size, making compact code beneficial. Of course, the fundamental reason they are needed is that main memories (*i.e.*, dynamic RAM today) remain slow compared to a (high performance) CPU-core.

4.3 DIFFERENCE BETWEEN RISC AND CISC

There is still considerable controversy among experts about which architecture is better. Some say that RISC is cheaper and faster and therefore the architecture of the future. Others note that by making the hardware simpler, RISC puts a greater burden on the software. Software needs to become more complex. Software developers need to write more lines for the same tasks. Therefore they argue that RISC is not the architecture of the future, since conventional CISC chips are becoming faster and cheaper anyway. RISC has now existed more than 10 years and hasn't been able to kick CISC out of the market. If we forget about the embedded market and mainly look at the market for PC's, workstations and servers I guess a least 75% of the processors are based on the CISC architecture. Most of them are the x86 standard (Intel, AMD, etc.), but even in the mainframe territory CISC is dominant via the IBM/390 chip. Looks like CISC is here to stay …

Is RISC then really not better? The answer isn't quite that simple. RISC and CISC architectures are becoming more and more alike. Many of today's RISC chips support just as many instructions as yesterday's CISC chips. The PowerPC 601, for example, supports *more* instructions than the Pentium. Yet the 601 is considered a RISC chip, while the Pentium is definitely CISC. Furthermore today's CISC chips use many techniques formerly associated with RISC chips.

4.4 SCALAR PROCESSOR

Scalar processors represent the simplest class of computer processors. A scalar processor processes one data item at a time (typical data items being integers or floating point numbers). In a vector processor, by contrast, a single instruction operates simultaneously on multiple data items. The difference is analogous to the difference between scalar and vector arithmetic.

4.5 SUPER SCALAR PROCESSOR

The evolution of microprocessors has reached the point where architectural concepts pioneered in vector processors and mainframe computers of the 1970s (most notably the CDC-6600 and Cray-1) are starting to appear in RISC processors. Early RISC machines were very simple single-chip processors. As VLSI technology improved more room became available on the chip. Rather than increase the complexity of the architecture, most designers decided to use this room on techniques to improve the execution of their current architecture. The two principle techniques are on-chip caches and instruction pipelines.

The latest step in this evolutionary process is the *superscalar processor*. The name means these processors are scalar processors that are capable of executing more than one instruction in each cycle. The keys to superscalar execution are an instruction fetching unit that can fetch more than one instruction at a time from cache; instruction decoding logic that can decide when instructions are independent and thus executed simultaneously; and sufficient execution units to be able to process several instructions at one time. Note that the execution units may be pipelined, *e.g.*, they may be floating point adders or multipliers, in which case the cycle time for each stage matches the cycle times on the fetching and decoding logic. In many systems the high level architecture is unchanged from earlier scalar designs. The superscalar designs use instruction level parallelism for improved implementation of these architectures.

Graphically the vector chaining is represented as shown in figure on next page:

A good example of a superscalar processor is the IBM RS/6000 [10]. There are three major subsystems in this processor: the instruction fetch unit, an integer processor, and a floating point processor. The instruction fetch unit is a 2-stage pipeline; during the first stage a packet of four instructions is fetched from an instruction cache, and in the second stage instructions are routed to the integer processor and/or floating point processor. An interesting feature of this instruction unit is that it executes branch instructions itself so that in a tight loop there is effectively no overhead from branching since the instruction unit executes branches while the data units are computing values. The integer unit is a four-stage pipeline. In addition to executing data processing instructions this unit does some preprocessing for the floating point unit. The floating point unit itself is six stages deep.

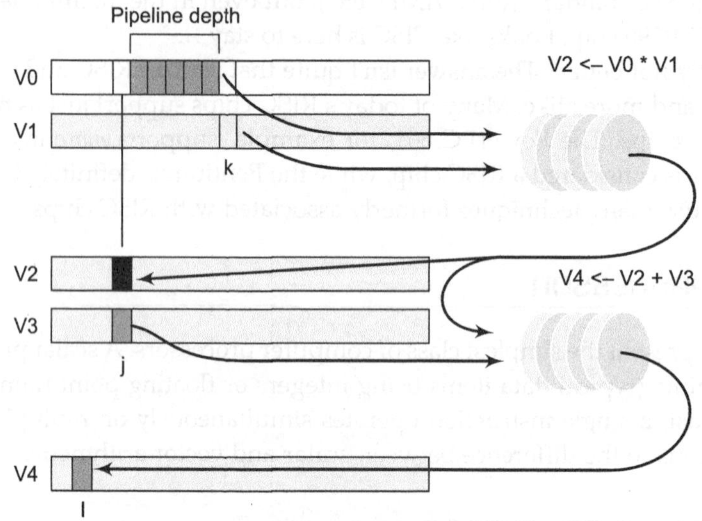

Vector Chaining

The following example from [10] shows the potential of this style of computing. This code from a computer graphics application rotates and displaces a set of (x, y) pairs by an angle q and displacement (x_d, y_d):

$$x_i' = x_i \cos \Theta - y_i \sin \Theta + x_d$$

$$y_i' = y_i \cos \Theta - x_i \sin \Theta + y_d$$

4.6 ARRAY PROCESSOR OR VECTOR PROCESSOR

A vector processor is a processor that can operate on entire vectors with one instruction, *i.e.*, the operands of some instructions specify complete vectors. For example, consider the following add instruction:

$$C = A + B$$

A **vector processor**, or **array processor**, is a CPU design wherein the instruction set includes operations that can perform mathematical operations on multiple data elements simultaneously. This is in contrast to a scalar processor, which handles one element at a time using multiple instructions. The vast majority of CPUs are scalar (or close to it). Vector processors were common in the scientific computing area, where they formed the basis of most supercomputers through the 1980s and into the 1990s, but general increases in performance and processor design saw the near-disappearance of the vector processor as a general-purpose CPU.

Today, most commodity CPU designs include single instructions for some vector processing on multiple (vectorised) data sets, typically known as SIMD (Single Instruction, Multiple Data), common examples include MMX, SSE, and AltiVec. Modern video game consoles and consumer computer-graphics hardware rely heavily on vector processing in their architecture. In 2000, IBM, Toshiba and Sony collaborated to create the Cell processor, consisting of one scalar processor and eight vector processors, which found use in the Sony PlayStation 3 among other applications.

Other CPU designs may include some multiple instructions for vector processing on multiple (vectorised) data sets, typically known as MIMD (**M**ultiple **I**nstruction, **M**ultiple **D**ata), such designs are very special and delicate breeds for dedicated purpose and these are not commonly marketed for general purpose applications.

The more advanced approach is not the active multiplicity of instructions in parallel but the active multiplicity in sequence, which led to the pipelining concept.

Description

In general terms, CPUs are able to manipulate one or two pieces of data at a time. For instance, many CPUs have an instruction that essentially says "add A to B and put the result in C," while others such as the MOS 6502 require two or three instructions to perform these types of operations.

The data for A, B and C could be—in theory at least—encoded directly into the instruction. However, things are rarely that simple. In general the data is rarely sent in raw form, and is instead "pointed to" by passing in an address to a memory location that holds the data. Decoding this address and getting the data out of the memory takes some time. As CPU speeds have increased, this *memory latency* has historically become a large impediment to performance.

In order to reduce the amount of time this takes, most modern CPUs use a technique known as instruction pipelining in which the instructions pass through several sub-units in turn. The first sub-unit reads the address and decodes it, the next "fetches" the values at those addresses, and the next does the math itself. With pipelining the "trick" is to start decoding the next instruction even before the first has left the CPU, in the fashion of an assembly line, so the address decoder is constantly in use. Any particular instruction takes the same amount of time to complete, a time known as the *latency*, but the CPU can process an entire batch of operations much faster than if it did so one at a time.

Vector processors take this concept one step further. Instead of pipelining just the instructions, they also pipeline the data itself. They are fed instructions that say not just to add A to B, but to add all of the numbers "from here to here" to all of the numbers "from there to there". Instead of constantly having to decode instructions and then fetch the data needed to complete them, it reads a single instruction from memory, and "knows" that the next address will be one larger than the last. This allows for significant savings in decoding time.

To illustrate what a difference this can make, consider the simple task of adding two groups of 10 numbers together. In a normal programming language you would write a "loop" that picked up each of the pairs of numbers in turn, and then added them. To the CPU, this would look something like this:

execute this loop 10 times
read the next instruction and decode it
fetch this number
fetch that number
add them
put the result here
end loop

But to a vector processor, this task looks considerably different:

read instruction and decode it
fetch these 10 numbers
fetch those 10 numbers
add them
put the results here

There are several savings inherent in this approach. For one, only two address translations are needed. Depending on the architecture, this can represent a significant savings by itself. Another saving is fetching and decoding the instruction itself, which has to be done only one time instead of ten. The code itself is also smaller, which can lead to more efficient memory use.

But more than that, the vector processor typically has some form of superscalar implementation, meaning there is not one part of the CPU adding up those 10 numbers, but perhaps two or four of them. Since the output of a vector command does not rely on the input from any other, those two (for instance) parts can each add five of the numbers, thereby completing the whole operation in half the time.

As mentioned earlier, the Cray implementations took this a step further, allowing several different types of operations to be carried out at the same time. Consider code that adds two numbers and then multiplies by a third; in the Cray, these would all be fetched at once, and both added and multiplied in a single operation. Using the pseudocode above, the Cray did:

read instruction and decode it
fetch these 10 numbers
fetch those 10 numbers
fetch another 10 numbers
add and multiply them
put the results here

The math operations thus completed far faster overall, the limiting factor being the time required to fetch the data from memory.

Not all problems can be attacked with this sort of solution. Adding these sorts of instructions necessarily adds complexity to the core CPU. That complexity typically makes *other* instructions run slower—*i.e.*, whenever it is **not** adding up many numbers in a row. The more complex instructions also add to the complexity of the decoders, which might slow down the decoding of the more common instructions such as normal adding.

In fact, vector processors work best only when there are large amounts of data to be worked on. For this reason, these sorts of CPUs were found primarily in supercomputers, as the supercomputers themselves were, in general, found in places such as weather prediction centres and physics labs, where huge amounts of data are "crunched".

4.7 MICROPROCESSOR WITH MMX TECHNOLOGY

In January of 1997 Intel released its newest, and presumably last, fifth generation processor, the Intel Pentium with MMX Technology. It is an evolutionary design, adding some enhancements and new capabilities to the "classic" Pentium. In most ways however, it is the same chip.

The chief (well-publicized) difference between the Pentium with MMX and the Pentium Classic is the inclusion of the MMX instruction set extensions. Running software that is designed for MMX on a Pentium MMX will result in a significant increase in performance over the regular Pentium; some claim up to 70%. Even running non-MMX software on the MMX chip results in an improvement of about 20% over a regular Pentium of the same clock speed, due to the following other improvements over the older chip:

- **Doubled Primary Cache:** The Pentium with MMX has 16 KB for each of the level 1 data and instruction caches, as opposed to 8 KB each for the regular Pentium.
- **Improved Cache Mapping:** The primary cache is now 4-way set associative instead of 2-way.

- **Deepening of Internal Pipelines:** Both of the internal integer executions units are increased from 5 to 6 stages.
- **Better Use of Internal Pipelines:** More types of instructions can be run in parallel down the two execution pipes than on the older Pentium, so more use is made of the second pipe.
- **Improved Branch Prediction Unit:** The branch prediction unit's accuracy is enhanced over the classic Pentium.
- **Improved Instruction Decoder:** The instruction decoder is more efficient than the Pentium's.

The Pentium with MMX is pin-compatible with the regular Pentium and goes in a Socket 7 just like the old chip. However, there is one important difference between the two: the Pentium with MMX uses split-rail voltage: 3.3 volts for the external voltage and 2.8 volts for the core. Not all motherboards with a socket 7 support this dual voltage setting, because before the Pentium with MMX came out no processors needed it, so some manufacturers skimped by not planning for the future. Intel is making OverDrives for motherboards that don't support the lower core voltage.

Warning: Some unscrupulous vendors are misleading buyers by telling them that the Pentium with MMX will run with 3.3 volts applied to both external *and* internal voltage. In many cases this will not instantly fry the chip; it may even work for a little while. Eventually however it will fail and you could void the warranty on the chip.

The MMX capability is implemented on the Pentium with MMX by sharing the registers used by the floating point unit. This has led to silly rumors that using MMX would cause a "major" performance penalty when switching between applications using MMX and those using floating point applications. This is a myth; aside from the fact that very few applications make extensive use of the floating point unit, a switch between MMX and non-MMX applications takes no more time than a switch between floating point and non-floating-point applications. And the amount of time for a switch is less than 100 cycles on a processor that is usually running at 200,000,000 cycles per second or more.

Originally, the Pentium with MMX was supposed to be available only in 166 and 200 MHz versions, after which Intel was going to stop production on Socket 7 chips entirely. Intel surprised the PC world a bit a few months after the introduction of the original 166 and 200 chips by announcing a Pentium with MMX 233, running at 66 MHz and 3.5x multiplier. They now say that *this* will be the last Pentium, but who knows?)

Since socket 7 boards don't support a 3.5x multiplier, Intel made the chip respond to a 1.5x signal as if it were 3.5xs. This is becoming a common trick, which I believe AMD used first in creating its 5x86-133 (which runs in 486 motherboards as 4x when configured as 2x). So to set up a Pentium with MMX 233, you jumper its system bus speed and multiplier (but *not* core voltage!) as if it were a Pentium 100.

Overall, the Pentium with MMX is a very good chip, and with Intel aggressively cutting prices it has become pretty much the mainstream processor of choice over the last year or so. In recent months, however, with new sixth-generation chips by AMD (the K6) and Cyrix (the 6x86MX) that will run in fifth-generation motherboards becoming commonplace, the high-end performance picture on fifth-generation motherboards has become much more cloudy. For its part, Intel has moved on to the Pentium II, and as a result the Pentium with MMX is being used on fewer and fewer new systems, in favour of the newer chip.

4.8 MICROPROCESSOR DEVELOPMENT SCHEMES

4.8.1 Changing the Focus: More Cores Instead of More Transistors

Microprocessor advancement accelerated rapidly in 1968 when three engineers from Fairchild Semiconductor – Robert Noyce, Andy Grove and Gordon Moore – founded Intel Corporation in Mountain View, California, to develop new technologies for silicon-based chips. In 1971, Intel developers successfully embedded a central processing unit, memory, input and output controls on the world's first single chip microprocessor, the Intel 4004 (U.S. Patent #3,821,715). Moore was well known in the industry for a 1965 article where he predicted the doubling of the number of transistors on microprocessors (an approximate measure of processing power) every 18 to 24 months. For more than 40 years, what became known as Moore's Law has accurately described the pace of advancements that have driven the semiconductor industry to reach over $200 billion in annual revenue today and serve as the foundation of the trillion-dollar electronics industry! But even Gordon Moore, as quoted recently by Techworld, says Moore's Law "can't continue forever." ("Moore's Law is dead, says Gordon Moore" by Manek Dubash, Techworld, April 13, 2005). Heat generation and circuit power requirements become more and more of a barrier as transistor density and processor clock speeds increase.

Either directly or indirectly, processor clock speed, expressed in Megahertz (MHz) or Gigahertz (GHz), was once the common reference point used to predict how a given processor would perform for a given application and where a processor ranked on performance comparisons. However, microprocessor designers today are more focused on other properties that enable processors to deliver higher performance. Among these are methods that leverage the latest silicon production processes and designs that minimize microprocessor footprint size, power consumption and heat generation. Designers are also concerned with microarchitecture optimization, multiprocessing parallelism, reliability, designed-in security features, memory structure efficiency and better synergy between the hardware and accompanying software tools, such as compilers and system and applications libraries. The more attention that a designer devotes to refining the efficiency of the software code rather than making the hardware responsible for dynamic optimization, the higher the ultimate system performance will be.

4.8.2 Replacing the x86 Paradigm

As Thomas Kuhn once observed ("The Structure of Scientific Revolution," Thomas Kuhn, 1962.), scientific advancement is not evolutionary, but is rather a "series of peaceful interludes punctuated by intellectually violent revolutions", and in those revolutions "one conceptual world view is replaced by another."

The conceptual framework of the Intel x86 or 80x86 microprocessor architecture was first introduced by Intel in 1978, and has since followed an unwavering path of advancement. Starting with an 8-bit instruction set, the x86 ISA grew to a 16- bit and then to a 32-bit instruction set. These labels (8-bit, 16-bit, 32-bit) designate the number of bits that each of the microprocessor's general-purpose registers (GPRs) can hold. The term "32-bit processor" translates to "a processor with GPRs that store 32-bit numbers." Similarly, a "32-bit instruction" is an instruction that operates on 32-bit numbers.

A 32-bit address space allows the CPU to directly address 4 GB of data. Though 4 GB once seemed gargantuan, the size requirements of memory-intensive applications such as multimedia programs or database query engines are often much higher. In response to this shortcoming, the 64-bit micro architecture has increased RAM addressability from 4 GB to a theoretical 18 million terabytes (1019 bytes). However, since the virtual address space of x86-64 is 48 bits and the physical address space is 40 bits, in reality the yield is only an approximate 256 terabytes. The 64-bit processor also has registers and arithmetic logic units that can manipulate larger instructions (64-bits worth) at each processing step. Since 64-bit processors can handle chunks of data and instructions twice as large as 32-bit processors, the 64-bit micro architecture should theoretically be able to process twice as much data per clock cycle as the 32-bit microprocessor. Unfortunately, that's not quite true. As Jonathan Stokes writes, "only applications that require and use 64-bit integers will see a performance increase on 64-bit hardware that is due solely to a 64-bit processor's wider registers and increased dynamic range."

4.8.3 Advancing to Multi-Core Designs

In order to advance beyond current x86 capabilities without adding more transistors, microprocessor designs are now headed down two paths. One path represents an embellishment of the traditional x86 design – multiple cores packaged on one board running parallel threads on each. The other path – Intel Itanium – represents one of those paradigm shifts described by Kuhn as a "revolution in which one conceptual world view is replaced by another."

Conceptually, multi-core architecture refers to a single processor package containing two or more processor "execution cores," or computational engines that deliver fully parallel execution of multiple software threads. The operating system treats each of its execution cores as a discrete processor, with all associated execution resources.

Multi-core processors thus deliver higher performance and greater efficiency without the heat problems and other disadvantages experienced by single core processors run at higher frequencies to squeeze out more performance. By multiplying the number of cores in the processor, it is possible to dramatically increase computing resources, higher multithreaded throughput, and the benefits of parallel computing.

Although multi-core technology was first discussed by Intel in 1989, the company released its first dual-core processor in April 2005 to mark the first step in its transition to multi-core computing. The company is now engaged in research on architectures that could include dozens or even hundreds of processors on a single die.

Intel has publicly committed itself to a vision of "moving beyond gigahertz" to deliver greater value, performance and functionality with multi-core with multicore architectures and a platform-centric approach. Central to this strategy, Intel and its industry partners allied through the Itanium Solutions Alliance (founded in September, 2005) are making billions of dollars in strategic technology investments to assure that the Intel Itanium processor becomes the platform of choice for mission critical enterprise systems and technical computing within four years. ISA founding sponsors include Bull, Fujitsu, Fujitsu Siemens Computers, Hitachi, HP, Intel, NEC, SGI and Unisys. Charter members include BEA, Microsoft, Novell, Oracle, Red Hat, SAP, SAS and Sybase. Over a dozen additional technology organizations have also joined the alliance.

4.8.4 Leveraging Software Tools for Optimal Performance

Some microprocessor designs of the past have been overly complex and have relied on out-of-order logic to reshuffle and optimize software instructions. Going forward, designers will continue to deliver better and better software tools, higher software optimization and better compilers.

Among semiconductor companies, Intel provides one of the strongest suites of software tools to support and enhance the performance of microprocessors such as the Intel Itanium. This includes an 18-month roadmap that Intel shares with its major customers showing how its software tools will advance and evolve over time. This enables Itanium developers to deliver higher performance to their users and customers.

Among the most important software tools from Intel are a family of compilers optimized for each of its microprocessor platforms, including Xeon and Itanium. According to Intel, some of these compilers work cross platform, often enabling one set of source code to be delivered for multiple platforms, occasionally requiring a simple recompile for each platform. Leveraging a compiler to enhance code performance for Itanium applications is a key benefit for Itanium users, as explained by technology writer Johan De Gelas. "The main philosophy behind Itanium is... that a compiler can statically schedule instructions much better than a hardware scheduler, which has to decide this dynamically in a few clock cycles... the compiler can search through thousands of instructions ahead while the hardware scheduler can check only a few tens of instructions for independent instructions. The compiler will make groups of instructions that can be issued simultaneously without dependencies or interlocks. These groups can be one or tens of instructions."

In addition to Itanium compilers, Intel offers a math kernel library (MKL) and an integrative performance primitive's library (IPP), that can significantly enhance Itanium code performance for scientific cluster applications, video image processing, audio processing and image recognition.

4.8.5 Optimizing Micro Architecture through Parallelism

As microprocessor technology continues to advance with new architectures such as the Intel Itanium, efficiency and performance optimization become even more critical. A new technique used to achieve this is multi-processing parallelism and a shift from instruction level parallelism to thread level parallelism.

The Intel Itanium processor family is designed around small footprint cores that are remarkably compact in terms of transistor count, especially when one considers the amount of processing work that they accomplish. Itanium has taken instruction level parallelism to a new level, and this can be used to leverage more processor cores and more threads per core to produce higher performance.

When Intel introduced Itanium in 2001, the company made a commitment to a design that takes a quantum leap ahead of instruction level parallelism. Instruction level parallelism is a process in which independent instructions (instructions not dependent on the outcome of one another) execute concurrently to utilize more of the available resources of a processor core and increase instruction throughput. The ability of the processor to work on more than one instruction at a time improves the cycles per instruction and increases the frequency from previous architectures. Contrast this with a processor equipped with thread-level parallelism that executes separate threads of code. This could be one thread running from an application and a second thread running from an operating system, or parallel threads running from within a single application.

In the past, microprocessor advancements have been based on improving the performance of a single thread. Since there was only one core per processor, advancements were achieved by adding more transistors and increasing the speed of each transistor to improve program performance. This, along with various pipelining and code strategies, made it possible for multiple instructions to be issued in parallel.

Moving forward, Itanium will leverage multiple cores and thread level parallelism to produce greater performance enhancements that would be possible through single cores and single thread performance boost. This move toward chip-level multiprocessing architectures with a large number of cores continues a decades-long trend at Intel, offering dramatically increased performance and power characteristics. This also presents significant challenges, including a need to make multi-core processors easy to program, which is accomplished, in part, using the software tools described above.

4.8.6 Combining Performance with Security

The dramatic performance advancements delivered by Itanium make it possible for additional security features to be incorporated into applications, enabling developers to avoid the trade-offs often made between performance and security.

Itanium provides four privilege levels that the operating system can leverage to provide a clean separation between what the user can access versus what the virtual machine can access. More importantly, Itanium provides a protection key scheme based on container logic. With the appropriate application support, a developer can compartmentalize everything contained in vast memory stores, and, in effect, put a security wrapper around each process. Itanium's ability to run high degrees of instructions per cycle means that developers can create security layers that run very efficiently to protect code stacks against a variety of attacks. There's also an application that takes strong advantage of the instructions per cycle to create an efficient security layer. In addition to that, there are floating point units unique to Itanium that provide very fast encryption The combination of all of these features – parallelism, floating point units, privilege levels, and the protection key scheme, when taken together and using the right application, provide micro architecture security that is second-to-none.

4.8.7 Further Advancing the Microprocessor: 2015 and Beyond

Considering the dramatic progress made in microprocessor design and architecture since 1965, it's risky to project what new technologies could become available in ten to fifteen years. And yet, a group of future-minded researchers are expressing optimism about the potential of tiny nanoelectronic components, organic molecules, carbon nanotubes and individual electrons that could serve as the underlying technology for new generation of microprocessors emerging around 2015.

There are limits to what can be accomplished with silicon, says Philip J. Kuekes, a physics researcher at Hewlett-Packard Laboratories quoted in the New York Times. Kuekes says H-P is currently working on molecular-scale nanotechnology switches that, it is hoped, will be able to overcome some of the present-day technological challenges discussed throughout this section.

EXERCISE

1. Explain the Architectural advancements of microprocessors.
2. What are RISC and CISC? Explain
3. What do you understand by "Multimedia extension" technology?
4. What do you mean by Pipelining? How does this occur?
5. Explain the concept of cache memory.
6. Explain the various generations of microprocessors.
7. Give the difference between third generation and fourth generation microprocessor.
8. What is the meaning of optimizing micro architecture through parallism?
9. Explain the concept of multi core processor design.
10. Explain the various microprocessor development schemes.

5

8085 Microprocessor Architecture

The **Intel 8085** is an 8-bit microprocessor introduced by Intel in 1977. It was binary-compatible with the more-famous Intel 8080 but required less supporting hardware, thus allowing simpler and less expensive microcomputer systems to be built.

The "5" in the model number came from the fact that the 8085 required only a +5-volt (V) power supply rather than the +5V, –5V and +12V supplies the 8080 needed. Both processors were sometimes used in computers running the CP/M operating system, and the 8085 later saw use as a microcontroller (much by virtue of its component count reducing feature). Both designs were eclipsed for desktop computers by the compatible but more capable Zilog Z80, which took over most of the CP/M computer market as well as taking a large share of the booming home computer market in the early-to-mid-1980s.

The 8085 had a very long life as a controller. Once designed into such products as the DECtape controller and the VT100 video terminal in the late 1970s, it continued to serve for new production throughout the life span of those products (generally many times longer than the new manufacture lifespan of desktop computers).

5.1 DESCRIPTION

The 8085 is a conventional von Neumann design based on the Intel 8080. Unlike the 8080 it had no state signals multiplexed onto the data bus, but the 8-bit data bus was instead multiplexed with the lower part of the 16-bit address bus (in order to limit the number of pins to 40). The processor was designed using nMOS circuitry and the later "H" versions were implemented in Intel's enhanced nMOS process called HMOS, originally developed for fast static RAM products. The 8085 used approximately 6,500 transistors.

The 8085 incorporated the functionality of the 8224 (clock generator) and the 8228 (system controller), increasing the level of integration. A downside compared to similar contemporary designs (such as the Z80) was the fact that the buses required demultiplexing, however, address latches in the Intel 8155, 8355, and 8755 memory chips allowed a direct interface, so an 8085 along with these chips was almost a complete system.

The 8085 had extensions to support new interrupts: It had three maskable interrupts (RST 7.5, RST 6.5 and RST 5.5), one Non-Maskable interrupt (TRAP), and one externally serviced interrupt (INTR). The RST n.5 interrupts refer to actual pins on the processor—a feature which permitted simple systems to avoid the cost of a separate interrupt controller.

Like the 8080, the 8085 could accommodate slower memories through externally generated wait states (pin 35, READY), and had provisions for Direct Memory Access (DMA) using HOLD and HLDA signals (pins 39 and 38). An improvement over the 8080 was that the 8085 can itself drive a piezoelectric crystal directly connected to it, and a built in clock generator generates the internal high amplitude two-phase clock signals at half the crystal frequency (a 6.14 MHz crystal would yield a 3.07 MHz clock for instance).

The functional block diagram or architechture of 8085 Microprocessor is very important as it gives the complete details about a Microprocessor Figure shows the Block diagram of a Microprocessor.

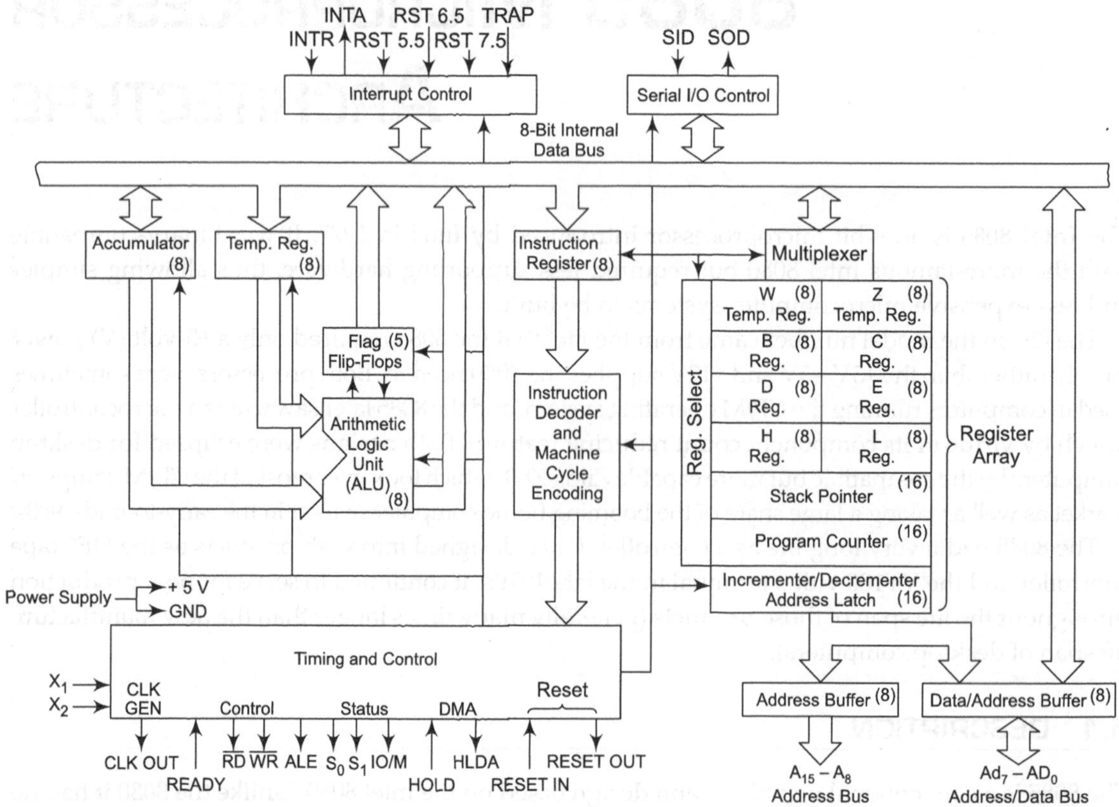

5.2 FEATURES OF 8085 MICROPROCESSOR

Before knowing about the 8085 architecture in detail, let us briefly discuss about the basic features of 8085 processor.

8085 microprocessor is an 8-bit microprocessor with a 40 pin dual in line package. The address and data bus are multiplexed in this processor which helps in providing more control signals. 8085 microprocessor has 1 Non-maskable interrupt and 3 maskable interrupts. It provides serial interfacing with serial input data (SID) and serial output data (SOD).

It has a set of registers for performing various operations. The various registers include

- Accumulator (register A)
- Registers: B, C, D, E, H and L
- Stack pointer
- Program Counter

- Temporary register
- Instruction register

5.2.1 Specific Features of Microprocessor

The following are the specific features of Microprocessor:

(*a*) **The Microprocessor's Power:** The power of a microprocessor is measured in bits. The more bits, the more information (data) the microprocessor is capable of bandying about and, therefore, the more powerful the microprocessor. Specifically, an 8-bit CPU can add/subtract/ multiply/etc. two 8-bit numbers, while a 32-bit ALU can manipulate 32-bit numbers. An 8-bit arithmetic logic unit (ALU) would have to execute four instructions to add two 32-bit numbers, while a 32-bit ALU can do it in one instruction.

In a way, the number of bits can be compared to lanes on a freeway: If you have only two lanes, too many cars congest traffic, and things slow down. A six-lane freeway, however, has plenty of room for lots of cars, and traffic flows smoothly. With more bits, just as with more lanes on a freeway, the microprocessor can do more powerful programming, operations and processes.

(*b*) **Microprocessor Speed:** Technically, microprocessor speed is measured in *hertz,* or cycles per second. Microprocessors are capable of doing *billions* of things in one second. Therefore, their speed is measured in gigahertz (GHz). A microprocessor that can do 1 billion things per second is rated at 1.0 GHz. A microprocessor that can do 3.66 billion things per second is rated at 3.66 GHz. This rate is much faster than 1.0 GHz. The higher the speed value, the faster the chip. And, naturally, the faster the chip, the more you can do in a given amount of time.

5.3 8085 BUS STRUCTURE

8085 Bus Structure is represented as follows:

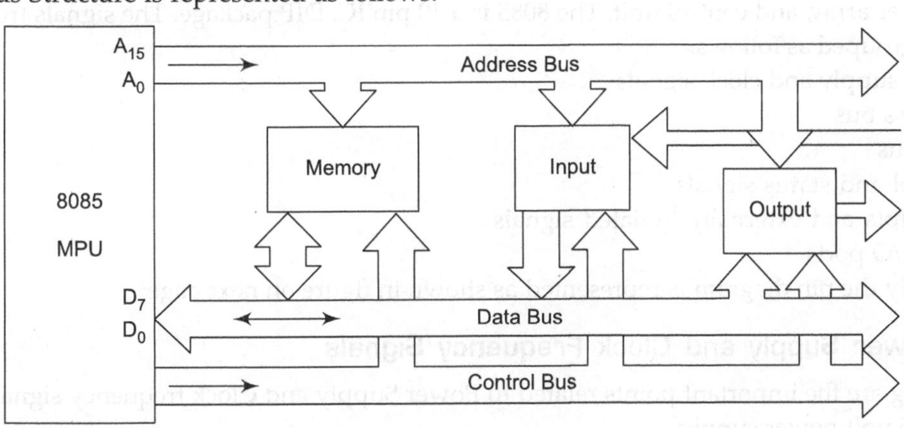

5.3.1 Address Bus

The following are the important points related to Address Bus:

- The address bus is a group of 16 lines generally identified as A0 to A15.
- The address bus is unidirectional: bits flow in one direction-from the MPU to peripheral devices.
- The MPU uses the address bus to perform the first function: identifying a peripheral or a memory location.

5.3.2 Data Bus

The following are the important points related to Data Bus:

- The data bus is a group of eight lines used for data flow.
- These lines are bi-directional—data flow is in both directions between the MPU and memory and peripheral devices.
- The MPU uses the data bus to perform the second function: transferring binary information.
- The eight data lines enable the MPU to manipulate 8-bit data ranging from 00 to FF (28 = 256 numbers).
- The largest number that can appear on the data bus is 11111111.

5.3.3 Control Bus

The following are the important points related to Control Bus:

- The control bus carries synchronization signals and provides timing signals.
- The MPU generates specific control signals for every operation it performs. These signals are used to identify a device type with which the MPU wants to communicate.

5.4 PIN DIAGRAM OF 8085 MICROPROCESSOR

The microprocessor is a clock-driven semiconductor device consisting of electronic logic circuits manufactured by using either a large-scale integration (LSI) or very-large-scale integration (VLSI) technique. The microprocessor is capable of performing various computing functions and making decisions to change the sequence of program execution. In large computers, a CPU implemented on one or more circuit boards performs these computing functions. The microprocessor is in many ways similar to the CPU, but includes the logic circuitry, including the control unit, on one chip. The microprocessor can be divided into three segments for the sake of clarity, arithmetic/logic unit (ALU), register array, and control unit. The 8085 is a 40 pin IC, DIP package. The signals from the pins can be grouped as follows:

- Power supply and clock signals
- Address bus
- Data bus
- Control and status signals
- Interrupts and externally initiated signals
- Serial I/O ports

Graphically the pin diagram is represented as shown in figure on next page:

5.4.1 Power Supply and Clock Frequency Signals

The following are the important points related to Power Supply and Clock frequency signals:

- Vcc + 5 volt power supply
- Vss Ground
- X1, X2: Crystal or R/C network or LC network connections to set the frequency of internal clock generator.
- The frequency is internally divided by two. Since the basic operating timing frequency is 3 MHz, a 6 MHz crystal is connected externally.
- CLK (output)-Clock Output is used as the system clock for peripheral and devices interfaced with the microprocessor.

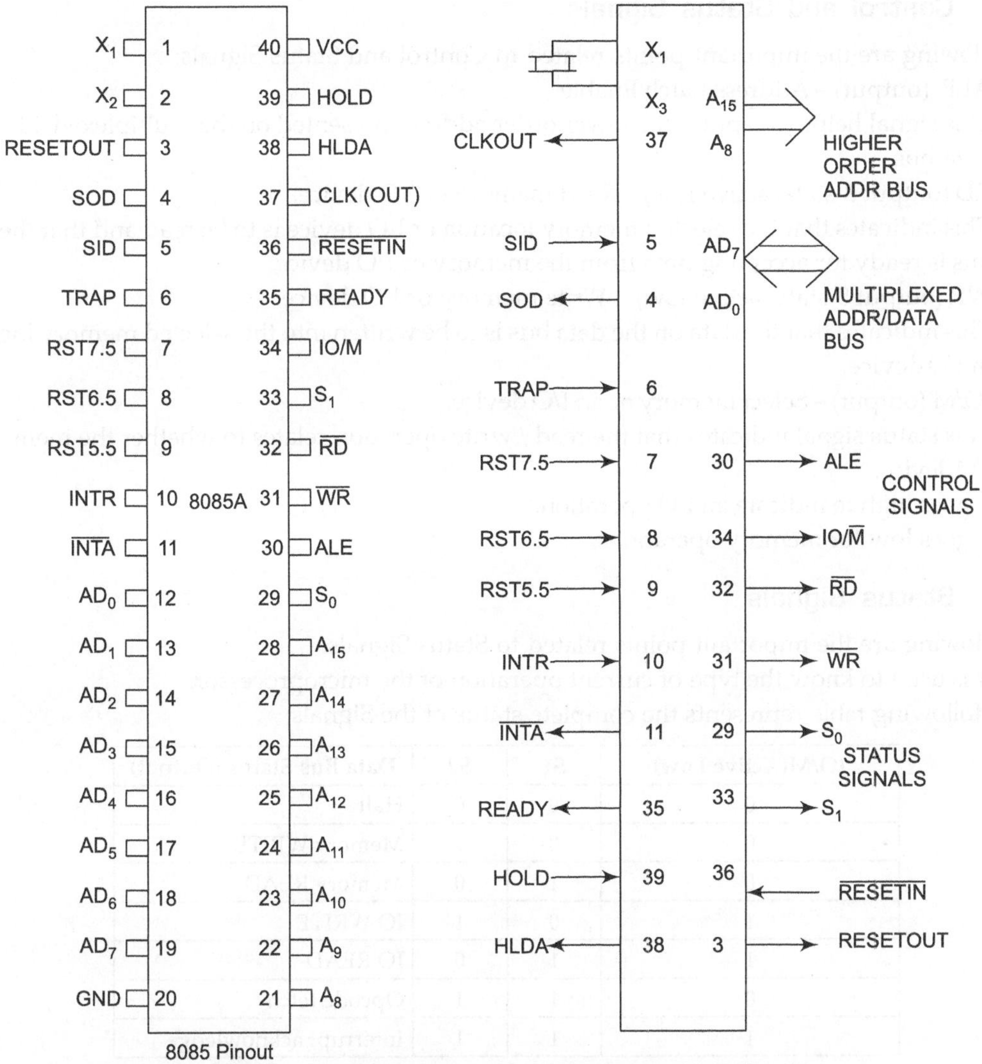

(a) - Pin Diagram of 8085 & (b) - Logical schematic of Pin diagram

5.4.2 Address Bus

The following are the important points related to Address Bus:

- A8 – A15 (output; 3-state)
- It carries the most significant 8 bits of the memory address or the 8 bits of the I/O address.

5.4.3 Multiplexed Address/Data Bus

The following are the important points related to Multiplexed Address/Data Bus:

- AD0 – AD7 (input/output; 3-state)
- These multiplexed set of lines are used to carry the lower order 8 bit address as well as data bus.
- During the opcode fetch operation, in the first clock cycle, the lines deliver the lower order address A0 – A7.
- In the subsequent IO/memory, read/write clock cycle, the lines are used as data bus.
- The CPU may read or write out data through these lines.

5.4.4 Control and Status Signals

The following are the important points related to Control and Status Signals:

- ALE (output) – Address Latch Enable.
- This signal helps to capture the lower order address presented on the multiplexed address / data bus.
- RD (output 3-state, active low) – Read memory or I/O device.
- This indicates that the selected memory location or I/O device is to be read and that the data bus is ready for accepting data from the memory or I/O device.
- WR (output 3-state, active low) – Write memory or I/O device.
- This indicates that the data on the data bus is to be written into the selected memory location or I/O device.
- IO/M (output) – Select memory or an I/O device.
- This status signal indicates that the read / write operation relates to whether the memory or I/O device.
- It goes high to indicate an I/O operation.
- It goes low for memory operations.

5.4.5 Status Signals

The following are the important points related to Status Signals:

- It is used to know the type of current operation of the microprocessor.

The following table represents the complete status of the Signals:

IO/M(Active Low)	S1	S2	Data Bus Status (Output)
0	0	0	Halt
0	0	1	Memory WRITE
0	1	0	Memory READ
1	0	1	IO WRITE
1	1	0	IO READ
0	1	1	Opcode fetch
1	1	1	Interrupt acknowledge

5.4.6 Interrupts and Externally Initiated Operations

The following are the important points related to Interrupts and externally initiated operations:

- They are the signals initiated by an external device to request the microprocessor to do a particular task or work.
- There are five hardware interrupts called,

TRAP
RST 7.5
RST 6.5 (inputs)
RST 5.5
INTR
INTA (active low output)

- On receipt of an interrupt, the microprocessor acknowledges the interrupt by the active low INTA (Interrupt Acknowledge) signal.

Reset In (input, active low)

- This signal is used to reset the microprocessor.
- The program counter inside the microprocessor is set to zero.
- The buses are tri-stated.

Reset out (Output)

- It indicates CPU is being reset.
- Used to reset all the connected devices when the microprocessor is reset.

5.4.7 Direct Memory Access (DMA)

Tri state devices are represented as follows:

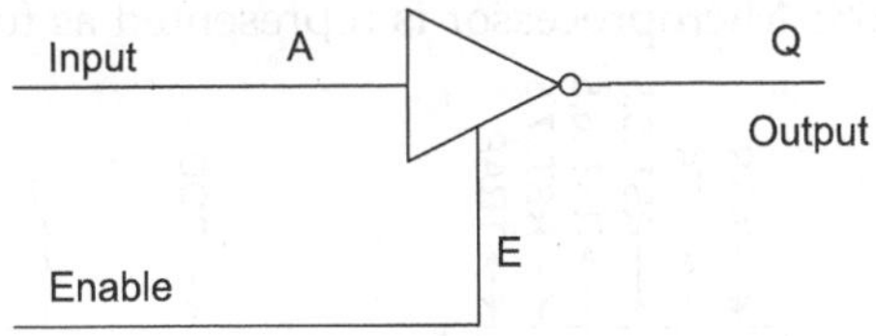

The following are the important points related to Direct Memory Access:

- 3 output states are high & low states and additionally a high impedance state.
- When enable E is high the gate is enabled and the output Q can be 1 or 0 (if A is 0, Q is 1, otherwise Q is 0). However, when E is low the gate is disabled and the output Q enters into a high impedance state.
- For both high and low states, the output Q draws a current from the input of the OR gate.
- When E is low, Q enters a high impedance state; high impedance means it is electrically isolated from the OR gate's input, though it is physically connected. Therefore, it does not draw any current from the OR gate's input.
- When 2 or more devices are connected to a common bus, to prevent the devices from interfering with each other, the tristate gates are used to disconnect all devices except the one that is communicating at a given instant.
- The CPU controls the data transfer operation between memory and I/O device. Direct Memory Access operation is used for large volume data transfer between memory and an I/O device directly.
- The CPU is disabled by tri-stating its buses and the transfer is effected directly by external control circuits.
- HOLD signal is generated by the DMA controller circuit. On receipt of this signal, the microprocessor acknowledges the request by sending out HLDA signal and leaves out the control of the buses. After the HLDA signal the DMA controller starts the direct transfer of data.

READY (input)

- Memory and I/O devices will have slower response compared to microprocessors.
- Before completing the present job such a slow peripheral may not be able to handle further data or control signal from CPU.
- The processor sets the READY signal after completing the present job to access the data.
- The microprocessor enters into WAIT state while the READY pin is disabled.

Single Bit Serial I/O ports

- SID (input) – Serial input data line
- SOD (output) – Serial output data line
- These signals are used for serial communication.

E	A	Q	State
1(high)	0	1	High
1	1	0	Low
0(low)	0	0	High impedance
0	1	0	High impedance

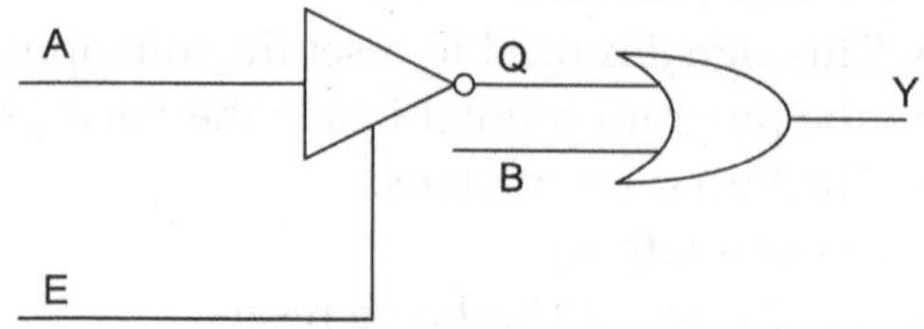

(a) Pin Diagram of 8085 (b) Logical schematic of Pin diagram

5.5 8085 INTERNAL ARCHITECTURE

The internal Architecture of 8085 Microprocessor is represented as follows:

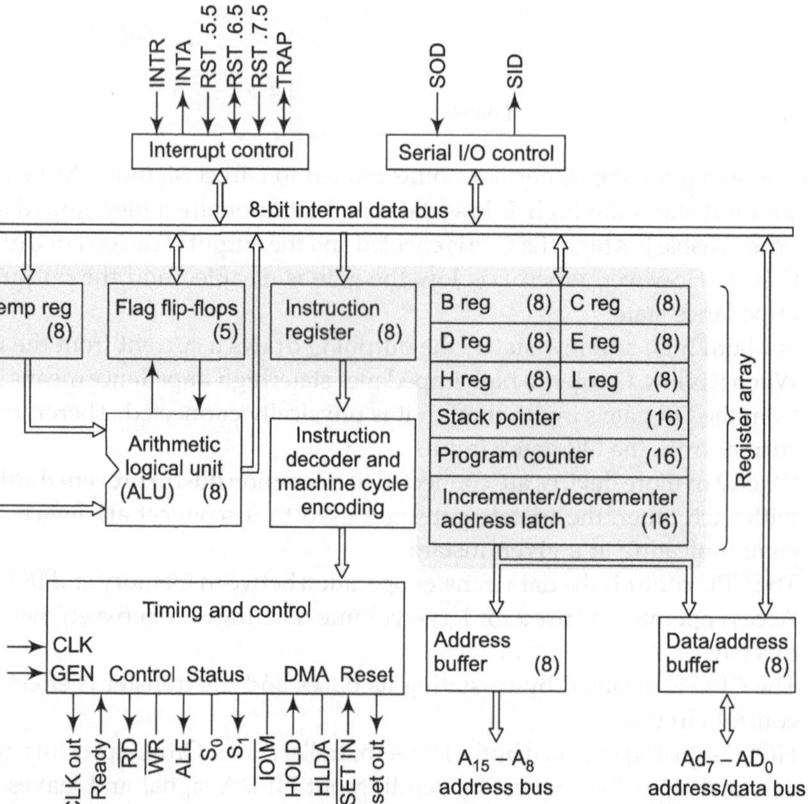

General Microprocessor Architecture

Following is the explanation of the above figure:

(*a*) **Control Unit:** Generates signals within microprocessor to carry out the instruction, which has been decoded. In reality causes certain connections between blocks of the microprocessor to be opened or closed, so that data goes where it is required, and so that ALU operations occur.

(*b*) **Arithmetic Logic Unit:** The ALU performs the actual numerical and logic operation such as 'add', 'subtract', 'AND', 'OR', etc. Uses data from memory and from Accumulator to perform arithmetic. Always stores result of operation in Accumulator.

(*c*) **Registers:** The 8085/8080A-programming model includes six registers, one accumulator, and one flag register, as shown in the above figure. In addition, it has two 16-bit registers: the stack pointer and the program counter. They are described briefly as follows.
The 8085/8080A has six general-purpose registers to store 8-bit data; these are identified as B, C, D, E,H, and L as shown in the figure. They can be combined as register pairs—BC, DE, and HL—to perform some 16-bit operations. The programmer can use these registers to store or copy data into the registers by using data copy instructions.

(*d*) **Accumulator:** The accumulator is an 8-bit register that is a part of arithmetic/logic unit (ALU). This register is used to store 8-bit data and to perform arithmetic and logical operations. The result of an operation is stored in the accumulator. The accumulator is also identified as register A.

(*e*) **Flags:** The ALU includes five flip-flops, which are set or reset after an operation according to data conditions of the result in the accumulator and other registers. They are called Zero (Z), Carry (CY), Sign (S), Parity (P), and Auxiliary Carry (AC) flags. The most commonly used flags are Zero, Carry, and Sign. The microprocessor uses these flags to test data conditions.
For example, after an addition of two numbers, if the sum in the accumulator is larger than eight bits, the flip-flop uses to indicate a carry — called the Carry flag (CY) — is set to one. When an arithmetic operation results in zero, the flip-flop called the Zero (Z) flag is set to one. The first Figure shows an 8-bit register, called the flag register, adjacent to the accumulator. However, it is not used as a register; five bit positions out of eight are used to store the outputs of the five flip-flops. The flags are stored in the 8-bit register so that the programmer can examine these flags (data conditions) by accessing the register through an instruction.
These flags have critical importance in the decision-making process of the microprocessor. The conditions (set or reset) of the flags are tested through the software instructions. For example, the instruction JC (Jump on Carry) is implemented to change the sequence of a program when CY flag is set. The thorough understanding of flag is essential in writing assembly language programs.

(*f*) **Program Counter (PC):** This 16-bit register deals with sequencing the execution of instructions. This register is a memory pointer. Memory locations have 16-bit addresses, and that is why this is a 16-bit register.
The microprocessor uses this register to sequence the execution of the instructions. The function of the program counter is to point to the memory address from which the next byte is to be fetched. When a byte (machine code) is being fetched, the program counter is incremented by one to point to the next memory location.

(*g*) **Stack Pointer (SP):** The stack pointer is also a 16-bit register used as a memory pointer. It points to a memory location in R/W memory, called the stack. The beginning of the stack is defined by loading 16-bit address in the stack pointer. The stack concept is explained in the chapter "Stack and Subroutines."

(*h*) **Instruction Register/Decoder:** Temporary store for the current instruction of a program. Latest instruction sent here from memory prior to execution. Decoder then takes instruction and 'decodes' or interprets the instruction. Decoded instruction is then passed to next stage.

(*i*) **Memory Address Register:** Holds address, received from PC, of next program instruction. Feeds the address bus with addresses of location of the program under execution.

(*j*) **Control Generator:** Generates signals within microprocessor to carry out the instruction which has been decoded. In reality causes certain connections between blocks of the microprocessor to be opened or closed, so that data goes where it is required, and so that ALU operations occur.

(*k*) **Register Selector:** This block controls the use of the register stack in the example. Just a logic circuit which switches between different registers in the set will receive instructions from Control Unit.

(*l*) **General Purpose Registers:** Microprocessor requires extra registers for versatility. Can be used to store additional data during a program. More complex processors may have a variety of differently named registers.

(*m*) **Microprogramming:** How the ìP does know what an instruction means, especially when it is only a binary number? The microprogram in a microprocessor/microcontroller is written by the chip designer and tells the microprocessor/microcontroller the meaning of each instruction, microprocessor/microcontroller can then carry out operation.

5.6 8085 TIMING DIAGRAM

Timing Diagram is a graphical representation. It represents the execution time taken by each instruction in a graphical format. The execution time is represented in T-states.

The timing diagram is the diagram which provides information about the various conditions of signals such as high/low, when a machine cycle is being executed. Without the knowledge of timing diagram it is not possible to match the peripheral devices to the microprocessors. These peripheral devices include memories, ports etc. Such devices can only be matched with microprocessors with the help of timing diagram.

Graphically it can be represented as follows:

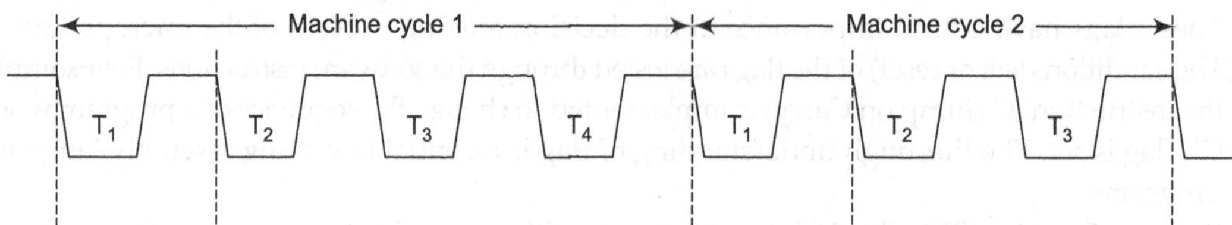

Before dealing with timing diagram, we have to make ourselves familiar with certain terms.

(*a*) **Instruction Cycle:** Assume that the microprocessor is executing an instruction. Instruction cycle is nothing but the time taken to complete the execution of that instruction by the microprocessor. The 8085 instruction cycle consists of 1 to 6 machine cycles.

(*b*) **Machine Cycle:** Machine cycle is nothing but the time required to complete one operation of accessing memory, I/O. It is also the time required to complete one operation of acknowledging an external request. One machine cycle may consist of 3 to 6 T-states.

(*c*) **T-state:** T-state is nothing but one subdivision of the operation performed in one clock period. These subdivisions are internal state of the microprocessor synchronized with system clock.

Now let us discuss the timing diagram for various signals that are associated with 8085 microprocessor.

5.6.1 Timing Diagram of Various Signals

(*a*) **Address Latch Enable:** Address latch enable is an active high signal, (*i.e.*) the latch becomes enabled when the signal is high. It is activated during the beginning of T1 state of each machine cycle and it remains active in the T1state. But in case of bus idle machine cycle it is not activated.

(*b*) **Data Bus (D0-D7):** While dealing with data bus, two types of data flow are possible. The data can be transferred from memory to microprocessor and vice versa.
This process occurs during the T2 and T3 states.
There are 2 cycles. One is Read machine cycle and the other is write machine cycle. In read machine cycle, the data will appear during the later part of T2 state, while in Write machine cycle the data will appear on the beginning of T2 state.
But for reading a data from memory or I/O device, first we need to select the required device. After selecting the device, the required data to be read or written is taken from the selected location and placed on data bus. A certain amount of time is required to perform this action. This time is called "access time". But for write cycle the access time is 0. This is because the data to be written is present on the registers of microprocessor and so it can put the data directly to data bus without any time delay.

(*c*) **Lower Byte Addresses (A0-A7):** The lower byte of address is available on the time multiplexed address/date bus during the T1 state of machine cycle, except the bus idle machine cycle.

(*d*) **Higher Byte Addresses (A8-A15):** The higher byte addresses (A8-A15) is available for T1, T2 and T3 states of each machine cycle, except the bus idle machine cycle.

(*e*) **IO/M′, S0, S1:** From the previous discussions about 8085 microprocessor, we very well know that IO/M′, S0, S1 are the status signals of the microprocessor. These status signals decide the type of machine cycle is to be executed. So they remain activated from the beginning T1 state of a particular machine cycle and remain till the end of that machine cycle.

(*f*) **RD′ and WR′:** These 2 signals RD′ and WR′ decides the direction of the data transfer.

RD′ is Active: When RD′ goes active, the data is transmitted from memory, I/O device or any other peripherals to the microprocessor.

WR′ is active: When WR′ goes active, the data is transmitted from microprocessor to the memory or any other peripheral devices.

Now there might be a question arising within your mind.

Can both the signals become active at same time? The answer for this question is NO. In 8085 microprocessor either RD′ goes high or WR′ goes high. Both cannot take place at same time.

The data transfer both RD′ and WR′ takes place during T2 and T3 states of machine cycle. So these signals are activated during the T2 and T3 states.

5.7 CLASSIFICATION OF MACHINE CYCLES

The timing diagram of a microprocessor consists of machine cycles and instruction cycles. There are different types of machine cycles and they are based on various factors. Read here to know about the machine cycle and its types in detail.

A machine cycle is the time required to complete one operation of accessing the memory, I/O or acknowledge an external signal or request. Usually machine cycle consists of 3 to 6 T-states. In this section let us discuss about their different types and how they are being classified.

Types of machine cycle:

There are various types of machine cycles which are classified based on

Status signals (IO/M′, S1 and S0)

Control Signals (RD′, WR′, INTA)

The different types of machine cycle available in 8085 microprocessor are:

- Opcode Fetch
- Memory Read

- Memory write
- I/O Read
- I/O Write
- INTR Acknowledge
- Bus Idle

The following table shows the information about all the cycles:

Machine Cycle	Status $IO/\overline{M}$	S1	S0	No. of Machine cycles	Control
Opcode Fetch	0	1	1	4	$\overline{RD}$ H0
Memory Read	0	1	0	3	$\overline{RD}$ H0
Memory Write	0	0	1	3	$\overline{WR}$ H0
I/O Read	1	1	0	3	$\overline{RD}$ H0
I/O Write	1	0	1	3	$\overline{WR}$ H0
INTR Acknowledge	1	1	1	3	$\overline{INTA}$ H0
Bus Idle	0	0	0	3	—

(a) Opcode Fetch Machine Cycle

The Opcode fetch cycle, fetches the instructions from memory and delivers it to the instruction register of the microprocessor. For any instruction cycle, Opcode fetch is the first machine cycle. We know that each machine cycle may have 3 to 6 T-states. This Opcode fetch machine cycle consists of 4 T-states.

T1 State

During the T1 state, the contents of the program counter are placed on the 16 bit address bus. The higher order 8 bits are transferred to address bus (A8-A15) and lower order 8 bits are transferred to multiplexed A/D (AD0-AD7) bus.

After the address bits are transferred, the ALE (address latch enable) signal goes high. As soon as ALE goes high, the memory latches the AD0-AD7 bus. At the middle of the T state the ALE goes low and the complete 16-bit address is made available for the Opcode fetch machine cycle.

T2 State

During the beginning of this state, the RD′ signal goes low to enable memory. It is during this state, the selected memory location is placed on D0-D7 of the Address/Data multiplexed bus.

T3 State

In the previous state the Opcode is placed in D0-D7 of the A/D bus. In this state of the cycle, the Opcode of the A/D bus is transferred to the instruction register of the microprocessor. Now the RD′ goes high after this action and thus disables the memory from A/D bus.

T4 State

In this state the Opcode which was fetched from the memory is decoded.

Thus the cycle completes after 4 T-states. This very well explains the Opcode fetch machine cycle. For better understanding of the concept, a diagram explaining Opcode fetch cycle is shown below.

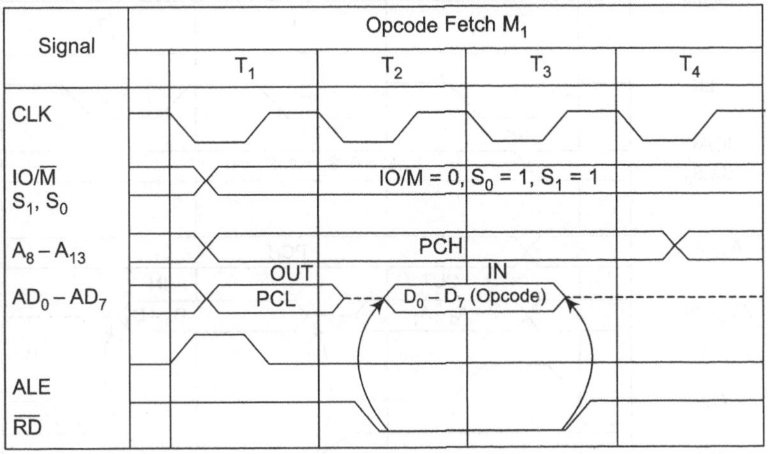

(b) Memory Read Machine Cycle

If the instruction is only one byte in length, then one machine cycle is enough to complete the process (Opcode fetch cycle). When the instruction has more than one byte of information to be processed, then the microprocessor may require more than one machine cycle to complete the process. The machine cycle in this case would require reading of address or data from memory or any other I/O devices. Hence these are known as memory read or I/O read machine cycles. These machine cycles have 3 T-states.

A simple example for memory read machine cycle is **MVI D, 24H.**

For the above example there are 2 machine cycles involved. One is the Opcode fetch cycle and the second one is the memory read cycle which transfers the operand 24H from the memory to the microprocessor.

T1 State

In this state the contents of the program counter is placed on the higher order address bus (A8-A15) and lower order address and data multiplexed (AD0-AD7) bus. ALE goes high so that the memory latches the (AD0-AD7) and then during the middle of the T1 state ALE goes low, so that complete 16-bit address are available.

The microprocessor then identifies the memory read machine cycle from the status signals IO/M′=0, S1=1, S0=0. This condition indicates the memory read cycle.

T2 State

Selected memory location is placed on the (D0-D7) of the A/D multiplexed bus.

T3 State

The data which was loaded on the previous state is transferred to the microprocessor. In the middle of the T3 state RD′ goes high and disables the memory read operation. The data which was obtained from the memory is then decoded.

The concept can be understood better with the aid of the diagram shown below.

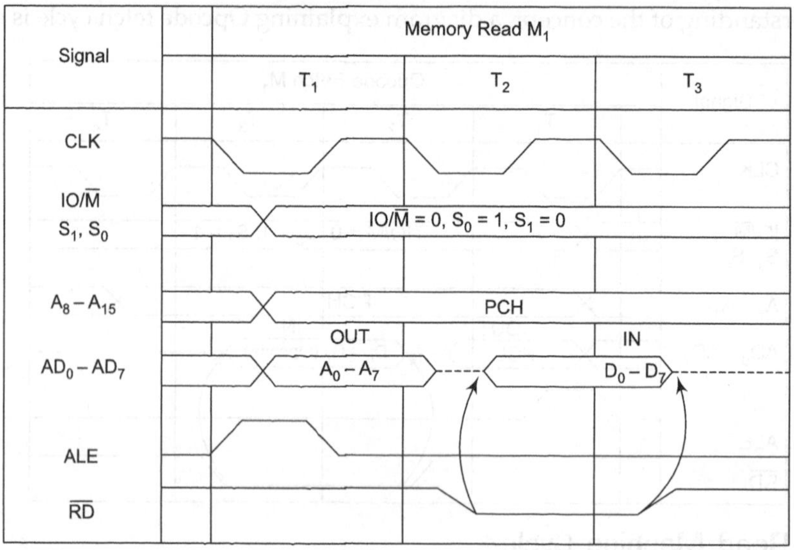

(c) I/O Read Machine Cycle

This machine cycle is very similar to memory read machine cycle. It is a 2 byte-I/O read instruction.

A simple example is IN 22H.

The first machine cycle is same as the memory read machine cycle, which is the Opcode fetch cycle. The second machine cycle is the I/O read machine cycle, where the content of port addresses (22H in this case) is transferred to the microprocessor.

The status signal for the I/O read machine cycle is different. The status signal values are IO/M'=1, S1=1, S0=0.

(d) Memory Write Machine Cycle

This cycle as the name specifies is used for sending data from the registers of the microprocessor to the memory or any other I/O devices.

A simple example for such an instruction is **MOV M, A.**

When the above instruction is executed the contents of the accumulator is moved to the specified memory location. For this operation to take place, we need two machine cycles. One is the Opcode fetch cycle and the second to transfer the contents of memory to the accumulator.

T1 State

This is the Opcode fetch cycle where the microprocessor places the 16 bit address on the higher order address bus (A8-A15) and lower order address and data multiplexed bus (AD0-AD7). The ALE goes high to latch the AD0-AD7 bus and during the middle of T1 state it goes low, so that the complete 16 bit address is available.

Now what the microprocessor has to do with this address and data? Should it write the data or read it? How will the microprocessor recognize about what operation should be performed?

The microprocessor can recognize it using the status signal IO/M', S0 and S1. For memory write cycle the values of status signals should be

IO/M'=0

S1=0

S0=1

This status information is always maintained throughout the machine cycle.

T2 State

At the beginning of this state WR′ goes low (as it is a memory write cycle), whereas in read cycle the RD′ goes low to enable memory. During this state the contents of the register is placed on the Data bus.

T3 State

The data which was placed on the data bus in the previous state is now transferred to the specific memory location. In the middle of this state the WR′ goes high and disables the memory.

Thus the data is transferred from the accumulator to specific memory location. This is the memory write machine cycle.

For better understanding the diagram of memory write machine cycle is shown below.

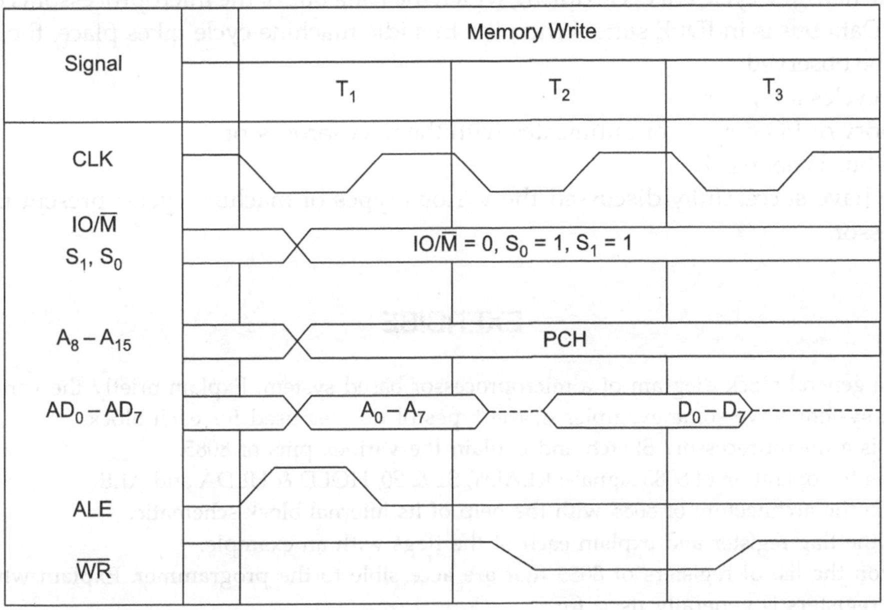

(e) I/O Write Machine Cycle

In the previous section, we saw memory read and I/O read. Both operations are similar to each other.

Similarly the memory write and I/O write operations are similar to each other.

Example of I/O write machine cycle is OUT 23H.

When the above instruction is executed, the data is transferred from accumulator to the specified location of the output device (in this case it is 23H). The microprocessor recognizes the I/O write machine cycle from the status signals IO/M′=1, S0=1, S1=0.

In the I/O write machine cycle, external latching using ALE is not necessary. This is because in this machine cycle, the address is duplicated and is available on the address bus till the end of the machine cycle.

(f) INTA Machine Cycle

Before dealing with the INTA machine cycle, first we must know what INTA is. INTA is the Interrupt acknowledge signal.

When microprocessor executes an instruction or a program, it checks the INTR during the execution of each instruction. When the INTR is low, the program execution continues. But when the INTR goes high, the microprocessor completes the execution of current instruction and sends an Interrupt acknowledge signal (INTA). The microprocessor now saves the address of the next instruction to be executed in the stack. After this step, the microprocessor finds the cause for which it was interrupted and then returns to the program which it was executing before the INTR signal. These operations come under the INTA machine cycle.

(g) Bus Idle Machine Cycle

When we discussed about the basics of timing diagram and timing diagram of various signals, we often came across one phrase "BUS IDLE MACHINE CYCLE".

So what exactly is this bus idle machine cycle?

A bus idle machine cycle is a cycle during which the Data bus of the microprocessor is not in use. That is the Data bus is in IDLE state. When the Bus idle machine cycle takes place, the following events can be observed:

No wait cycles are possible

No memory or I/O device communicates with the microprocessor.

The data bus is not used.

Thus we have successfully discussed the various types of machine cycles present in an 8085 microprocessor.

EXERCISE

1. Give a general block diagram of a microprocessor based system. Explain briefly the various blocks of the system. Give some examples of the types of devices used for each block.
2. What is a microprocessor? Sketch and explain the various pins of 8085.
3. Explain the operation of 8085 signals: READY, S1 & S0, HOLD & HLDA and ALE.
4. Explain the architecture of 8085 with the help of its internal block schematic.
5. Write the flag register and explain each of the flags with an example.
6. Mention the list of registers of 8085 that are accessible to the programmer. Explain what each of these registers is generally used for.
7. Explain the purpose of the following signals of 8085: AD0-AD7, /RESET IN, RESET OUT, and RST 7.5.
8. Explain with schematic diagram how separate address, data signals can be generated from 8085 common address-data lines.
9. Explain the schematic which decodes IO//M, /RD, /WR into two active low signals /MERD, /MEWR, /IORD and /IOWR.
10. List the status signals in 8085. Explain their functions.
11. List the control signals in 8085. Explain their functions.
12. Explain, how the microprocessor works.
13. Explain the operations of microprocessor.
14. Define Instruction cycle, Machine cycle and T-States with examples.
15. Explain the Bus Structure of 8085.
16. Explain the pin diagram of 8085 with details.
17. What is ALE? Explain the functions of ALE in 8085.
18. Explain timing diagram for opcode fetch cycle.
19. Describe the sequence of events that may occur during the different T states in the opcode Fetch machine cycle of 8085.
20. List out the maskable and non-maskable interrupts available in 8085.

6

8085 Microprocessor Instruction Set

An **instruction** is a binary pattern designed inside a microprocessor to perform a specific function. The entire group of instructions, called the **instruction set**, determines what functions the microprocessor can perform. These instructions can be classified into the following five functional categories: data transfer (copy) operations, arithmetic operations, logical operations, branching operations, and machine-control operations.

6.1 INSTRUCTION FORMAT

An **instruction** is a command to the microprocessor to perform a given task on a specified data. Each instruction has two parts: one is task to be performed, called the **operation code** (opcode), and the second is the data to be operated on, called the **operand.** The operand (or data) can be specified in various ways. It may include 8-bit (or 16-bit) data, an internal register, a memory location, or 8-bit (or 16-bit) address. In some instructions, the operand is implicit.

6.1.1 Instruction Word Size

The 8085 instruction set is classified into the following three groups according to word size:

1. One-word or 1-byte instructions
2. Two-word or 2-byte instructions
3. Three-word or 3-byte instructions

In the 8085, "byte" and "word" are synonymous because it is an 8-bit microprocessor. However, instructions are commonly referred to in terms of bytes rather than words.

6.1.1.1 One-Byte Instructions

A 1-byte instruction includes the opcode and operand in the same byte. Operand(s) are internal register and are coded into the instruction.

For Example:

Task	Op Code	Operand	Binary Code	Hex Code
Copy the contents of the accumulator in the register C.	MOV	C,A	0100 1111	4FH
Add the contents of register B to the contents of the accumulator.	ADD	B	1000 0000	80H
Invert (complement) each bit in the accumulator.	CMA		0010 1111	2FH

These instructions are 1-byte instructions performing three different tasks. In the first instruction, both operand registers are specified. In the second instruction, the operand B is specified and the accumulator is assumed. Similarly, in the third instruction, the accumulator is assumed to be the implicit operand. These instructions are stored in 8-bit binary format in memory; each requires one memory location.

Example 1. MOV rd, rs

rd <-- rs copies contents of rs into rd.

Coded as 01 ddd sss where ddd is a code for one of the 7 general registers which is the destination of the data, sss is the code of the source register.

Example 2. MOV A, B

Coded as 01111000 = 78H = 170 octal (octal was used extensively in instruction design of such processors).

Example 3. ADD r

A <-- A + r

6.1.1.2 Two-Byte Instructions

In a two-byte instruction, the first byte specifies the operation code and the second byte specifies the operand. Source operand is a data byte immediately following the opcode.

For Example:

Task	Op Code	Operand	Binary Code	Hex Code	
Load an 8-bit data byte in the accumulator.	MVI	A, Data	0011 1110	3E	First Byte
			DATA	Data	Second Byte

Assume that the data byte is 32H. The assembly language instruction is written as

Mnemonics	Hex Code
MVI A, 32H	3E 32H

The instruction would require two memory locations to store in memory.

Example 1. MVI r, data

r <-- data

Example 2. MVI A, 30H coded as 3EH 30H as two contiguous bytes. This is an example of immediate addressing.

Example 3. ADI data

A <-- A + data

OUT port

Where port is an 8-bit device address. (Port) <-- A. Since the byte is not the data but points directly to where it is located, this is called direct addressing.

6.1.1.3 Three-Byte Instructions

In a three-byte instruction, the first byte specifies the opcode, and the following two bytes specify the 16-bit address. Note that the second byte is the low-order address and the third byte is the high-order address.

opcode + data byte + data byte

For example:

Task	Op Code	Operand	Binary Code	Hex Code	
Transfer the program sequence to the memory location 2085H.	JMP	2085H	1100 0011	C3	First Byte
			1000 0101	85	Second Byte
			0010 0000	20	Third Byte

This instruction would require three memory locations to store in memory.

Three byte instructions – opcode + data byte + data byte

Example 1. LXI rp, data16

rp is one of the pairs of registers BC, DE, HL used as 16-bit registers. The two data bytes are 16-bit data in L H order of significance.

rp <-- data16

Example 2. LXI H, 0520H coded as 21H 20H 50H in three bytes. This is also immediate addressing.

Example 3. LDA addr

A <-- (addr) Addr is a 16-bit address in L H order. Example: LDA 2134H coded as 3AH 34H 21H. This is also an example of direct addressing.

6.2 ADDRESSING MODES

The instructions MOV B, A or MVI A, 82H are to copy data from a source into a destination. In these instructions the source can be a register, an input port, or an 8-bit number (00H to FFH). Similarly, a destination can be a register or an output port. The sources and destination are operands. The various formats for specifying operands are called the ADDRESSING MODES. For 8085, they are:

1. Immediate Addressing
2. Direct Addressing
3. Register Addressing
4. Register Indirect Addressing
5. Implied Addressing.

6.2.1 Immediate Addressing Mode

In immediate addressing mode, the data is specified in the instruction itself. The data will be a part of the program instruction.

Example. MVI B, 3EH – Move the data 3EH given in the instruction to B register; LXI SP, 2700H.

6.2.2 Direct Addressing Mode

In direct addressing mode, the address of the data is specified in the instruction. The data will be in memory. In this addressing mode, the program instructions and data can be stored in different memory.

Direct Addressing Mode is used to accept data from outside devices to store in the accumulator or send the data stored in the accumulator to the outside device. Accept the data from the port 00H and store them into the accumulator or Send the data from the accumulator to the port 01H.

Example. LDA 1050H – Load the data available in memory location 1050H in to accumulator; SHLD 3000H.

6.2.3 Register Addressing Mode

In register addressing mode, the instruction specifies the name of the register in which the data is available.

Example. MOV A, B – Move the content of B register to A register; SPHL; ADD C.

6.2.4 Register Indirect Addressing

In register indirect addressing mode, the instruction specifies the name of the register in which the address of the data is available. Here the data will be in memory and the address will be in the register pair.

Example. MOV A, M – The memory data addressed by H L pair is moved to A register. LDAX B.

6.2.5 Implied Addressing

In implied addressing mode, the instruction itself specifies the data to be operated.

Example. Complement the content of accumulator; RAL

6.3 INSTRUCTION SET CLASSIFICATION

An **instruction** is a binary pattern designed inside a microprocessor to perform a specific function. The entire group of instructions, called the **instruction set**, determines what functions the microprocessor can perform. These instructions can be classified into the following five functional categories: data transfer (copy) operations, arithmetic operations, logical operations, branching operations, and machine-control operations.

6.3.1 Data Transfer (Copy) Operations

This group of instructions copy data from a location called a source to another location called a destination, without modifying the contents of the source. In technical manuals, the term *data transfer* is used for this copying function. However, the term *transfer* is misleading; it creates the impression that the contents of the source are destroyed when, in fact, the contents are retained without any modification. The various types of data transfer (copy) are listed below together with examples of each type:

6.3.2 Arithmetic Operations

These instructions perform arithmetic operations such as addition, subtraction, increment, and decrement.

(*a*) **Addition:** Any 8-bit number, or the contents of a register or the contents of a memory location can be added to the contents of the accumulator and the sum is stored in the accumulator. No two other 8-bit registers can be added directly (e.g., the contents of register B cannot be added directly to the contents of the register C). The instruction DAD is an exception; it adds 16-bit data directly in register pairs.

(*b*) **Subtraction:** Any 8-bit number, or the contents of a register, or the contents of a memory location can be subtracted from the contents of the accumulator and the results stored in the accumulator. The subtraction is performed in 2's compliment, and the results if negative, are expressed in 2's complement. No two other registers can be subtracted directly.

(*c*) **Increment/Decrement:** The 8-bit contents of a register or a memory location can be incremented or decrement by 1. Similarly, the 16-bit contents of a register pair (such as BC) can be incremented or decrement by 1. These increment and decrement operations differ from addition and subtraction in an important way; *i.e.*, they can be performed in any one of the registers or in a memory location.

The following table represents the complete explanation of data transfer instructions:

Data Transfer Instructions

Opcode	Operand	Explanation of Instruction	Description
MOV	**Rd, Rs** **M, Rs** **Rd, M**	Copy from source (Rs) to destination(Rd)	This instruction copies the contents of the source register into the destination register; the contents of the source register are not altered. If one of the operands is a memory location, its location is specified by the contents of the HL registers. **Example: MOV B, C or MOV B, M**
MVI	**Rd, data** **M, data**	Move immediate 8-bit	The 8-bit data is stored in the destination register or memory. If the operand is a memory location, its location is specified by the contents of the HL registers. **Example: MVI B, 57H or MVI M, 57H**
LDA	**16-bit address**	Load accumulator	The contents of a memory location, specified by a 16-bit address in the operand, are copied to the accumulator. The contents of the source are not altered. **Example: LDA 2034H**
LDAX	**B/D Reg. pair**	Load accumulator indirect	The contents of the designated register pair point to a memory location. This instruction copies the contents of that memory location into the accumulator. The contents of either the register pair or the memory location are not altered. **Example: LDAX B**
LXI	**Reg. pair, 16-bit data**	Load register pair immediate	The instruction loads 16-bit data in the register pair designated in the operand. **Example: LXI H, 2034H or LXI H, XYZ**
LHLD	**16-bit address**	Load H and L registers direct	The instruction copies the contents of the memory location pointed out by the 16-bit address into register L and copies the contents of the next memory location into register H. The contents of source memory locations are not altered. **Example: LHLD 2040H**
STA	**16-bit address**	16-bit address	The contents of the accumulator are copied into the memory location specified by the operand. This is a 3-byte instruction, the second byte specifies the low-order address and the third byte specifies the high-order address. **Example: STA 4350H**
STAX	**Reg. pair**	Store accumulator indirect	The contents of the accumulator are copied into the memory location specified by the contents of the operand (register pair). The contents of the accumulator are not altered. **Example: STAX B**
SHLD	**16-bit address**	Store H and L registers direct	The contents of register L are stored into the memory location specified by the 16-bit address in the operand and the contents of H register are stored into the next memory location by incrementing the operand. The

			contents of registers HL are not altered. This is a 3-byte instruction, the second byte specifies the low-order address and the third byte specifies the high-order address. **Example: SHLD 2470H**
XCHG	**none**	Exchange H and L with D and E	The contents of register H are exchanged with the contents of register D, and the contents of register L are exchanged with the contents of register E. **Example: XCHG**
SPHL	**none**	Copy H and L registers to the stack pointer	The instruction loads the contents of the H and L registers into the stack pointer register, the contents of the H register provide the high-order address and the contents of the L register provide the low-order address. The contents of the H and L registers are not altered. **Example: SPHL**
XTHL	**none**	Exchange H and L with top of stack	The contents of the L register are exchanged with the stack location pointed out by the contents of the stack pointer register. The contents of the H register are exchanged with the next stack location (SP+1); however, the contents of the stack pointer register are not altered. **Example: XTHL**
PUSH	**Reg. pair**	Push register pair onto stack	The contents of the register pair designated in the operand are copied onto the stack in the following sequence. The stack pointer register is decremented and the contents of the high order register (B, D, H, A) are copied into that location. The stack pointer register is decremented again and the contents of the low-order register (C, E, L, flags) are copied to that location. **Example: PUSH B or PUSH A**
POP	**Reg. pair**	Pop off stack to register pair	The contents of the memory location pointed out by the stack pointer register are copied to the low-order register (C, E, L, status flags) of the operand. The stack pointer is incremented by 1 and the contents of that memory location are copied to the high-order register (B, D, H, A) of the operand. The stack pointer register is again incremented by 1. **Example: POP H or POP A**
OUT	**8-bit port address**	Output data from accumulator to a port with 8-bit address	The contents of the accumulator are copied into the I/O port specified by the operand. **Example: OUT F8H**
IN	**8-bit port address**	Input data to accumulator from a port with 8-bit address	The contents of the input port designated in the operand are read and loaded into the accumulator. **Example: IN 8CH**

The following table represents the complete explanation of Arithmetic instructions:

Arithmetic Instructions

Opcode	Operand	Explanation of Instruction	Description
ADD	**R** **M**	Add register or memory, to accumulator	The contents of the operand (register or memory) are added to the contents of the accumulator and the result is stored in the accumulator. If the operand is a memory location, its location is specified by the contents of the HL registers. All flags are modified to reflect the result of the addition. **Example: ADD B or ADD M**
ADC	**R** **M**	Add register to accumulator with carry	The contents of the operand (register or memory) and M the Carry flag are added to the contents of the accumulator and the result is stored in the accumulator. If the operand is a memory location, its location is specified by the contents of the HL registers. All flags are modified to reflect the result of the addition. **Example: ADC B or ADC M**
ADI	**8-bit data**	Add immediate to accumulator	The 8-bit data (operand) is added to the contents of the accumulator and the result is stored in the accumulator. All flags are modified to reflect the result of the addition. **Example: ADI 45H**
ACI	**8-bit data**	Add immediate to accumulator with carry	The 8-bit data (operand) and the Carry flag are added to the contents of the accumulator and the result is stored in the accumulator. All flags are modified to reflect the result of the addition. **Example: ACI 45H**
LXI	**Reg. pair, 16-bit data**	Load register pair immediate	The instruction loads 16-bit data in the register pair designated in the operand. **Example: LXI H, 2034H or LXI H, XYZ**
DAD	**Reg. pair**	Add register pair to H and L registers	The 16-bit contents of the specified register pair are added to the contents of the HL register and the sum is stored in the HL register. The contents of the source register pair are not altered. If the result is larger than 16 bits, the CY flag is set. No other flags are affected. **Example: DAD H**
SUB	**R** **M**	Subtract register or memory from accumulator	The contents of the operand (register or memory) are subtracted from the contents of the accumulator, and the result is stored in the accumulator. If the operand is a memory location, its location is specified by the contents of the HL registers. All flags are modified to reflect the result of the subtraction. **Example: SUB B or SUB M**
SBB	**R** **M**	Subtract source and borrow from accumulator	The contents of the operand (register or memory) and M the Borrow flag are subtracted from the contents of the accumulator and the result is placed in the accumulator. If the operand is a memory location, its location is specified by the contents of the HL registers. All flags are modified to reflect the result of the subtraction. **Example: SBB B or SBB M**

SUI	**8-bit data**	Subtract immediate from accumulator	The 8-bit data (operand) is subtracted from the contents of the accumulator and the result is stored in the accumulator. All flags are modified to reflect the result of the subtraction. **Example: SUI 45H**
SBI	**8-bit data**	Subtract immediate from accumulator with borrow	The contents of register H are exchanged with the contents of register D, and the contents of register L are exchanged with the contents of register E. **Example: XCHG**
INR	**R** **M**	Increment register or memory by 1	The contents of the designated register or memory) are incremented by 1 and the result is stored in the same place. If the operand is a memory location, its location is specified by the contents of the HL registers. **Example: INR B or INR M**
INX	**R**	Increment register pair by 1	The contents of the designated register pair are incremented by 1 and the result is stored in the same place. **Example: INX H**
DCR	**R** **M**	Decrement register or memory by 1	The contents of the designated register or memory are M decremented by 1 and the result is stored in the same place. If the operand is a memory location, its location is specified by the contents of the HL registers. **Example: DCR B or DCR M**
DCX	**R**	Decrement register pair by 1	The contents of the designated register pair are decremented by 1 and the result is stored in the same place. **Example: DCX H**
DAA	**none**	Decimal adjust accumulator	The contents of the accumulator are changed from a binary value to two 4-bit binary coded decimal (BCD) digits. This is the only instruction that uses the auxiliary flag to perform the binary to BCD conversion, and the conversion procedure is described below. S, Z, AC, P, CY flags are altered to reflect the results of the operation. If the value of the low-order 4-bits in the accumulator is greater than 9 or if AC flag is set, the instruction adds 6 to the low-order four bits. If the value of the high-order 4-bits in the accumulator is greater than 9 or if the Carry flag is set, the instruction adds 6 to the high-order four bits. **Example: DAA**

6.3.3 Logical Operations

These instructions perform various logical operations with the contents of the accumulator.

(*a*) **AND, OR Exclusive-OR:** Any 8-bit number, or the contents of a register, or of a memory location can be logically ANDed, ORed, or Exclusive-ORed with the contents of the accumulator. The results are stored in the accumulator.

(*b*) **Rotate:** Each bit in the accumulator can be shifted either left or right to the next position.

(*c*) **Compare:** Any 8-bit number or the contents of a register, or a memory location can be compared for equality, greater than, or less than, with the contents of the accumulator.

(*d*) **Complement:** The contents of the accumulator can be complemented. All 0s are replaced by 1s and all 1s are replaced by 0s.

6.3.4 Branching Operations

This group of instructions alters the sequence of program execution either conditionally or unconditionally.

(*a*) **Jump:** Conditional jumps are an important aspect of the decision-making process in the programming. These instructions test for a certain conditions (e.g., Zero or Carry flag) and alter the program sequence when the condition is met. In addition, the instruction set includes an instruction called *unconditional jump*.

(*b*) **Call, Return, and Restart:** These instructions change the sequence of a program either by calling a subroutine or returning from a subroutine. The conditional Call and Return instructions also can test condition flags.

6.3.5 Machine Control Operations

These instructions control machine functions such as Halt, Interrupt, or do nothing.

6.4 8085 BASED ASSEMBLY LANGUAGE PROGRAMS

In this section, we will discuss some programs based on Assembly Language.

Program 1. Write an assembly program to add two numbers.

Solution:

```
MVI D, 8BH
MVI C, 6FH
MOV A, C
ADD D
OUT PORT1
HLT
```

Program 2. Write an assembly program to multiply a number by 8.

Solution:

```
MVI A, 30H
RRC
RRC
RRC
OUT PORT1
HLT
```

The following table represents the complete explanation of Logical instructions:

Logical Instructions

Opcode	Operand	Explanation of Instruction	Description

CMP	**R**
	M
		Compare register or memory with accumulator	The contents of the operand (register or memory) are M compared with the contents of the accumulator. Both contents are preserved . The result of the comparison is shown by setting the flags of the PSW as follows:
			if (A) < (reg/mem): carry flag is set
			if (A) = (reg/mem): zero flag is set
			if (A) > (reg/mem): carry and zero flags are reset.
			Example: CMP B or CMP M

CPI	**8-bit data**	Compare immediate with accumulator	The second byte (8-bit data) is compared with the contents of the accumulator. The values being compared remain unchanged. The result of the comparison is shown by setting the flags of the PSW as follows:
			if (A) < data: carry flag is set
			if (A) = data: zero flag is set
			if (A) > data: carry and zero flags are reset.
			Example: CPI 89H

ANA	**R** **M**	Logical AND register or memory with accumulator	The contents of the accumulator are logically ANDed with M the contents of the operand (register or memory), and the result is placed in the accumulator. If the operand is a memory location, its address is specified by the contents of HL registers. S, Z, P are modified to reflect the result of the operation. CY is reset. AC is set. **Example: ANA B or ANA M**
ANI	**8-bit data**	Logical AND immediate with accumulator	The contents of the accumulator are logically ANDed with the8-bit data (operand) and the result is placed in the accumulator. S, Z, P are modified to reflect the result of the operation. CY is reset. AC is set. **Example: ANI 86H**
XRA	**R** **M**	Exclusive OR register or memory with accumulator	The contents of the accumulator are Exclusive ORed with M the contents of the operand (register or memory), and the result is placed in the accumulator. If the operand is a memory location, its address is specified by the contents of HL registers. S, Z, P are modified to reflect the result of the operation. CY and AC are reset. **Example: XRA B or XRA M**
XRI	**8-bit data**	Exclusive OR immediate with accumulator	The contents of the accumulator are Exclusive ORed with the 8-bit data (operand) and the result is placed in the accumulator. S, Z, P are modified to reflect the result of the operation. CY and AC are reset. **Example: XRI 86H**
ORA	**R** **M**	Logical OR register or	The contents of the accumulator are logically ORed with M the contents of the operand (register or memory), and the result is placed in the accumulator. If the operand is a memory location, its address is specified by

		memory with accumulator	the contents of HL registers. S, Z, P are modified to reflect the result of the operation. CY and AC are reset. **Example: ORA B or ORA M**
ORI	**8-bit data**	Logical OR immediate with accumulator	The contents of the accumulator are logically ORed with the 8-bit data (operand) and the result is placed in the accumulator. S, Z, P are modified to reflect the result of the operation. CY and AC are reset. **Example: ORI 86H**
RLC	**none**	Rotate accumulator left	Each binary bit of the accumulator is rotated left by one position. Bit D_7 is placed in the position of D_0 as well as in the Carry flag. CY is modified according to bit D_7. S, Z, P, AC are not affected. **Example: RLC**
RRC	**none**	Rotate accumulator right	Each binary bit of the accumulator is rotated right by one position. Bit D_0 is placed in the position of D_7 as well as in the Carry flag. CY is modified according to bit D_0. S, Z, P, AC are not affected. **Example: RRC**
RAL	**none**	Rotate accumulator left through carry	Each binary bit of the accumulator is rotated left by one position through the Carry flag. Bit D_7 is placed in the Carry flag, and the Carry flag is placed in the least significant position D_0. CY is modified according to bit D_7. S, Z, P, AC are not affected. **Example: RAL**
RAR	**none**	Rotate accumulator right through carry	Each binary bit of the accumulator is rotated right by one position through the Carry flag. Bit D_0 is placed in the Carry flag, and the Carry flag is placed in the most significant position D_7. CY is modified according to bit D_0. S, Z, P, AC are not affected. **Example: RAR**
CMA	**none**	Complement accumulator	The contents of the accumulator are complemented. No flags are affected. **Example: CMA**
CMC	**none**	Complement carry	The Carry flag is complemented. No other flags are affected. **Example: CMC**
STC	**none**	Set Carry	Set Carry **Example: STC**

The following table represents the complete explanation of Branching instructions:

Branching Instructions

<table>
<tr><th>Opcode</th><th>Operand</th><th>Explanation of Instruction</th><th>Description</th></tr>
<tr><td>JMP</td><td>16-bit address</td><td>Jump unconditionally</td><td>The program sequence is transferred to the memory location specified by the 16-bit address given in the operand.
Example: JMP 2034H or JMP XYZ</td></tr>
<tr><td><table>
<tr><th>Opcode</th><th>Description</th><th>Flag Status</th></tr>
<tr><td>JC</td><td>Jump on Carry</td><td>CY = 1</td></tr>
<tr><td>JNC</td><td>Jump on no Carry</td><td>CY = 0</td></tr>
<tr><td>JP</td><td>Jump on positive</td><td>S = 0</td></tr>
<tr><td>JM</td><td>Jump on minus</td><td>S = 1</td></tr>
<tr><td>JZ</td><td>Jump on zero</td><td>Z = 1</td></tr>
<tr><td>JNZ</td><td>Jump on no zero</td><td>Z = 0</td></tr>
<tr><td>JPE</td><td>Jump on parity even</td><td>P = 1</td></tr>
<tr><td>JPO</td><td>Jump on parity odd</td><td>P = 0</td></tr>
</table></td><td>16-bit address</td><td>Jump conditionally</td><td>The program sequence is transferred to the memory location specified by the 16-bit address given in the operand based on the specified flag of the PSW as described below.
Example: JZ 2034H or JZ XYZ</td></tr>
<tr><td><table>
<tr><th>Opcode</th><th>Description</th><th>Flag Status</th></tr>
<tr><td>CC</td><td>Call on Carry</td><td>CY = 1</td></tr>
<tr><td>CNC</td><td>Call on no Carry</td><td>CY = 0</td></tr>
<tr><td>CP</td><td>Call on positive</td><td>S = 0</td></tr>
<tr><td>CM</td><td>Call on minus</td><td>S = 1</td></tr>
<tr><td>CZ</td><td>Call on zero</td><td>Z = 1</td></tr>
<tr><td>CNZ</td><td>Call on no zero</td><td>Z = 0</td></tr>
<tr><td>CPE</td><td>Call on parity even</td><td>P = 1</td></tr>
<tr><td>CPO</td><td>Call on parity odd</td><td>P = 0</td></tr>
</table></td><td>16-bit address</td><td>Unconditional subroutine call</td><td>The program sequence is transferred to the memory location specified by the 16-bit address given in the operand. Before the transfer, the address of the next instruction after CALL (the contents of the program counter) is pushed onto the stack.
Example: CALL 2034H or CALL XYZ</td></tr>
<tr><td>RET</td><td>none</td><td>Return from subroutine unconditionally</td><td>The program sequence is transferred from the subroutine to the calling program. The two bytes from the top of the stack are copied into the program counter, and program execution begins at the new address.
Example: RET</td></tr>
<tr><td><table>
<tr><th>Opcode</th><th>Description</th><th>Flag Status</th></tr>
<tr><td>RC</td><td>Return on Carry</td><td>CY = 1</td></tr>
</table></td><td>none</td><td>Return from subroutine conditionally</td><td>The program sequence is transferred from the subroutine to the calling program based on the specified flag of the PSW as described below. The</td></tr>
</table>

<table>
<tr>
<td colspan="2"><table><tr><td>RNC</td><td>Return on no Carry</td><td>CY = 0</td></tr><tr><td>RP</td><td>Return on positive</td><td>S = 0</td></tr><tr><td>RM</td><td>Return on minus</td><td>S = 1</td></tr><tr><td>RZ</td><td>Return on zero</td><td>Z = 1</td></tr><tr><td>RNZ</td><td>Return on no zero</td><td>Z = 0</td></tr><tr><td>RPE</td><td>Return on parity even</td><td>P = 1</td></tr><tr><td>RPO</td><td>Return on parity odd</td><td>P = 0</td></tr></table></td>
<td></td>
<td>Two bytes from the top of the stack are copied into the program counter, and program execution begins at the new address.
Example: RZ</td>
</tr>
<tr>
<td>PCHL</td>
<td>none</td>
<td>Load program counter with HL contents</td>
<td>The contents of registers H and L are copied into the program counter. The contents of H are placed as the high-order byte and the contents of L as the low-order byte.
Example: PCHL</td>
</tr>
<tr>
<td>RST</td>
<td>0-7</td>
<td>Restart</td>
<td>The RST instruction is equivalent to a 1-byte call instruction to one of eight memory locations depending upon the number. The instructions are generally used in conjunction with interrupts and inserted using external hardware. However these can be used as software instructions in a program to transfer program execution to one of the eight locations. The addresses are:
<table><tr><th>Instruction</th><th>Restart Address</th></tr><tr><td>RST 0</td><td>0000H</td></tr><tr><td>RST 1</td><td>0008H</td></tr><tr><td>RST 2</td><td>0010H</td></tr><tr><td>RST 3</td><td>0018H</td></tr><tr><td>RST 4</td><td>0020H</td></tr><tr><td>RST 5</td><td>0028H</td></tr><tr><td>RST 6</td><td>0030H</td></tr><tr><td>RST 7</td><td>0038H</td></tr></table>
The 8085 has four additional interrupts and these interrupts generate RST instructions internally and thus do not require any external hardware. These instructions and their Restart addresses are:</td>
</tr>
</table>

			Interrupts	Restart Address
			TRAP	0024H
			RSR 5.5	002CH
			RST 6.5	0034H
			RST 7.5	003CH

The following table represents the complete explanation of Control instructions:

Control Instructions

Opcode	Operand	Explanation of Instruction	Description
NOP	**none**	No operation	No operation is performed. The instruction is fetched and decoded. However no operation is executed. **Example: NOP**
HLT	**none**	Halt and enter wait state	The CPU finishes executing the current instruction and halts any further execution. An interrupt or reset is necessary to exit from the halt state. **Example: HLT**
DI	**none**	Disable interrupts	The interrupt enable flip-flop is reset and all the interrupts except the TRAP are disabled. No flags are affected. **Example: DI**
EI	**none**	Enable interrupts	The interrupt enable flip-flop is set and all interrupts are enabled. No flags are affected. After a system reset or the acknowledgement of an interrupt, the interrupt enable flip-flop is reset, thus disabling the interrupts. This instruction is necessary to reenable the interrupts (except TRAP). **Example: EI**
RIM	**none**	Read interrupt mas	This is a multipurpose instruction used to read the status of interrupts 7.5, 6.5, 5.5 and read serial data input bit. The instruction loads eight bits in the accumulator with the following interpretations. **Example: RIM**

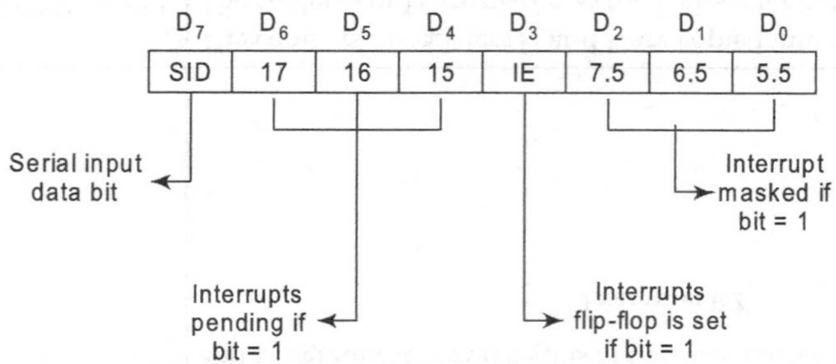

<table>
<tr>
<td>SIM</td>
<td>none</td>
<td>Set interrupt mask</td>
<td>This is a multipurpose instruction and used to implement the 8085 interrupts 7.5, 6.5, 5.5, and serial data output. The instruction interprets the accumulator contents as follows.

Example: SIM

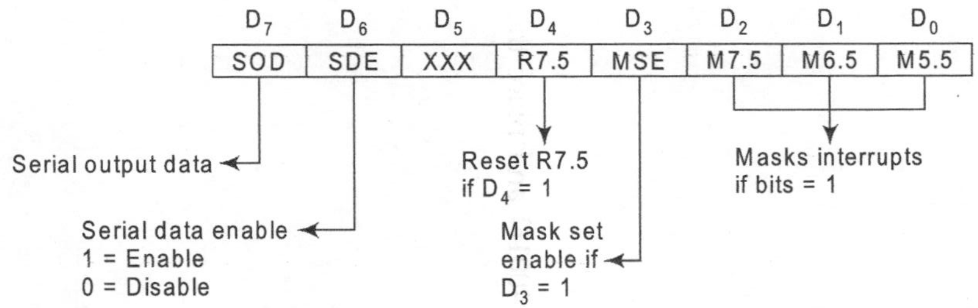

• SOD—Serial Output Data: Bit D7 of the accumulator is latched into the SOD output line and made available in a serial peripheral if bit $D_6 = 1$.
• SDE—Serial Data Enable: If this but = 1, it enables the serial output. To implement serial output, this bit needs to be enabled.
• XXX—Don't Care
• R7.5—Reset RST 7.5: If this bit = 1, RST 7.5 flip-flop is reset. This is an additional control to reset RST 7.5.
• MSE—Mask Set Enable: If this bit is high, it enables the functions of bits D_2, D_1, D_0. This is a master control over all the interrupt masking bits. If this bit is low, bits D_2, D_1, and D_9 do not have any effect on the masks.
• M7.5—D_2 = 0, RST 7.5 is enabled.
= 1, RST 7.5 is masked or disabled.
• M6.5—D_1 = 0, RST 6.5 is enabled.
= 1, RST 6.5 is masked or disabled.
• M5.5—D_0 = 0, RST 5.5 is enabled.
= 1, RST 5.5 is masked or disabled.</td>
</tr>
</table>

Program 3. Write an assembly program to find greatest between two numbers?

Solution:

```
MVI B, 30H
MVI C, 40H
MOV A, B
CMP C
JZ EQU
JC GRT
OUT PORT1
HLT
EQU: MVI A, 01H
OUT PORT1
HLT
GRT: MOV A, C
OUT PORT1
HLT
```

Program 4. Add the contents of memory locations 4000H and 4001H and place the result in memory location 4002H.

Solution:

The sample code is as follows:

(4000H) = 14H

(4001H) = 89H

Result = 14H + 89H = 9DH

The Source Program is as follows:

```
LXI H 4000H: HL points 4000H
MOV A, M: Get first operand
INX H: HL points 4001H
ADD M: Add second operand
INX H: HL points 4002H
MOV M, A : Store result at 4002H
HLT: Terminate program execution
```

The Flow Chart is as follows

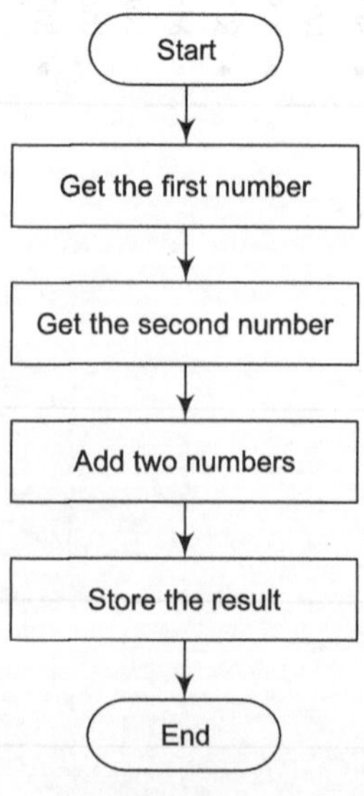

Note: In Sample Code, direct addressing instructions are used, whereas in Source Program, indirect addressing instructions are used.

Program 5. Add the 16-bit number in memory locations 4000H and 4001H to the 16-bit number in memory locations 4002H and 4003H. The most significant eight bits of the two numbers to be added are in memory locations 4001H and 4003H. Store the result in memory locations 4004H and 4005H with the most significant byte in memory location 4005H.

Solution:

The flow chart is as follows:

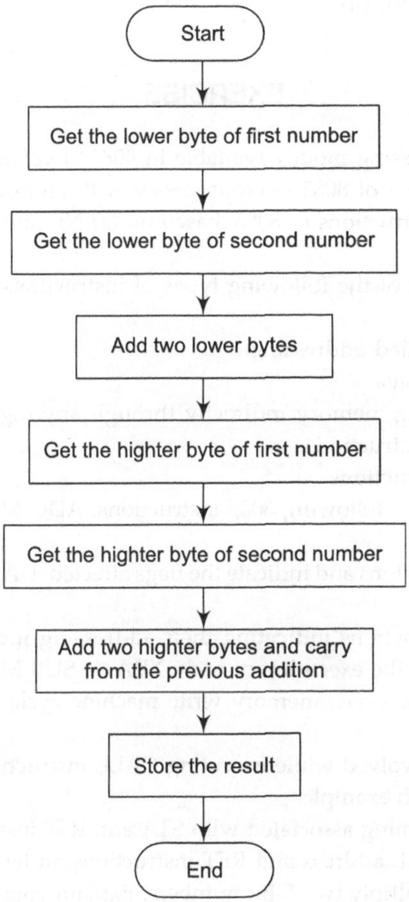

The sample code is as follows:

(4000H) = 15H

(4001H) = 1CH

(4002H) = B7H

(4003H) = 5AH

***Result*= 1C15 + 5AB7H = 76CCH**

(4004H) = CCH

(4005H) = 76H

The Source Program is as follows:

LHLD 4000H: Get first I6-bit number in HL

XCHG: Save first I6-bit number in DE

LHLD 4002H : Get second I6-bit number in HL
MOV A, E: Get lower byte of the first number
ADD L: Add lower byte of the second number
MOV L, A: Store result in L register
MOV A, D: Get higher byte of the first number
ADC H: Add higher byte of the second number with CARRY
MOV H, A : Store result in H register
SHLD 4004H: Store I6-bit result in memory locations 4004H and 4005H.
HLT: Terminate program execution

EXERCISE

1. What are the various addressing modes available in 8085? Explain with examples.
2. Explain the instruction format of 8085 microprocessor with an example.
3. How do you classify the instructions of 8085, based on (a) Number of bytes (b) Addressing modes? Explain with examples.
4. Give two examples for each of the following types of instructions and indicate what each of these instructions do:
   (*a*) Instructions with implied addressing.
   (*b*) ALU type of instructions.
   (*c*) Instructions addressing memory indirectly through any register pair.
   (*d*) Conditional branch instructions.
   (*e*) Interrupt related instructions.
5. Explain fully the action of the following 8085 instructions: ADC M, STAX D, RLC, LHLD 9898, and TRAP.
6. Explain the following instructions and indicate the flags affected: DAA, DAD D, XCHG, HALT & MOV M, A.
7. Explain the following instructions indicating their addressing modes, flags affected, number and names of machine cycles on the execution of each: XRA A, SUB M, & DCXSP.
8. Draw opcodes fetch machine cycle, memory write machine cycle and I/O read machine cycle and explain.
9. Explain the various steps involved while executing CALL instruction with an example.
10. Explain DAA instruction with example.
11. Explain the functions and timing associated with STA and RST instructions.
12. Explain the working of CALL address and RET instructions, in terms of machine cycles.
13. Write an ALP for 8085 to multiply two 8-bit numbers. Explain your program with only comments.
14. Write a program in 8085 to convert the binary number stored in location XX20h to BCD. Store the possible three BCD digits in unpacked manner from location XX50h.
15. What is a stack? On what principle does it work? Is it necessary to intialize stack pointer while a writing program? Explain with example.
16. With suitable programming example clearly explain passing the parameter through registers, through memory and through stack.
17. Sketch and explain briefly the timing diagram of the instruction MVI A, 32H, which is stored from address 3000h.
18. Draw the timing diagram for the instruction STA 9000h.
19. Explain any five logical instructions with their function, addressing modes and flags affected.
20. Explain various logical and bit manipulation instructions.

7

16 BIT MICROPROCESSOR 8086 ARCHITECTURE

The **8086** is a 16-bit microprocessor chip designed by Intel and introduced to the market in 1978, which gave rise to the x86 architecture. The Intel 8088, released in 1979, was a slightly modified chip with an external 8-bit data bus (allowing the use of cheaper and fewer supporting logic chips), and is notable as the processor used in the original IBM PC.

Intel 8086 microprocessor is a first member of x86 families of processors. Advertised as a "source-code compatible" with Intel 8080 and Intel 8085 processors, the 8086 was not object code compatible with them. The 8086 has complete 16-bit architecture - 16-bit internal registers, 16-bit data bus, and 20-bit address bus (1 MB of physical memory). Because the processor has 16-bit index registers and memory pointers, it can effectively address only 64 KB of memory. To address memory beyond 64 KB the CPU uses segment registers — these registers specify memory locations for code, stack, data and extra data 64 KB segments. The segments can be positioned anywhere in memory, and, if necessary, user programs can change their position. This addressing method has one big advantage — it is very easy to write memory-independent code when the size of code, stack and data is smaller than 64 KB each. The complexity of the code and programming increases, sometimes significantly, when the size of stack, data and/code is larger than 64 KB. To support different variations of this awkward memory addressing scheme many 8086 compilers included 6 different memory models: tiny, small, compact, medium, large and huge. 64 KB direct addressing limitation was eliminated with the introduction of the 32-bit protected mode in Intel 80386 processor.

7.1 REGISTER ORGANIZATION

The 8086 microprocessor has a total of fourteen registers that are accessible to the programmer. It is divided into four groups. They are as follows:

- Four General purpose registers
- Four Index/Pointer registers
- Four Segment registers
- Two Other registers

7.1.1 General Purpose Registers

The following figure represents all the general purpose registers:

General Purpose Registers

		15 — 0	
Accumulator	AX		Multiply, divide, I/O
Base	BX		Pointer to base address (data)
Count	CX		Count for loops, shifts
Data	DX		Multiply, divide, I/O

The explanation of the above figure is as follows:

Accumulator register consists of two 8-bit registers AL and AH, which can be combined together and used as a 16-bit register AX. AL in this case contains the low order byte of the word, and AH contains the high-order byte. Accumulator can be used for I/O operations and string manipulation.

Base register consists of two 8-bit registers BL and BH, which can be combined together and used as a 16-bit register BX. BL in this case contains the low-order byte of the word, and BH contains the high-order byte. BX register usually contains a data pointer used for based, based indexed or register indirect addressing.

Count register consists of two 8-bit registers CL and CH, which can be combined together and used as a 16-bit register CX. When combined, CL register contains the low order byte of the word, and CH contains the high-order byte. Count register can be used in Loop, shift/rotate instructions and as a counter in string manipulation

Data register consists of two 8-bit registers DL and DH, which can be combined together and used as a 16-bit register DX. When combined, DL register contains the low order byte of the word, and DH contains the high-order byte. Data register can be used as a port number in I/O operations. In integer 32-bit multiply and divide instruction the DX register contains high-order word of the initial or resulting number.

7.1.2 Index or Pointer Registers

These registers can also be called as Special Purpose registers and graphically can be represented as follows:

Pointer and Index Registers

		15 — 0	
Stack Pointer	SP		Pointer to top of stack
Base Pointer	BP		Pointer to base address (stack)
Source Index	SI		Source string/index pointer
Destination Index	DI		Destination string/index pointer
		15 — 0	

The explanation of the above figure is as follows:

Stack Pointer (SP) is a 16-bit register pointing to program stack, *i.e.*, it is used to hold the address of the top of stack. The stack is maintained as a LIFO with its bottom at the start of the stack segment (specified by the SS segment register).Unlike the SP register, the BP can be used to specify the offset of other program segments.

Base Pointer (BP) is a 16-bit register pointing to data in stack segment. It is usually used by subroutines to locate variables that were passed on the stack by a calling program. BP register is usually used for based, based indexed or register indirect addressing.

Source Index (SI) is a 16-bit register. SI is used for indexed, based indexed and register indirect addressing, as well as a source data address in string manipulation instructions. Used in conjunction with the DS register to point to data locations in the data segment.

Destination Index (DI) is a 16-bit register. Used in conjunction with the ES register in string operations. DI is used for indexed, based indexed and register indirect addressing, as well as a destination data addresses in string manipulation instructions. In short, Destination Index and SI Source Index registers are used to hold address.

7.1.3 Segment Registers

Most of the registers contain data/instruction offsets within 64 KB memory segment. There are four different 64 KB segments for instructions, stack, data and extra data. To specify where in 1 MB of processor memory these 4 segments are located the processor uses four segment registers.

Graphically these registers are represented as follows:

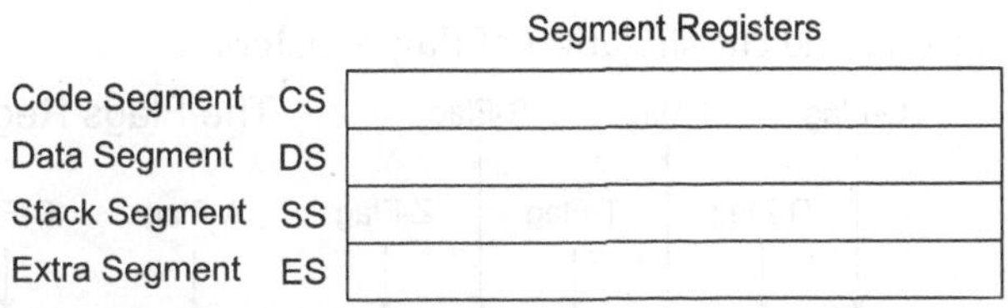

The explanation of the above figure is as follows:

Code segment (CS) is a 16-bit register containing address of 64 KB segment with processor instructions. The processor uses CS segment for all accesses to instructions referenced by instruction pointer (IP) register. CS register cannot be changed directly. The CS register is automatically updated during far jump, far call and far return instructions.

Stack segment (SS) is a 16-bit register containing address of 64 KB segment with program stack. By default, the processor assumes that all data referenced by the stack pointer (SP) and base pointer (BP) registers is located in the stack segment. SS register can be changed directly using POP instruction.

Data segment (DS) is a 16-bit register containing address of 64 KB segment with program data. By default, the processor assumes that all data referenced by general registers (AX, BX, CX, DX) and index register (SI, DI) is located in the data segment. DS register can be changed directly using POP and LDS instructions.

Extra segment (ES) used to hold the starting address of Extra segment. Extra segment is provided for programs that need to access a second data segment. Segment registers cannot be used in arithmetic operations.

7.1.4 Other Registers

Other registers consist of the following registers:

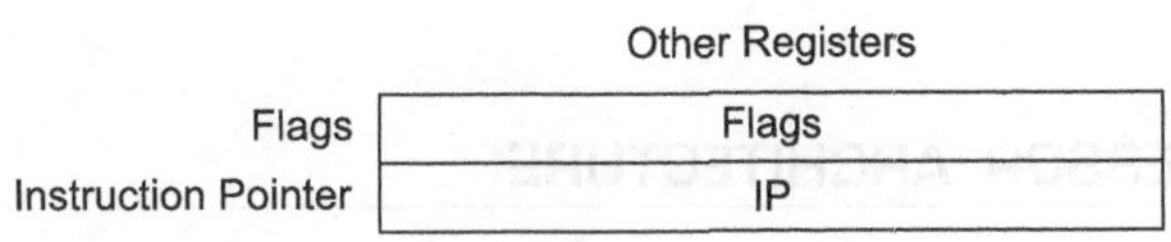

The explanation of the above figure is as follows:

Instruction Pointer (IP) is a 16-bit register. This is a crucially important register which is used to control which instruction the CPU executes. The ip, or program counter, is used to store the memory location of the next instruction to be executed. The CPU checks the program counter to ascertain which instruction to carry out next. It then updates the program counter to point to the next instruction. Thus the program counter will always point to the next instruction to be executed.

Flag Register contains a group of status bits called flags that indicate the status of the CPU or the result of arithmetic operations. There are two types of flags:

1. The **status flags** which reflect the result of executing an instruction. The programmer cannot set/reset these flags directly.
2. The **control flags** enable or disable certain CPU operations. The programmer can set/reset these bits to control the CPU's operation.

Nine individual bits of the status register are used as control flags (3 of them) and status flags (6 of them).The remaining 7 are not used.

A flag can only take on the values 0 and 1. We say a flag is set if it has the value 1.The status flags are used to record specific characteristics of arithmetic and of logical instructions.

Classification of Flag Registers

The following figure represents the classification of flag registers:

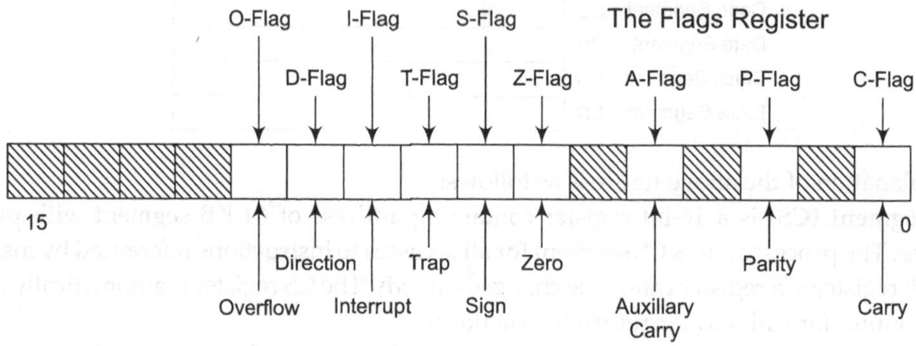

(a) Control Flags

There are three control flags:

1. **The Direction Flag (D):** Affects the direction of moving data blocks by such instructions as MOVS, CMPS and SCAS. The flag values are 0 = up and 1 = down and can be set/reset by the STD (set D) and CLD (clear D) instructions.
2. **The Interrupt Flag (I):** Dictates whether or not system interrupts can occur. Interrupts are actions initiated by hardware block such as input devices that will interrupt the normal execution of programs. The flag values are 0 = disable interrupts or 1 = enable interrupts and can be manipulated by the CLI (clear I) and STI (set I) instructions.
3. **The Trap Flag (T):** Determines whether or not the CPU is halted after the execution of each instruction. When this flag is set (*i.e.*, = 1), the programmer can single step through his program to debug any errors. When this flag = 0 this feature is off. This flag can be set by the INT 3 instruction.

7.2 80X86 PROCESSOR ARCHITECTURE

Processor Architecture

The following figure is a model for the 8086 microprocessor. It is organized as two separate processors,

- The *bus interface unit (BIU)* and
- The *execution unit (EU)*.

Processor model for the 8086 microprocessor. A separate execution unit (EU) and bus interface unit (BIU) are provided

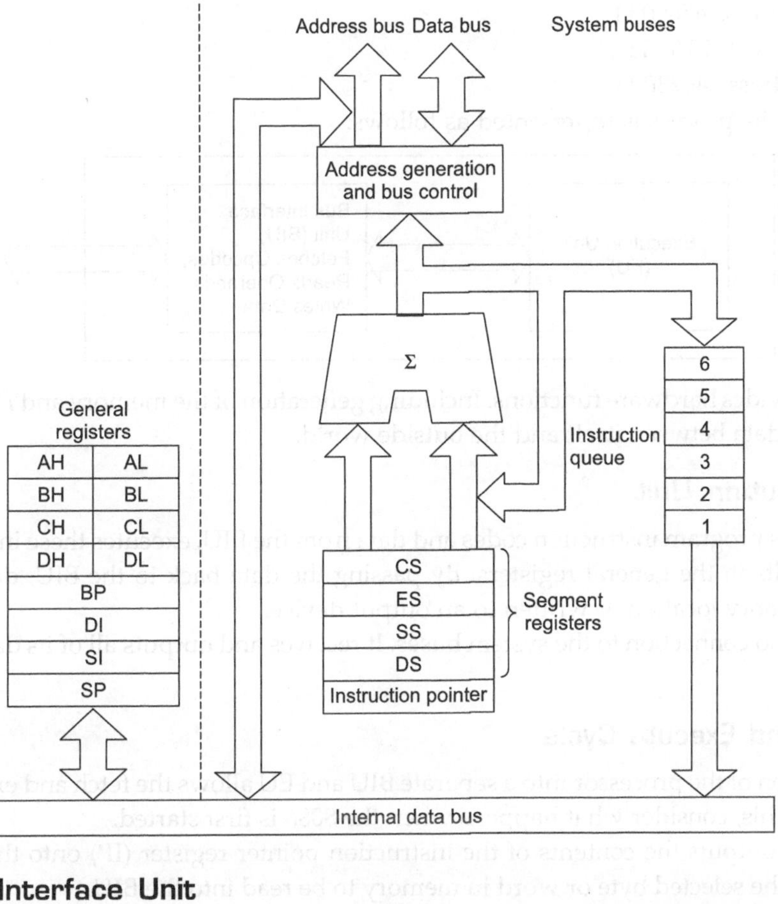

7.2.1 Bus Interface Unit

The Bus Interface Unit (BIU) generates the 20-bit physical memory address and provides the interface with external memory (ROM/RAM). As mentioned earlier, 8086 has a single memory interface. To speed up the execution, 6-bytes of instruction are fetched in advance and kept in a 6-byte Instruction Queue while other instructions are being executed in the Execution Unit (EU). Hence after the execution of an instruction, the next instruction is directly fetched from the instruction queue without having to wait for the external memory to send the instruction. This is called pipe-lining and is helpful for speeding up the overall execution process.

8086's BIU produces the 20-bit physical memory address by combining a 16-bit segment address with a 16-bit offset address. There are four 16-bit segment registers, viz., the code segment (CS), the stack segment (SS), the extra segment (ES), and the data segment (DS). These segment registers hold the corresponding 16-bit segment addresses. A segment address is the upper 16-bits of the starting address of that segment. The lower 4-bits of the starting address of a segment is always zero. The offset address is held by another 16-bit register. The physical 20-bit address is calculated by shifting the segment address 4-bit left and then adding that to the offset address.

For Example:
Code segment Register CS holds the segment address which is 4569 H
Instruction pointer IP holds the offset address which is 10A0 H
The physical 20-bit address is calculated as follows.
Segment address: 45690 H
Offset address : + 10A0 H
Physical address: 46730 H
Graphically the process is represented as follows:

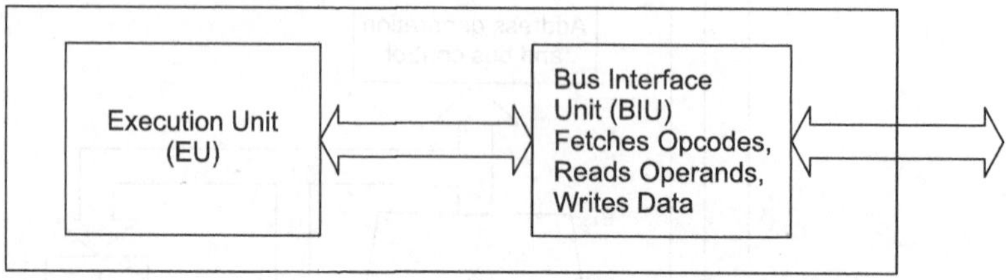

The BIU provides hardware functions. Including generation of the memory and I/0 addresses for the transfer of data between itself and the outside world.

7.2.2 Execution Unit

The EU receives program instruction codes and data from the BIU, executes these instructions, and stores the results in the general registers. By passing the data back to the BIU, data can also be stored in a memory location or written to an output device.

The EU has no connection to the system buses. It receives and outputs all of its data through the BIU.

The Fetch and Execute Cycle

The organization of the processor into a separate BIU and EU allows the fetch and execute cycles to *overlap.* To see this, consider what happens when the 8086 is first started.

1. The BIU outputs the contents of the instruction pointer register (IP) onto the address bus, causing the selected byte or word in memory to be read into the BIU.
2. Register IP is incremented by one to prepare for the next instruction fetch.
3. Once inside the BIU, the instruction is passed to the *queue:* a first-in/first-out storage register sometimes likened to a *pipeline.*
4. Assuming that the queue is initially empty, the EU immediately draws this instruction from the queue and begins execution.
5. While the EU is executing this instruction, the BIU proceeds to fetch a new instruction. Depending on the execution time of the first instruction, the BIU may fill the queue with several new instructions before the EU is ready to draw its next instruction.
6. The cycle continues, with the BIU filling the queue with instructions and the EU fetching and executing these instructions.

The BIU is programmed to fetch a new instruction whenever the queue has room for **two** additional bytes. The advantage to this pipelined architecture is that the EU can execute instructions (almost) continually instead of having to wait for the BIU to fetch a new instruction. This is shown schematically in the following figure.

(a) The nonpipelined microprocessor follows a sequential fetch and execute cycle.
(b) The 8086's pipelined architecture allows the EU to execute instructions without the delays associated with instruction fetching.

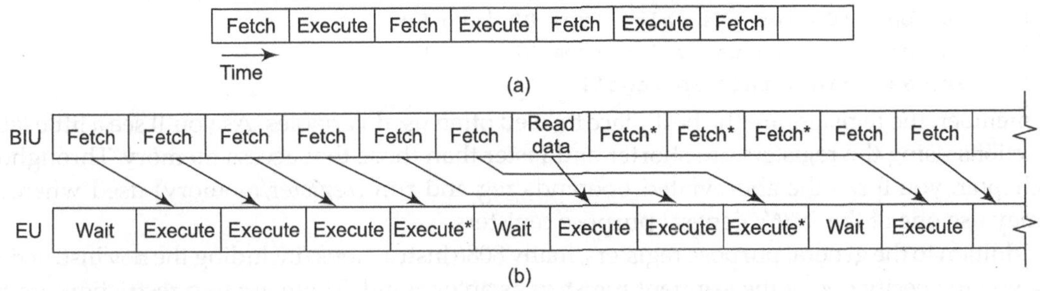

* These bytes are discarded.
* This instruction requires a request for data not in the queue.
* Jump instruction occurs

There are three conditions that will cause the EU to enter a "wait" mode:

- When an instruction requires access to a memory location. The BIU must suspend fetching instructions and output the address of this memory location. After waiting for the memory access, the EU can resume executing instruction codes from the queue, and the BIU can resume filling the queue.
- When the instruction to be executed is *a jump* instruction. In this case, control is to be transferred to a new address. The EU must wait while the instruction at the jump address is fetched. Any bytes presently in the queue must be discarded (they are overwritten).
- During the execution of slow-executing instructions.

7.3 THE 80X86 ADDRESSING MODES

The 80x86 processors let you access memory in many different ways. The 80x86 memory addressing modes provide flexible access to memory, allowing you to easily access variables, arrays, records, pointers, and other complex data types. Mastery of the 80x86 addressing modes is the first step towards mastering 80x86 assembly language.

When Intel designed the original 8086 processor, they provided it with a flexible, though limited, set of memory addressing modes. Intel added several new addressing modes when it introduced the 80386 microprocessor. Note that the 80386 retained all the modes of the previous processors; the new modes are just an added bonus. If you need to write code that works on 80286 and earlier processors, you will not be able to take advantage of these new modes. However, if you intend to run your code on 80386sx or higher processors, you can use these new modes. Since many programmers still need to write programs that run on 80286 and earlier machines, it's important to separate the discussion of these two sets of addressing modes to avoid confusing them.

7.3.1 Register Addressing Mode

Most 8086 instructions can operate on the 8086's general purpose register set. By specifying the name of the register as an operand to the instruction, you may access the contents of that register. Consider the 8086 mov (move) instruction:

```
mov    destination, source
```

This instruction copies the data from the source operand to the destination operand. The eight and 16 bit registers are certainly valid operands for this instruction. The only restriction is that both operands must be the same size. Now let's look at some actual 8086 mov instructions:

```
mov    ax, bx   ;Copies the value from BX into AX
mov    dl, al   ;Copies the value from AL into DL
mov    si, dx   ;Copies the value from DX into SI
mov    sp, bp   ;Copies the value from BP into SP
mov    dh, cl   ;Copies the value from CL into DH
mov    ax, ax   ;Yes, this is legal!
```

Remember, the registers are the best place to keep often used variables. As you'll see a little later, instructions using the registers are shorter and faster than those that access memory. Throughout this chapter, you'll see the abbreviated operands reg and r/m (register/memory) used wherever you may use one of the 8086's general purpose registers.

In addition to the general purpose registers, many 8086 instructions (including the `mov` instruction) allow you to specify one of the segment registers as an operand. There are two restrictions on the use of the segment registers with the `mov` instruction. First of all, you may not specify `cs` as the destination operand; second, only one of the operands can be a segment register. You cannot move data from one segment register to another with a single `mov` instruction. To copy the value of `cs` to `ds`, you'd have to use some sequence like:

```
mov    ax, cs
mov    ds, ax
```

You should never use the segment registers as data registers to hold arbitrary values. They should only contain segment addresses. But more on that, later. Throughout this text you'll see the abbreviated operand sreg used wherever segment register operands are allowed (or required).

7.3.2 8086 Memory Addressing Modes

The 8086 provides 17 different ways to access memory. This may seem like quite a bit at first, but fortunately most of the address modes are simple variants of one another so they're very easy to learn. And learn them you should! The key to good assembly language programming is the proper use of memory addressing modes.

The addressing modes provided by the 8086 family include displacement-only, base, displacement plus base, base plus indexed, and displacement plus base plus indexed. Variations on these five forms provide the 17 different addressing modes on the 8086.

7.3.3 The Displacement Only Addressing Mode

The most common addressing mode, and the one that's easiest to understand, is the displacement-only (or direct) addressing mode. The displacement-only addressing mode consists of a 16-bit constant that specifies the address of the target location. The instruction mov al,ds:[8088h] loads the al register with a copy of the byte at memory location 8088h. Likewise, the instruction mov ds:[1234h],dl stores the value in the dl register to memory location 1234h:

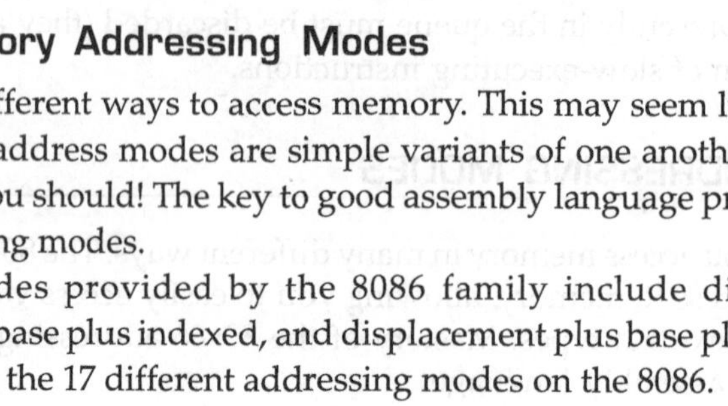

The displacement-only addressing mode is perfect for accessing simple variables. Of course, you'd probably prefer using names like "I" or "J" rather than "DS:[1234h]" or "DS:[8088h]". Well, fear not, you'll soon see it's possible to do just that.

Intel named this the displacement-only addressing mode because a 16 bit constant (displacement) follows the mov opcode in memory. In that respect it is quite similar to the direct addressing mode on the x86 processors (see the previous chapter). There are some minor differences, however. First of all, a displacement is exactly that- some distance from some other point. On the x86, a direct address can be thought of as a displacement from address zero. On the 80x86 processors, this displacement is an offset from the beginning of a segment (the data segment in this example). Don't worry if this doesn't make a lot of sense right now. You'll get an opportunity to study segments to your heart's content a little later in this chapter. For now, you can think of the displacement-only addressing mode as a direct addressing mode. The examples in this chapter will typically access bytes in memory. Don't forget, however, that you can also access words on the 8086 processors:

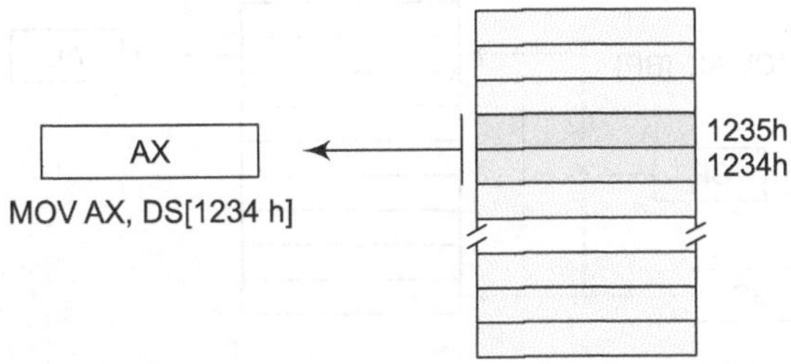

By default, all displacement-only values provide offsets into the data segment. If you want to provide an offset into a different segment, you must use a segment override prefix before your address. For example, to access location 1234h in the extra segment (es) you would use an instruction of the form mov ax,es:[1234h]. Likewise, to access this location in the code segment you would use the instruction mov ax, cs:[1234h]. The ds: prefix in the previous examples is not a segment override. The CPU uses the data segment register by default. These specific examples require ds: because of MASM's syntactical limitations.

7.3.4 The Register Indirect Addressing Modes

The 80x86 CPUs let you access memory indirectly through a register using the register indirect addressing modes. There are four forms of this addressing mode on the 8086, best demonstrated by the following instructions:

```
mov    al, [bx]
mov    al, [bp]
mov    al, [si]
mov    al, [di]
```

As with the x86 `[bx]` addressing mode, these four addressing modes reference the byte at the offset found in the `bx, bp, si,` or `di` register, respectively. The `[bx], [si],` and `[di]` modes use the ds segment by default. The `[bp]` addressing mode uses the stack segment `(ss)` by default.

You can use the segment override prefix symbols if you wish to access data in different segments. The following instructions demonstrate the use of these overrides:

```
mov    al, cs:[bx]
mov    al, ds:[bp]
mov    al, ss:[si]
mov    al, es:[di]
```

Intel refers to `[bx]` and `[bp]` as base addressing modes and `bx` and `bp` as base registers (in fact, `bp` stands for base pointer). Intel refers to the `[si]` and `[di]` addressing modes as indexed addressing modes (`si` stands for source index, `di` stands for destination index). However, these addressing modes are functionally equivalent. This text will call these forms register indirect modes to be consistent.

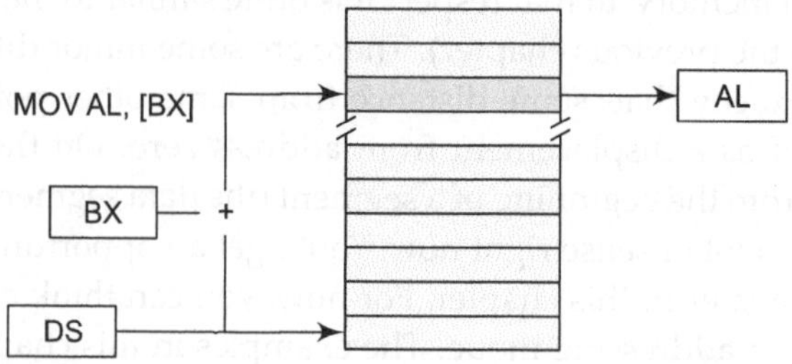

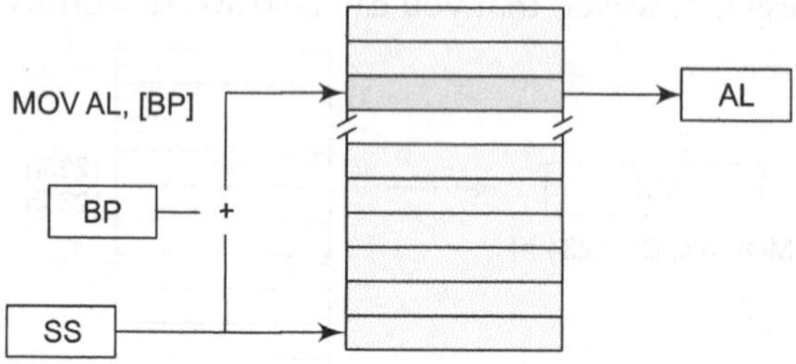

Note: The `[si]` and `[di]` addressing modes work exactly the same way, just substitute `si` and `di` for `bx` above.

7.3.5 Indexed Addressing Modes

The indexed addressing modes use the following syntax:

```
mov    al, disp[bx]
mov    al, disp[bp]
mov    al, disp[si]
mov    al, disp[di]
```

If bx contains 1000h, then the instruction mov cl,20h[bx] will load cl from memory location ds:1020h. Likewise, if bp contains 2020h, mov dh,1000h[bp] will load dh from location ss:3020.

The offsets generated by these addressing modes are the sum of the constant and the specified register. The addressing modes involving bx, si, and di all use the data segment, the disp[bp] addressing mode uses the stack segment by default. As with the register indirect addressing modes, you can use the segment override prefixes to specify a different segment:

```
mov    al, ss:disp[bx]
mov    al, es:disp[bp]
mov    al, cs:disp[si]
mov    al, ss:disp[di]
```

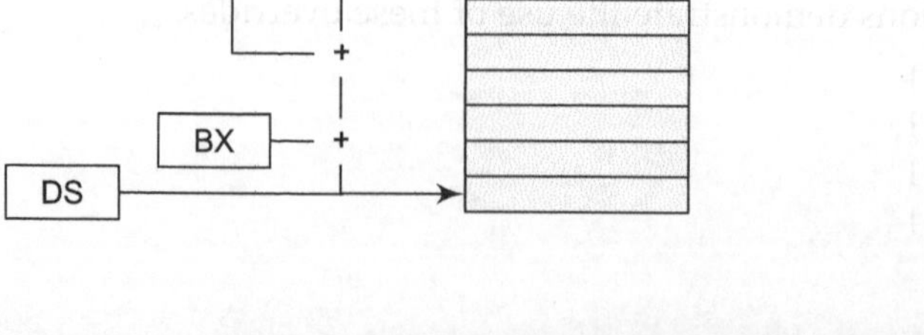

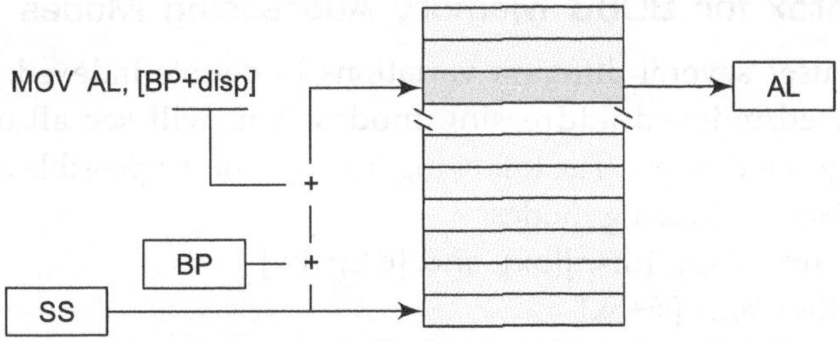

You may substitute si or di in the figure above to obtain the [si+disp] and [di+disp] addressing modes.

Note that Intel still refers to these addressing modes as based addressing and indexed addressing. Intel's literature does not differentiate between these modes with or without the constant. If you look at how the hardware works, this is a reasonable definition. From the programmer's point of view, however, these addressing modes are useful for entirely different things. Which is why this text uses different terms to describe them. Unfortunately, there is very little consensus on the use of these terms in the 80x86 world.

7.3.6 Based Indexed Addressing Modes

These addressing modes are a slight modification of the base/indexed addressing modes with the addition of an eight bit or sixteen bit constant. The following are some examples of these addressing modes:

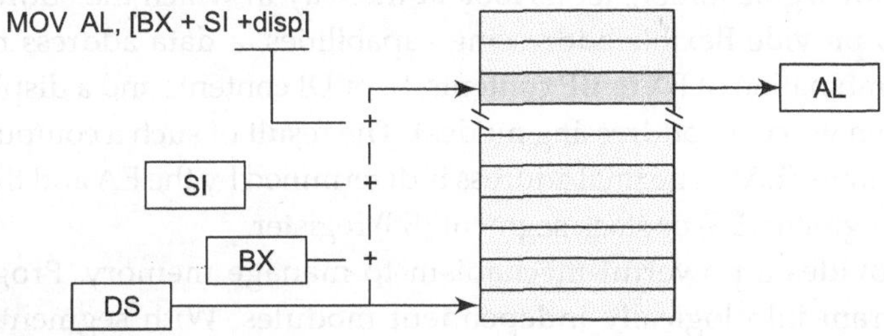


```
mov    al, disp[bx][si]
mov    al, disp[bx+di]
mov    al, [bp+si+disp]
mov    al, [bp][di][disp]
```

You may substitute di in the figure above to produce the [bx+di+disp] addressing mode.

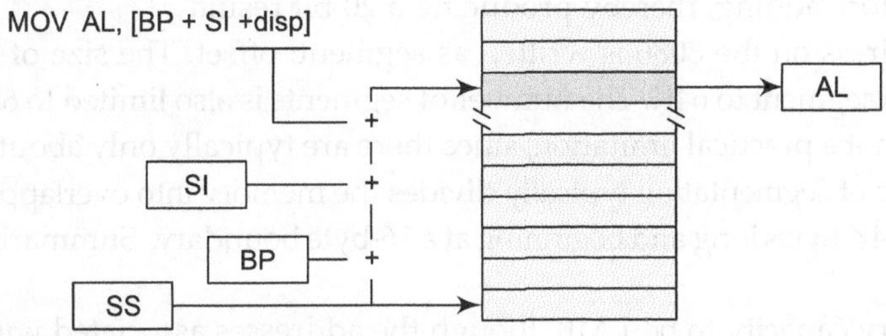

You may substitute di in the figure above to produce the [bp+di+disp] addressing mode.

Suppose bp contains 1000h, bx contains 2000h, si contains 120h, and di contains 5. Then mov al,10h[bx+si] loads al from address DS:2130; mov ch,125h[bp+di] loads ch from location SS:112A; and mov bx,cs:2[bx][di] loads bx from location CS:2007.

7.3.7 MASM Syntax for 8086 Memory Addressing Modes

Microsoft's assembler uses several different variations to denote indexed, based/indexed, and displacement plus based/indexed addressing modes. You will see all of these forms used interchangeably throughout this text. The following list some of the possible combinations that are legal for the various 80x86 addressing modes:

disp[bx], [bx][disp], [bx+disp], [disp][bx], and [disp+bx]

[bx][si], [bx+si], [si][bx], and [si+bx]

disp[bx][si], disp[bx+si], [disp+bx+si], [disp+bx][si], disp[si][bx], [disp+si][bx], [disp+si+bx], [si+disp+bx], [bx+disp+si], etc.

MASM treats the "[ ]" symbols just like the "+" operator. This operator is commutative, just like the "+" operator. Of course, this discussion applies to all the 8086 addressing modes, not just those involving BX and SI. You may substitute any legal registers in the addressing modes above.

7.4 MEMORY SEGMENTATION

The 8086 is a 16-bit processor, with 16-bit registers and hence the address space which can be accessed is 2^{16} bits (*i.e.*, 64 kilobytes). But with the clever use of segmentation, the address space can be expanded to 1 Megabyte. Though 1 Megabyte seems very small in today's context, in those days, it was a lot of memory (especially since the cost of making physical memory was exhorbitant).

Before we start with segmentation, let us look at the way in which the address computation is done in the 8086: To provide flexible addressing capabilities, a data address may be formed by adding together a combination of BX or BP contents, SI or DI contents and a displacement. (This we will see in detail, when we cover addressing modes). The result of such a computation is called the *offset* or the *effective address* (EA). The final address is determined by the EA and the appropriate data segment (DS), extra segment (ES) or stack segment (SS) register.

Segmentation provides a powerful mechanism to manage memory. Programmers tend to partition their program into logically independent modules. With segmentation, this logical partitioning can be extended to keep these modules in physically independent sections (called *segments*). Segmentation also allows two programs to share memory.

The effective address is a 16-bit value. However, to address 1 MB, there are 20 bits in the *physical address*. Where do these extra 4 bits come from? They are obtained from the segment register (to which the EA is added). The addition is carried out by appending four 0 bits to the value of the segment register, before adding, thereby producing a 20-bit result.

Therefore, an address on the 8086 is written as segment: offset. The size of the offset (16-bits) limits the size of each segment to 64K. The number of segments is also limited to 65,536 segments per program, but this is not a practical limitation, since there are typically only about a dozen segments in a program. The use of segmentation typically divides the memory into overlapping segments, with each segment being 64K bytes long and beginning at a 16-byte boundary. Summarizing segmentation, the advantages are:

- Allows memory capacity to be 1 MB, though the addresses associated with each instruction is only 16 bits.
- Allows the instruction, data or stack segment of a program to be more that 64K, by using multiple code segments (or data/stack segments).
- Facilitate the use of separate memory areas for a program, its data, its stack, etc.

One of the reasons why segmentation is not a very popular concept among programmers is that it is very difficult to access more than 256K of memory at once. This limitation is because there are only segment registers. Accessing more than 256K requires quite a bit of bookkeeping.

7.4.1 Segment Registers

The 8086 has 4 special segment registers, viz. cs, ds, es and ss. These are 16-bit registers and deal with selecting blocks (segments) of the memory.

- The cs *(Code Segment)* register points at the segment containing the currently executing machine instructions. Since you can change the value of the cs register, you can switch to a new code segment when you want to execute the code located there.
- The ds *(Data Segment)* register generally points at global variables for the program. You can change the value of the ds register to access additional data in other segments.
- The es *(Extra Segment)* register is an extra segment register. 8086 programs often use this segment register to gain access to segments when it is difficult or impossible to modify the other segment registers.
- The ss *(Stack Segment)* register points at the segment containing the 8086 *stack.* The stack is where the 8086 stores important machine state information, subroutine return addresses, procedure parameters, and local variables. In general, you do not modify the stack segment register because too many things in the system depend upon it.

7.5 OPERATING MODE

In this section, we will discuss the Minimum mode 8086 system and Maximum mode 8086 system.

7.5.1 Minimum Mode 8086 System & Timing

Minimum mode is one of two different hardware modes of the Intel 8086 and 8088 processors (CPU). The other is maximum mode. Mode selection is accomplished by how the chip is hard-wired in the circuit. Specifically, pin #33 (MN/MX) is used to select the mode, depending on whether it is wired to voltage or to ground. Changing the state of pin #33 changes the function of certain other pins. Mode cannot be changed by software.

Minimum Mode Interface

The following are the important points related to Minimum Mode interface:

- When the Minimum mode operation is selected, the 8086 provides all control signals needed to implement the memory and I/O interface.
- The minimum mode signal can be divided into the following basic groups: address/data bus, status, control, interrupt and DMA.
- **Address/Data Bus:** these lines serve two functions. As an address bus is 20 bits long and consists of signal lines A0 through A19. A19 represents the MSB and A0 LSB. A 20-bit address gives the 8086 a 1Mbyte memory address space. More over it has an independent I/O address space which is 64K bytes in length.
- The 16 data bus lines D0 through D15 are actually multiplexed with address lines A0 through A15 respectively. By multiplexed we mean that the bus work as an address bus during first machine cycle and as a data bus during next machine cycles. D15 is the MSB and D0 LSB.
- When acting as a data bus, they carry read/write data for memory, input/output data for I/O devices, and interrupt type codes from an interrupt controller.

Block Diagram of the Minimum Mode 8086 MPU is represented as follows:

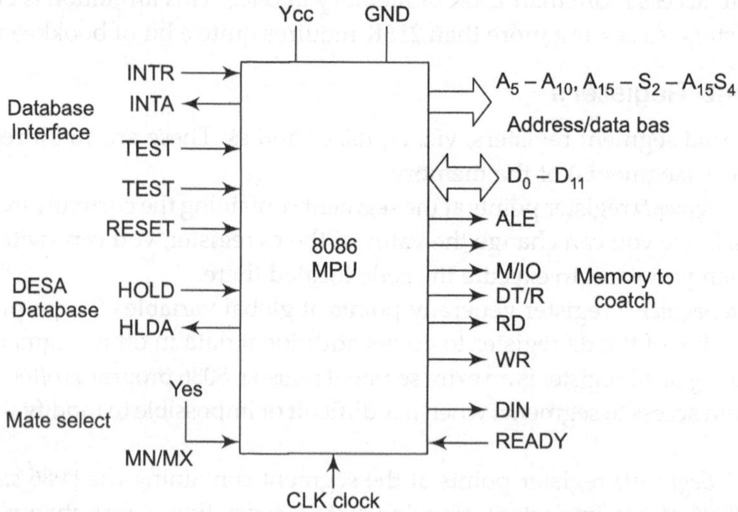

Status Signal

The four most significant address lines A19 through A16 are also multiplexed but in this case with status signals S6 through S3. These status bits are output on the bus at the same time that data are transferred over the other bus lines.

The following are the important points related to status signal:

- Bit S4 and S3 together from a 2-bit binary code that identifies which of the 8086 internal segment registers are used to generate the physical address that was output on the address bus during the current bus cycle.
- Code S4S3 = 00 identifies a register known as ***extra segment register*** as the source of the segment address
- Status line S5 reflects the status of another internal characteristic of the 8086. It is the logic level of the internal enable flag. The last status bit S6 is always at the logic 0 level.

The following table represents the information related to segment registers:

S_4	S_3	**Segment Register**
0	0	Extra
0	1	Stack
1	0	Code/none
1	1	Date

7.5.2 Maximum Mode 8086 System

Maximum mode is one of the two hardware modes available to the Intel 8086 and 8088 processors (CPU). The other is minimum mode.

Maximum mode is for large applications such as multiprocessing. The mode is hard-wired into the circuit and cannot be changed by software. Specifically, pin #33 (MN/MX) is either wired to voltage or to ground to determine the mode. Changing the state of pin #33 changes the function of certain other pins, most of which have to do with how the CPU handles the (local) bus. The IBM PC and PC/XT use an Intel 8088 running in maximum mode.

Maximum Mode Interface

The following are the important points related to Maximum Mode Interface:

- When the 8086 is set for the maximum-mode configuration, it provides signals for implementing a multiprocessor/coprocessor system environment.
- By multiprocessor environment we mean that one microprocessor exists in the system and that each processor is executing its own program.
- Usually in this type of system environment, there are some system resources that are common to all processors.
- They are called as ***global resources***. There are also other resources that are assigned to specific processors. These are known as ***local or private resources***.
- Coprocessor also means that there is a second processor in the system. In this two processor does not access the bus at the same time.
- One passes the control of the system bus to the other and then may suspend its operation.
- In the maximum-mode 8086 system, facilities are provided for implementing allocation of global resources and passing bus control to other microprocessor or coprocessor.

8086 Maximum mode Block Diagram is represented as follows:

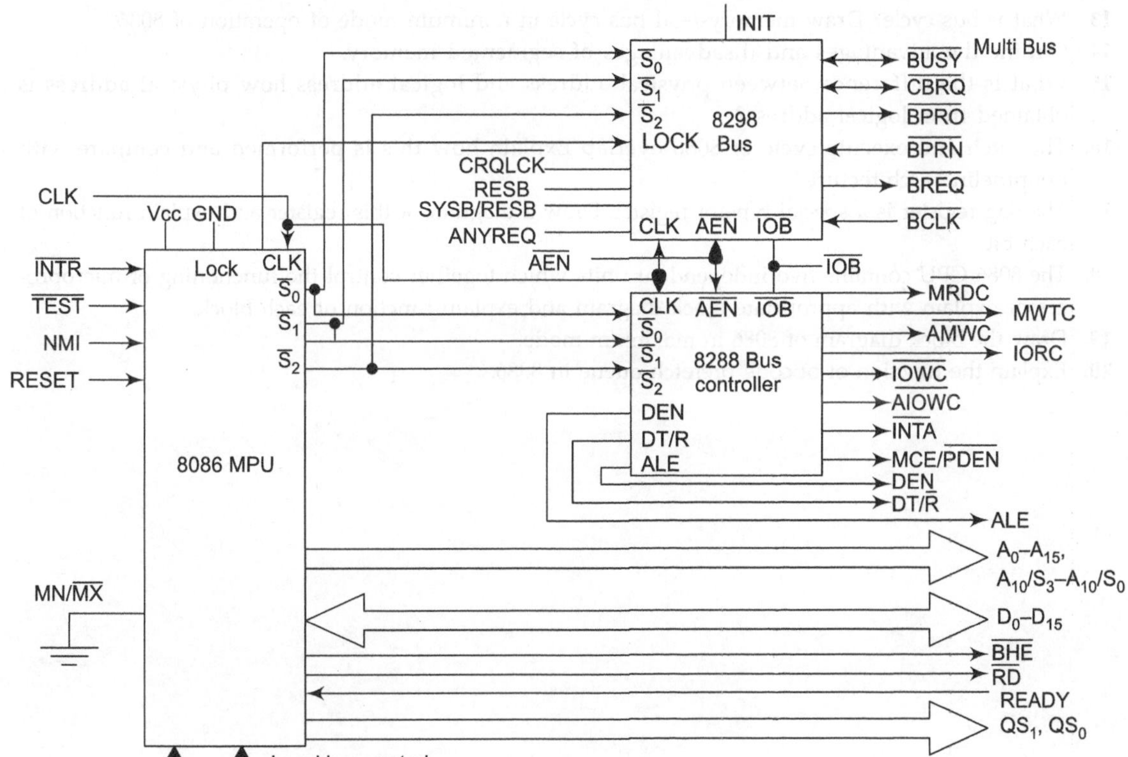

8086 Maximum mode Block Diagram

EXERCISE

1. What are differences between 8086 and 8085 processor?
2. What do you mean by pipelined architecture of CPU?
3. In respect of 8085 or 8086 processors, state any two states signals and any two control signals. Explain how these signals may be used in the system?
4. Describe the Minimum Mode Configuration System of 8086 with a neat block schematic.
5. Draw the register organization of 8086 and explain typical applications of each register.
6. With a neat block diagram explain the Internal Architecture of 8086/8088.
7. Explain conditional and control flags of 8086.
8. Explain why memory is divided into segments in 8086? What are its advantages?
9. Explain the Minimum and Maximum mode of operation of 8086 with details of each pin.
10. Discuss 8086 based system configured in Maximum mode with a neat block diagram.
11. Discuss 8086 based system configured in Minimum mode with a neat block diagram.
12. State the function of the following signals of 8086:

    (*i*) RQ/GT (*ii*) LOCK (*iii*) DT/R (*iv*) MN/MX
    (*v*) QS0, QS1 (*vi*) TEST (*vii*) BHE (*viii*) RESET
    (*ix*) IO/M (*x*) READY
13. What is bus cycle? Draw memory-read bus cycle in minimum mode of operation of 8086.
14. Outline the advantages and disadvantages of segmented memory.
15. What is the difference between physical address and logical address how physical address is obtained from logical address?
16. The fetch and execute cycle of 8086 overlap Explain how this is performed and compare with nonpipeline architecture?
17. The flag register is a special type of register. Draw a diagram of this register and explain function of each bit.
18. The 8086 CPU contains two independent units which together control the functioning of microprocessor explain with appropriate block diagram and explain function of each block.
19. Draw the block diagram of 8086 in maximum mode.
20. Explain the function of opcode prefetch queue in 8086.

8

8086 Instruction Set

This chapter discusses the 80x86 real mode instruction set. Like any programming language, there are going to be several instructions you use all the time, some you use occasionally, and some you will rarely, if ever, use. This chapter organizes its presentation by instruction class rather than importance. Since beginning assembly language programmers do not have to learn the entire instruction set in order to write meaningful assembly language programs, you will probably not have to learn how every instruction operates. The following list describes the instructions this chapter discusses. A "•" symbol marks the important instructions in each group. If you learn only these instructions, you will probably be able to write any assembly language program you want. There are many additional instructions, especially on the 80386 and later processors. These additional instructions make assembly language programming easier, but you do not need to know them to begin writing programs. 80x86 instructions can be (roughly) divided into eight different classes:

1. **Data movement instructions**
   mov, lea, les , push, pop, pushf, popf
2. **Conversions**
   cbw, cwd, xlat
3. **Arithmetic instructions**
   add, inc sub, dec, cmp, neg, mul, imul, div, idiv
4. **Logical, shift, rotate, and bit instructions**
   and, or, xor, not, shl, shr, rcl, rcr
5. **I/O instructions**
   in, out
6. **String instructions**
   movs, stos, lods
7. **Program flow control instructions**
   jmp, call, ret, conditional jumps
8. **Miscellaneous instructions**
   clc, stc, cmc

The following sections describe all the instructions in these groups and how they operate. At one time a text such as this one would recommend against using the extended 80386 instruction set. After all, programs that use such instructions will not run properly on 80286 and earlier processors. Using these additional instructions could limit the number of machines your code would run on. However, the 80386 processor is on the verge of disappearing as this text is being written. You can safely assume that most systems will contain an 80386sx or later processor. This text often uses the 80386 instruction set in various example programs. Keep in mind, though, that this is only for convenience. There is no program that appears in this text that could not be recoded using only 8088 assembly language instructions.

8.1 THE PROCESSOR STATUS REGISTER (FLAGS)

The flags register maintains the current operating mode of the CPU and some instruction state information. The following figure shows the layout of the flags register.

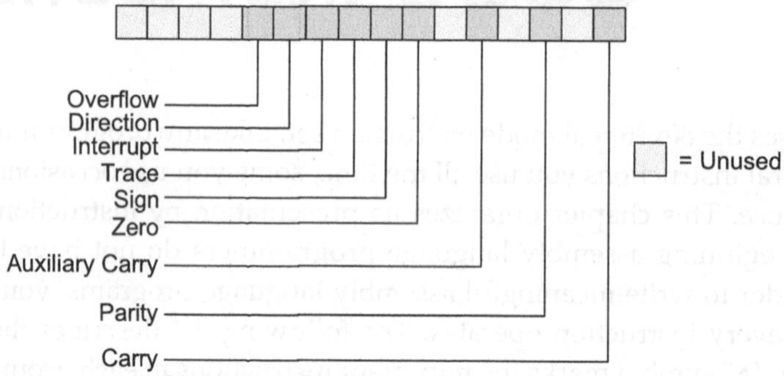

80x86 Flags Register

The carry, parity, zero, sign, and overflow flags are special because you can test their status (zero or one) with the Setcc and conditional jump instructions. The 80x86 uses these bits, the *condition codes*, to make decisions during program execution.

Various arithmetic, logical, and miscellaneous instructions affect the *overflow flag*. After an arithmetic operation, this flag contains a one if the result does not fit in the signed destination operand. For example, if you attempt to add the 16 bit signed numbers 7FFFh and 0001h the result is too large so the CPU sets the overflow flag. If the result of the arithmetic operation does not produce a signed overflow, then the CPU clears this flag.

Since the logical operations generally apply to unsigned values, the 80x86 logical instructions simply clear the overflow flag. Other 80x86 instructions leave the overflow flag containing an arbitrary value.

The 80x86 string instructions use the *direction flag*. When the direction flag is clear, the 80x86 processes string elements from low addresses to high addresses; when set, the CPU processes strings in the opposite direction.

The *interrupt enable/disable flag* controls the 80x86's ability to respond to external events known as interrupt requests. Some programs contain certain instruction sequences that the CPU must not interrupt. The interrupt enable/disable flag turns interrupts on or off to guarantee that the CPU does not interrupt those critical sections of code.

The *trace flag* enables or disables the 80x86 trace modes. Debuggers (such as CodeView) use this bit to enable or disable the single step/trace operation. When set, the CPU interrupts each instruction and passes control to the debugger software, allowing the debugger to *single step* through the application. If the trace bit is clear, then the 80x86 executes instructions without the interruption. The 80x86 CPUs do not provide any instructions that directly manipulate this flag. To set or clear the trace flag, you must:

- Push the flags onto the 80x86 stacks,
- Pop the value into another register,
- Tweak the trace flag value,

- Push the result onto the stack, and then
- Pop the flags off the stack.

If the result of some computation is negative, the 80x86 sets the *sign flag*. You can test this flag after an arithmetic operation to check for a negative result. Remember, a value is negative if its H.O. bit is one. Therefore, operations on unsigned values will set the sign flag if the result has a one in the H.O. position. Various instructions set the *zero flag*

When they generate a zero result. You'll often use this flag to see if two values are equal (*e.g.*, after subtracting two numbers, they are equal if the result is zero). This flag is also useful after various logical operations to see if a specific bit in a register or memory location contains zero or one. The *auxiliary carry flag* supports special binary coded decimal (BCD) operations. Since most programs don't deal with BCD numbers, you'll rarely use this flag and even then you'll not access it directly. The 80x86 CPUs do not provide any instructions that let you directly test, set, or clear this flag. Only the add, adc, sub, sbb, mul, imul, div, idiv, and BCD instructions manipulate this flag.

The *parity flag* is set according to the parity of the L.O. eight bits of any data operation. If an operation produces an even number of one bit, the CPU sets this flag. It clears this flag if the operation yields an odd number of one bit. This flag is useful in certain data communications programs, however, Intel provided it mainly to provide some compatibility with the older 8080 m P. The *carry flag* has several purposes. First, it denotes an unsigned overflow (much like the overflow flag detects a signed overflow). You will also use it during multiprecision arithmetic and logical operations. Certain bit test, set, clears, and invert instructions on the 80386 directly affect this flag. Finally, since you can easily clear, set, invert, and test it, it is useful for various boolean operations. The carry flag has many purposes and knowing when to use it, and for what purpose, can confuse beginning assembly language programmers. Fortunately, for any given instruction, the meaning of the carry flag is clear. The use of these flags will become readily apparent in the coming sections and chapters. This section is mainly a formal introduction to the individual flags in the register rather than an attempt to explain the exact function of each flag.

8.2 INSTRUCTION ENCODINGS

The 80x86 uses a binary encoding for each machine operation. While it is important to have a general understanding of how the 80x86 encodes instructions, it is not important that you memorize the encodings for all the instructions in the instruction set. If you were to write an assembler or disassemble (debugger), you would definitely need to know the exact encodings. For general assembly language programming, however, you won't need to know the exact encodings. However, as you become more experienced with assembly language you will probably want to study the encodings of the 80x86 instruction set. Certainly you should be aware of such terms as *opcode, mod-reg-r/m byte, displacement value,* and so on. Although you do not need to memorize the parameters for each instruction, it is always a good idea to know the lengths and cycle times for instructions you use regularly since this will help you write better programs. This chapter presents a higher level view of each instruction and assumes that you don't care how the machine treats bits in memory.

8.3 DATA TRANSFER INSTRUCTIONS

These instructions are used to transfer data from a source operand to a destination operand. The source operand in most of the cases remains unchanged. The operand can be a literal, a memory location, a register, or even an I/O port address, as the case may be. Let us discuss these instructions with the following table:

Mnemonic	Description	Example
MOV des, src	des ← src; Both the operands should be byte or word, src operand can be register, memory location or an immediate operand des can be register or memory operand. Restriction: Both source and destination cannot be memory operands at the same time.	MOV CX.037AH ; CX register is initialized : with immediate value ; 037AH. MOV AX.BX ; AX ← BX
PUSH operand	Pushes, the operand into a stack. SP ← SP – 2; value [TOS] ← operand. Initialise stack segment register, and the stack pointer properly before using this instruction. No flags are effected by this instruction. The operand can be a general purpose register, a segment register, or a memory location. Please note it is a word stack and memory address is a byte address, thus, you decrement by 2. Also you decrement as SP is initialised to maximum offset and condition of stackful is a zero offset (so it is a reversed stack)	PUSH BX ; decrement stack pointer ; by; two, and copy BX to ; stack. ; decrement stack pointer ; by two, and copy ; BX to stack
POP des	POP a word from stack. The des can be a general-purpose register, a segment register [except for CS register), or a memory location. Steps are: des ← value [TOS] SP ← SP + 2	POP AX ; Copy content for top ; of stack to AX.
XCHG des, sic	Used to exchange bytes or words of sic and des. It requires at least one of the operands to be a register operand. The other can be a register or memory operand. Thus, the instruction cannot exchange two memory locations directly. Both the operands should be either byte type or word type. The segment registers cannot be used as operands for this instruction.	XCHG DX,AX ; Exchange word in DX ; with word m AX
XLAT	Translate a byte in AL using a table stored in the memory. The instruction replaces the AL register with a byte from the lookup table. Tins instruction is a complex instruction.	Example is available in Unit 3.
IN accumulator, port address	It transfers a byte or word from specified port to accumulator register. In case an 8-bit port is supplied as an operand then the data byte read from that part will be transferred to AL register. If a 16-bit port is read then the AX will get 16 bit word that was read. The port address can be an immediate operand, or contained in DX register. This instruction does not change any flags.	IN AL.028h ; read a byte from port ; 028h to AL register
OUT port address. Accumulator	It transfers a byte or word from accumulator register to specified port. This instruction is used to output on devices like the monitor or the printer.	
LEA register. source	Load "effective address" (refer to this term in block 2, Unit 1 in addressing modes) of operand into specified 16 — bit register. Since, an address is an offset in a segment and maximum can be of 16 bits, therefore, the register can only	LEA BX, PRICES ; Assume PRICES is ; an array in the data ; segment. The

	be a 16-bit register. LEA instruction does not change any flags. The instruction is very useful for array processing.	; instruction loads the ; offset of the first byte of ; PRICES directly into ; the BX register.
LDS des-reg	It loads data segment register and other specified register by using consecutive memory locations.	LDS SI. DATA ; DS ← content of memory ; location DATA & ; DATA + 1 ; SI ← content of ; memory locations ; DATA + 2 & DATA + 3
LES des-reg	It loads ES register and other specified register by using consecutive memory locations. This instruction is used exactly like the LDS except in this case ES & other specified registers are initialized.	
LAHF	Copies the lower byte of flag register to AH. The instruction does not change any flags and has no operands.	
SAHF	Copies the value of AH register to low byte of flag register. This instruction is just the opposite of LAHF instruction. This instruction has no operands.	
PUSHF	Pushes flag register to top of stack. SP ← SP – 2: stack [SP] ← Flag Register.	
POPF	Pops the stack top to Flag register. Flag register ← stack [SP] SP ← SP + 2	

8.4 ARITHMETIC INSTRUCTIONS

The following table shows the various Arithmetic Instructions used in 8086 environment.

Mnemonic	Description	Example

ADD	Adds byte to byte, or word to word. The source may be an immediate operand, a register or a memory location. The rules for operands are the same as that of MOV instruction. To add a byte to a word, first copy the byte to a word location, then fill up the upper byte of the word with zeros. This instruction effects the following flags: AF, CF, OF, PF, SF, ZF.	ADD AL, 76H
		; Add the number 74H to
		: AL register, and store the
		; result back in AL
		ADD DX,BX
		; Add the contents of DX
		; to BX and store the
		; result in:
		DX, BX remains
		; unaffected.

ADC des, src	Add byte – byte – carry flag, or word – word – carry flag. It adds the two operands with the carry flag. Rest all the details are the same as that of ADD instruction.	
INC des	It increments specified byte or word operand by one. The operand can be register or a memory location. It can effect	INC BX ; ADD 1 to the contents of

	AF, SF, ZF, PF, and OF flags. It does not affect the carry flag, that is, if you increment a byte operand having OFFH then it will result in 0 value in register and no carry flag.	; BX register INC BL ; Add 1 to the contents of ; BL register
AAA	ASCII adjusts after addition. The data entered from the terminal is usually in ASCII format. In ASCII 0-9 are represented by codes 30–39. This instruction allows you to add the ASCII codes instead of first converting them to decimal digit using masking of upper nibble. AAA instruction is men used to ensure that the result is the correct unpacked BCD.	ADD AL,BL ; AL=00110101, ASCII 05 ; BL=00111001, ASCII 09 ; after addition ; AL = 01101110, that is, ; 6EH- incorrect ; temporary result AAA ; AL = 00000100. ; Unpacked BCD for 04 ; carry = 1. medicates ; the result is 14
DAA	Decimal (BCD) adjust after addition. This is used to make sure that the result of adding two packed BCD numbers is adjusted to be a correct BCD number. DAA only works on AL register.	; AL = 0101 1001(59 ; BCD) ; BL = 00110101(35 ; BCD) ADD AL. BL ; AL= 10001101 or ; 8Eh (incorrect BCD) DAA ; AL = 1001 0100 ; = 94 BCD : Correct.
SUB des, src	Subtract byte from byte, or word from word, (des ← des – src). For subtraction the carry flag functions as a borrow flag, that is, if the number in the source is greater than the number in the destination the borrow flag is to set 1. Other details are equivalent to that of the ADD instruction.	SUB AX, 3427h ; Subtract 3427h from AX ; register, and store the ; result back m AX
SBB des. src	Subtract operands involving previous carry if any. The instruction is similar to SUB, except that it allows us to subtract two multibyte numbers, because any borrow produced by subtracting less-significant byte can be included in the result using this instruction.	SBB AL.CH ; subtract the contents ; of CH and CF from AL ; and store the result ; back in AL.
DEC src	Decrement specified byte or specified word by one. Rules regarding the operands and the flags that are affected are same as INC: instruction. Please note that if the contents of the operand is equal to zero then after decrementing the contents it becomes OFFH or OFFFFH, as the case may be. The carry flag in this case is not affected.	DECBP ; Decrement the contents ; of BP ; register by one.
NEG src	Negate - creates 2's complement of a given number, this changes the sign of a number. However, please note that if you apply this instruction on operand having value -128 (byte operand) or-32768 (word operand) it will result in	NEGAL ; Replace the number in ; AL with it's 2's ; complement

	overflow condition. The overflow (OF) flag will be set to indicate that operation could not be done.	
CMP des, src	It compares two specified byte operands or two specified word operands. The source and destination operands can be an immediate number, a register or a memory location. But, both the operands cannot be memory locations at the same time. The comparison is done simply by internally subtracting the source operand from the destination operand. The value of source and the destination, operands is not changed, but the flags are set to indicate the results of the comparison.	CMP CX, BX ; Compare the CX register ; with the BX register ; In the example above, ; the CF, ZF. and the SF flags ; will be set as follows. ; CX=BX0 1 0; result of ; subtraction is zero ; CX>BX 0 0 0; no borrow ;required therefore, CF=0 ; CX<BX 1 0 1 ; subtraction require ; borrow, so CF=1
AAS	ASCII adjust after subtraction. This instruction is similar to AAA. (ASCII adjust after addition) instruction. The AAS instruction works on the AL register only. It updates the AF and CF flags, but the OF, PF, SF and the ZF flags remain undefined.	; AL = 00110101 ASCII 5 ; BL=0011 1001 ASCII 9 SUB AL,BL ; (5-9) result: ; AL=1111 1100 = – 4 in ; 2's complement, CF = 1 AAS :result: ; AL = 0000 0100 = ; BCD 04. ; CF = 1 borrow needed.
DAS	Decimal adjust after subtraction. This instruction is used after subtracting two packed BCD numbers to make sure the result is the packed BCD. DAS only works on the AL register. The DAS instruction updates the AF, CF, SF, PF and ZF flags. The overflow (OF) is undefined after DAS.	; AL=86 BCD ; BH=57 BCD SUB AL,BH ; AL=2Fh, CF =0 DAS ; Results in AL = 29 BCD

MUL src	This is an unsigned multiplication instruction that multiplies two bytes to produce a word operand or two words to produce a double word such as: AX ← AL* src (byte multiplication src is also byte) DX or AX ← AX * src (word multiplication is two word). This instruction assumes one of the operand in AL (byte) or AX (word): the src operand can be register or memory operand. If the most significant word of the result is zero then, the CF and the OF flags are both made zero. The AF, SF, PF, ZF flags are not defined after the MUL instruction. If you want to multiply a byte with a word, then fast convert byte to a word operand.	MOV AX,Q5; AX=Q5
		MOV CX,02; CX=02
		MUL CX
		; results in DX=0
		; AX=0Ah

AAM	ASCII adjust after multiplication. Please note that two ASCII numbers cannot be multiplied directly. To multiply first convert the ASCII number to numeric digits by	; AL=0000 0101 unpacked ; BCD 05 ; BH=0000 1001 unpacked

	masking off the upper nibble of each byte. This leaves unpacked BCD in the register. AAM instruction is used to adjust the product to two unpacked BCD digits in AX after the multiplication has been performed. AAM defined by the instruction while the CF, OF and the AF flags are left undefined.	; BCD 09 MUL BH ; AX=AL * BH=002Dh AAM ; AX=00000100 00000101 ; BCD 45 : Correct result
DIV src	This instruction divides unsigned word by byte, or unsigned double word by word. For dividing a word by a byte, the word is stored in AX register, divisor the src operand and the result is obtained in AH : remainder AL: quotient. It can be represented as: AH: Remainder AL: Quotient } ← AX/ src Similarly for double word division by a word we have DX: Remainder AX: Quotient } ← DX : A X/ src A division by zero result in run time error. The divisor src can be either in a register or a memory operand.	: AX = 37D7h= 14295 ; decimal :BH = 97h = 151 decimal DIV BH ; AX/BH quotient ; AL = 5Eh = 94 ; decimal KemainderAH = ; 65h=101 ; decimal
IDIV	Divide signed word by byte or signed double word by word. For this division the operand requirement, the general format of the instruction etc. are all same as the DIV instruction. IDIV instruction leaves all flags undefined.	; AL= 11001010 = -26h = ; - 38 decimal ; CH = 00000011 =-3h = ; 3 decimal ; According to the operand ; rules to divide by a byte ; the number should be ; present in a word ; register, *i.e.*, AX. So, first ; convert the operand in ; AL to word operand. This ; can be done by sign ; extending the ; AL register, ; this makes AX ; 11111111 11001010. ; (Sign extension can also ; be done with the help of ; an instruction discussed ; later)IDIV CH ; AX CH ; AL= 11110100= -OCH ; = -12 Decimal ; AH= 11111110 =-02H = ; -02 Decimal ; Although, the quotient is ; actually closer to -13 ; (-12.66667) than -12, but

		; 8086 truncates the result ; to give -12.
AAD	ASCII adjust after division. The BCD numbers are First unpacked, by masking off the upper nibble of each byte. Then ADD instruction is used to convert the unpacked BCD digits in AL and AH registers to adjust them to equivalent binary prior to division. Such division will result in unpacked BCD quotient and remainder. The PF, SF. ZF flags are updated while the AF, CF, and the OF flags are left undefined.	; AX= 0607 unpacked ; BCD for 6 ; and 7 CH = 09h AAD ; adjust to binary before ; division AX= 0043 = ; 043h = 67 Decimal DIV CH ; Divide AX by unpacked ; BCD in CH ; AL = 07 unpacked BCD ; AH = 04 unpacked BCD ; PF = SF = ZF = 0
CBW	Fill upper-byte or word with copies of sign bit of lower bit. This is called sign extension of byte to word. This instruction does not change any flags. This operation is done with AL register in die result being stored in AX.	AL = 10011011 = -155 ; decimal AH = 00000000 CBW ; convert signed ; byte in AL to signed ; word in AX = 11111111 ; 10011011 = -155 decimal
CWD	Fill upper word or double word with sign bit of lower word. This instruction is an extension of the previous instruction. This instruction results in sign extension of AX register to DX: AX double word, masking off the upper nibble of each byte. Then ADD instruction is used to convert the unpacked BCD digits in AL and AH registers to adjust them to equivalent binary prior to division Such division will result in unpacked BCD quotient and remainder. The PF, SF. ZF flags are updated while the AT, CF, and the OF flags are left undefined.	; DX : 0000 0000 0000 0000 ; AX: 1111 0000 0101 0001 CWD ; DX:AX = 1111 1111 1111 ; 1111: 11110000 01010001 ; and 7 CH = 09h AAD ; adjust to binary before ; division AX = 0043 = ; 043h = 67 Decimal DIV CH ; Divide AX by unpacked ; BCD in CH ; AL = 07 unpacked BCD ; AH = 04 unpacked BCD ; PF = SF = ZF = 0
CBW	Fill upper-byte or word with copies of sign bit of lower bit. This is called sign extension of byte to word. This instruction does not change any flags. This operation is done with AL register in the result being stored in AX.	; AL= 10011011 = -155 ; decimal AH = 00000000 CBW ; convert signed ; byte in AL to signed ; word in AX = 1111 1111 ; 10011011 =155 decimal
CWD	Fill upper word or double word with sign bit of lower word. This instruction is an extension of the previous instruction. This instruction results in sign extension of AX register to DX: AX double word.	; DX : 0000 0000 0000 0000 ; AX: 1111 0000 0101 0001 CWD ; DX:AX=1111 1111 1111 ; 1111: 1111 0000 0101 0001

8.5 BIT MANIPULATION INSTRUCTIONS

These instructions are used at the bit level. These instructions can be used for testing a zero bit, set or reset a bit and to shift bits across registers. Let us look into some such basic instructions.

Mnemonic	Description	Example
NOT des	Complements each bit to produce l's complement of the specified byte or word operand. The operand can be a register or a memory operand.	; BX = 0011 1010 0001 0000 NOT BX ; BX = 1100 0101 1110 1111
AND des, src	Bitwise AND of two byte or word operands. The result is des ← des AND src. The source can be an immediate operand a register, or a memory operand. The destination can be a register or a memory operand. Both operands cannot be memory operands at the same time. The CF and the OF flags are both zero after the AND operation. PF, SF and ZF area updated. AF is left undefined.	:BH = 0011 1010 before AND BH, OFh ; BH = 0000 1010 ; after the AND operation
OR des, src	OR each corresponding bits of the byte or word operands. The other operands rules are same as AND. des ← des OR. src	; BH = 0011 1010 before ORBH, OFh ; BH = 0011 1111 after
XOR des, src	XOR each corresponding bit in a byte or word operands rules are two same as AXD and OR. des ← Des + src	; BX = 0011 1101 0110 1001 ; CX = 0000 0000 1111 1111 XOR BX,CX ; BX = 0011 1101 1001 0110 ; Please note, that the bits ; in the lower byte are ; inverted.
TEST des, src	AXD the operands to update flags, but do not change operands value. It can be used to set and test conditions. CF and OF are both set to zero, PF, SF and ZF are all updated, AF is left undefined after the operation.	: AL = 0101 0001 TEST AL, 80h. ; This instruction would ; test if the MSB bit of the ; AL register is zero or ; one. After the TEST ; operation ZF will be set ; to 1 if the MSB of AL ; is zero.
SHLSAL des, count	Shift bits of word or byte left, by count. It puts zero(s) in LSB(s). MSB is shifted into the carry flag. If more than one bits are shifted left, then the CF gets the most recently moved MSB. If the number of bits desired to be shifted is only 1, then the immediate number. 1 can be written as one of the operands. However, if the number of bits desired to be shifted is more than one, then the second operand is put in CL register.	SAL BX, 01 ; if CF = 0 ; BX = 1000 1001 ; result: CF = 1 ; BX = 0001 0010
SHR des, count	It shifts bits of a byte or word to register put zero in MSB. LSB is moved into CF.	SHR BX,01 ; if CF = 0 ; BX= 1000 1001 ; result: CF = 1

		; BX= 0100 0100 MOY CL, 02 SHR BX, CL ; with same BX, the ; result would be ; CF = 0 ; BX= 0010 0100
SAR des, count	Shift bits of word or byte right, but it retains the value of new MSB to that of old MSB. This is also called arithmetic shift operation, as it does not change the MSB, which is sign bit of a number.	; AL=0001 1101 = +29 ; decimal. CF = 0 SAR AL, 01 ; AL = 0000 1110 = +14 ; decimal CF = 1 ; OF = PF = SF = ZF = 0 ; BH = 11110011 = – 13 ; decimal SAR BH,01 ; BH= 1111 1001 = –7 ; decimal, CF = 1 ; OF = ZF = 0 ; PF = SF = 1
ROL des, count	Rotate bits of word or byte left. MSB is transferred to LSB and also to CF. Diagrammatically, it can be represented as: 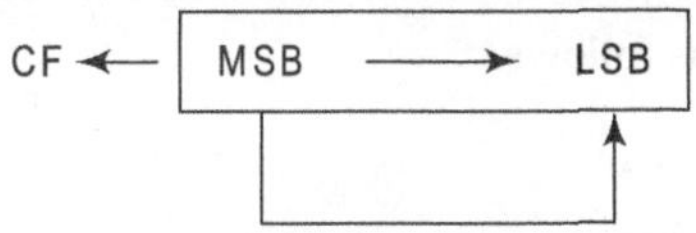 The operation is called rotate as it circulates bits. The operands can be register or memory operand.	
ROR des. count	Rotate bits of word or byte right, LSB is transferred to MSB and also to CF. The same can be represented diagrammatically as follows: 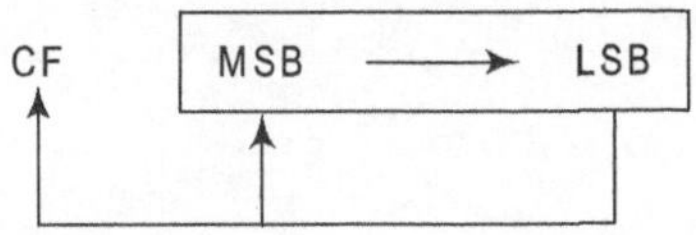	; CF = 0. ; BX = 0011 1011 0111 0101 ROR BX, 1 ; results ; CF = 1. ; BX = 1001 1101 1011 1010
RCL des. count	Rotate bits of words or byte left, MSB to CF and CF to LSB. The operation is circular and involves carry flag in rotation. 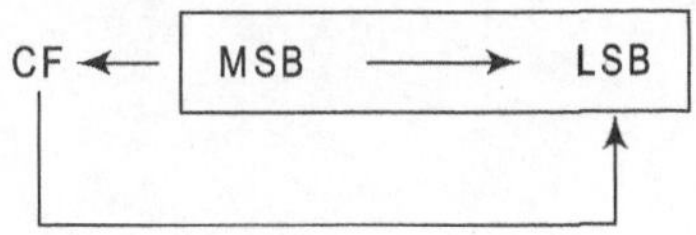	
RCR des. count	Rotate bits, of word or byte right. LSB to CF and CF to MSB. This instruction rotates left. 	

8.6 PROGRAM EXECUTION TRANSFER INSTRUCTIONS

These instructions are the ones that cause change in the sequence of execution of instruction. This change can be through a condition or sometimes may be unconditional. The conditions are represented by flags. For example, an instruction may be jump to an address if zero flag is set, that is the last ALU operation has resulted in zero value. These instructions are often used after a compare instruction, or some arithmetic instructions that are used to set the flags, for example, ADD or SUB. LOOP is also a conditional branch instruction and is taken till loop variable is below a certain count. Please note that a "/" is used to separate two mnemonics which represent the same instruction.

Mnemonic	Description	Example
CALL proc 1	This function results in a procedure/function call. The return address is saved on the stack. There are two basic types of CALLS. NEAR or Intra-Segment calls: if the call is made to a procedure in the same segment as the calling program. FAR or Inter segment call: if the call is made to a procedure in the segment, other than the calling program. The saved return address for NEAR procedure call is just the IP. For FAR Procedure call IP and CS are saved as return address. A procedure can also be called indirectly, by first initialize-ing some 16-bit register, or some other memory location with the new addresses as follows.	CALL proc1 CALL proc2 The new instruction address is determined by name declaration proc1 is a near procedure, thus, only IP is involved, proc2 involves new CS: IP pair. On call to proc 1 stack ← IP IP ← address offset of proc1 on call to proc2 Stack [top] ← CS Stack [top] ← IP CS ← code segment of proc2 IP ← address offset of proc2 Here we assume that proc1 is defined within the same segment as the calling procedure, while proc2 is defined in another segment. As far as the calling program is concerned, both the procedures have been called in the same manner. But while declaring these procedures, we declare proc1 as NEAR procedure and proc 2 as FAR procedure, as follows: proc1 PROC: NEAR proc 2 PROC FAR LEA BX

		proc1 ; initialize BX with the ; offset of the procedure ; procl CALL BX ; CALL prod indirectly : using BX register
RET number	It returns the control from procedure to calling program. Every CALL should be a RET instruction. A RET instruction, causes return from NEAR or FAR procedure call. For return from near procedure the values of the instruction pointer is restored from stack. While for far procedure the CS:IP pair get is restored. RET instruction can also be followed by a number.	RET 6 ; In this case. 8086 ; increments the stack ; pointer by this number ; after popping off the IP ; (for new) or IP and CS ; registers (for far) from ; the stack. This cancels ; the local parameters, or ; temporary parameters : created by the ; programmer. RET ; instruction does not ; affect any flags.
JMP Label	Unconditionally go to specified address and get next instruction from the label specified. The label assigns the instruction to which jump has to take place within the program, or it could be a register that has been initialised with the offset value. JMP can be a NEAR JMP or a FAR jump, just like CALL.	JMP CONTINUE ; CONTINUE is the ; label given to the ; instruction where the ; control needs to be ; transferred JMPBX ; initialize BX with the ; offset of the instruction, ; where the control needs ; to be transferred.

Conditional Jump	All the conditional jumps follow some conditional statement, or any instruction that affects the flag.	MOV CX. 05
		MOV BX, 04
		CMP CX, BX
		; this instruction will set
		; various flags like the ZF,
		; and the CF
		JE LABEL1
		; conditional jump can
		; now be applied, which
		; checks for the ZF. and if
		; it is set implying CX =
		; BX, it makes
		; a jump to LABEL1,
		; otherwise the control
		; simply falls

		; through to next ; instruction ; in the above example as ; CX is not equal to BX ; the jump will not take ; place and the next ; instruction to conditional ; jump instruction will be ; executed. However if ; JNE (Jump if not equal : to) or JA (Jump if above), ; or JAE (Jump above or ; equal) jump instructions ; if applied instead of JE, ; will cause the ; conditional jump to ; occur.
	All the conditional jump instructions which are given below are self explanatory.	
JA/JNBE	Jump if above / Jump if not below nor equal	
JAE/JNB	Jump if above or equal/ Jump if not below	
JB/JNAE	Jump if below/ Jump if not above nor equal	
JBE/JNA	Jump if below or equal/ Jump if not above	
JC	Jump if carry flag set	
JE/JZ	Jump if equal / Jump if zero flag is set	
JNC	Jump if not carry	
JNE/JNZ	Jump if not equal / Jump if zero flag is not set	
JO	Jump if overflow flag is set	
JNO	Jump if overflow flag is not set	
JP/JPE	Jump if parity flag is set / Jump if parity even	
JNP/JPO	Jump if not parity / Jump if parity odd	
JG/JNLE	Jump if greater than / Jump if not less than nor equal	
JA/JNL	Jump if above / Jump if not less than	
JL/JNGE	Jump if less than / Jump if not .greater than nor equal	
JLE/JNG	Jump if less than or equal to / Jump if not greater than	
JS	Jump if sign flag is set	
JXS	Jump if sign flag is not set	
LOOP label	This is a looping instruction of assembly. The number of times the looping is required is placed in CX register. Each iteration decrements CX register by one implicitly, and the Zero Flag is checked to check whether to loop again. If the zero flag is not set (CX is zero) greater than the control goes back to the specified label in the instruction, or else the control falls through to the next instruction. The LOOP	; Let us assume we want ; to add 07 to AL register, ; three times. MOV CX03 ; count of iterations LI: ADDAL.07 LOOP LI ; loop back to L1,

	instruction expects the label destination at offset of – 128 to +127 from the loop instruction offset.	; until CX ; becomes equal to zero ; Loop affects no flags.
LOOPE/ LOOPZ label	Loop through a sequence of instructions while zero flag = 1 and CX is not equal to zero. There are two ways to exit out of the loop, firstly when the count in the CX register becomes equal to zero, or when the quantities that are being compared become unequal.	Let us assume we have an array of 20 bytes. We want to see if all the elements of that array are equal to 0FFh or not. To scan 20 elements of the array, we loop 20 times. And we come out of the loop, when either the count of iterations has become equal to 20, or in other words CX register has decremented to zero, which means all the elements of the array are equal to 0FFh, or an element in the array is found which is not equal to 0FFh. In this case, the CX register may still be greater than zero, when the control comes out. This can be coded as follows: (Please note here that you might not understand everything at this place, that is because you are still not familiar with the various addressing modes. Just concentrate on the LOOPE instruction): MOV BX7 OFFSET ARRAY ; Point BX at the start ; of the ARRAY DEC: BX ; put number of ; array elements in CX MOVCX, 10 LI: INC BX ; point to ; next element in array CMP [BX]-70FFh ; compare array element

		; with 0FFh LOOPE LI ; When the control comes ; out of the loop, it has : either scanned all the : elements and found ; them to be all equal to ; 0FFh, or it is pointing to ; the first non-0FFh. ; element in the array.
LOOPNE LOOPNZ label	This instruction causes Loop through a sequence of instructions while zero flag = 0 and CX is not equal to zero. This instruction is just the opposite of the previous instruction m its functionality.	
JCXZ label	Jump to specified address if CX = 0. This instruction will cause a jump, if the value of CX register is zero. Otherwise it will proceed with the next instruction in sequence.	This instruction is useful when you want to check whether CX is zero even prior to entering into a loop. Please note that LOOP instruction executes the loop at least once before decrementing and checking the value of CX register. Thus. CX=0 will execute the loop once and decrement the CX register making it 0FFFFh, which is non zero. This "will cause FFFFh times execution of loop. To avoid such type of conditions you can proceed as follows: JCXZ SKIP_LOOP ; if CX is already 07 skip ; loop L1: SUB [BX], 07h INC BX LOOP L1 ; loop until CX=0 SKIP LOOP:

In addition to these instructions, there are other interrupt handling instructions also, which too transfer the control of the program to some specified location.

8.7 STRING INSTRUCTIONS

These are a very strong set of 8086 instructions as these instructions process strings, in a compact manner, thus, reducing the size of the program by a considerable amount. "String" in assembly is just a sequentially stored bytes or words. A string often consists of ASCII character codes. A subscript B following the instruction indicates that the string of bytes is to be acted upon, while "W" indicates that it is the string of words that is being acted upon.

Mnemonic	Description	Example
REP	This is an instruction prefix. It causes repetition of the following instruction till CX becomes zero. REP. It is not an instruction, but it is an instruction prefix that causes the CX register to be decremented. This prefix causes the string instruction to be repeated, until CX becomes equal to zero.	REP MOVSB STR1, STR2 The above example copies byte by byte contents. The CX register is initialized to contain the length of source string REP repeats the operation MOVSB that copies the source string byte to destination byte. This operation is repeated until the CX register becomes equal to zero.
REPE/REPZ	It repeats the instruction following until CX =0 or ZF is not equal to one. REPE/REPZ may be used with the compare string instruction or the scan string instruction. REPE causes the string instruction to be repeated, till compared bytes or words are equal, and CX is not yet decremented to zero.	
REPNE/REPNZ	It repeats instruction following it until CX =0 or ZF is equal to 1. This comparison here is just inverse of REPE except for CX, which is checked to be equal to zero.	

MOVS/MOVSB/ MOVSW	It causes moving of byte or word from one string to another. This instruction assumes that:
	• Source string is in Data segment
	• Destination string is in extra data segment
	• SI stores offset of source string in extra segment
	• DI stores offset of destination string is in data segment
	• CX contains the count of operation
	A single byte transfer requires;
	• One byte transfer from source string to destination
	• Increment of SI and DI to next byte
	• Decrement count register that is CX register
		Assumes both data and extra segment start at address 1000 in the memory. Source string starts at offset 20h and the destination string starts at offset 3Oh Length of the source string is 10 bytes. To copy the source string to the destination string, proceed as follows:
		MOV AX, 1000h
		MOV DS, AX
		; initialize data segment
		; and
		MOV ES, AX
		; extra segment
		MOV SI, 20h
		MOV DI, 30h
		; load offset of start of
		; source string to SI
		; Load offset of start of

		; destination string to DI MOV CX, 10 ; load length of string to ; CX as counter REP MOVSB ; Decrement CX and ; MOVSB until ; CX = 0 ; after move SI will be one ; greater than offset of last ; byte in source string, DI ; will be one greater than ; offset of last destination ; string CX will be equal ; to zero.
CMPS/CMPSB/ CMPSW	It compares two string bytes or words. The source string and the destination strings should be present in data segment and the extra segment respectively. SI and DI are used as in the previous instruction. CX is used if more than one bytes or words are to be compared, however for such a case appropriate repeating prefix like REP, PEPE etc. need to be used.	MOV CX,10 MOV SIOFFSET SRC_STR ; offset of source ; string in SI MOV DI. OFFSET DES_STR ; offset of destination ; string in DI REPECMPSB ; Repeat the comparison ; of string bytes until ; end of string or until ; compared bytes are not ; equal.

SCAS/SCASB/ SCASW	It scans a string. Compare a string byte with byte in AL or a string word with a word in AX. The instruction does not change the operands in AL (AX) or the operand in the string. The string to be scanned must be present in the extra segment, and the offset of the string must be contained in the DI register. You can use CX if operation is to be repeated using REP prefixes.	MOV AL, 0Dh
		; Byte to be scanned
		; for in AL
		MOV DI. OFF SET
		DES_STR
		MOV CX,10
		REPNE SCAS DES_STR
		; Compare byte
		; inDES_STR with byte in
		; AL register Scanning is
		; repeated while the bytes
		; are not equal and it is not
		; end of string. If a
		; carriage return 0Dh is
		; found ZF = DI will point
		; at the next byte after the
		; carnage return. If a
		; carnage return is not

		; found then, ZF = 0 and ; CX = 0. SCASB or ; SCASW can be used to ; explicitly state whether ; the byte comparison or ; the word comparison is ; required.
LODS.LODSB/ LODSW	It loads string byte into AL or a string word into AX. The string byte is assumed to be pointed to by SI register. After the Load, the SI pointer is automatically adjusted to point to the next byte or word as the case may be. This instruction does not affect any flag.	MOV SI OFFSET SRC–STR LODS SRC–STR ; LODSB or LODSW can ; be used to indicate to the ; assembler, explicitly. ; whether it is the byte ; that is required to be ; loaded or the word.
STOS/STOSB/ STOSW	It stores byte from AL or word from AX into the string present in the extra segment with offset given by DI. After the copy. DI is automatically adjusted to point to the next byte or word as per the instruction. No flags are affected.	MOVDI. OFFSET DES-SIR STOSB DES_STR

8.8 PROCESSOR CONTROL INSTRUCTIONS

The objectives of these instructions are to control the processor. This raises two questions:

How can you control processor, as this is the job of control unit?

How much control of processor is actually allowed?

Well, 8086 only allows you to control certain control flags that causes the processing in a certain direction, processor synchronization if more than one processors are attached through LOCK instruction for buses etc.

Note: Please note that these instructions may not be very clear to you right now. Thus, some of these instructions have been discussed in more detail in later units. You must refer to further readings for more details on these instructions.

Mnemonic	Description	Example
STC	It sets carry flag to 1.	
CLC	It clears the carry flag to 0.	
CMC	It complements the state of the carry flag from 0 to 1 or 1 to 0 as the case may be.	CMC; Invert the carry flag
STD	It sets the direction flag to 1. The string instruction moves either forward (increment SI, DI) or backward (decrement SI, DI) based on this flag value. STD instruction does not affect any other flag. The set direction flag causes strings to move from right to left.	
CLD	This is opposite to STD, the string operation occurs in the reverse direction.	CLD ; Clear the direction flag ; so that the string ; pointers auto-increment

		MOV AX, 1000h
		MOV DS, AX
		; Initialize data segment
		; and extra segment
		MOV ES, AX
		MOV SI, 20h
		; Load offset of start of
		; source string to SI
		MOV DI, 30h
		; Load offset of start of
		; destination string to DI
		MOV CX, 10
		; Load length of string to
		; CX as counter
		REP MOVSB
		; Decrement CX and
		; increment
		; SI and DI to point to next
		; byte, then MOVSB until
		; CX = 0

There are many process control instructions other than these; you may please refer to further reading for such instructions. These instructions include instructions for setting and closing interrupt flag, halting the computer, LOCK (locking the bus), NOP etc.

8.9 ADDRESSING MODES

The basic set of operands in 8086 may reside in register, memory and immediate operand. How can these operands be accessed through various addressing modes? The answer to the question above is given in the following sub-section. Large number of addressing modes helps in addressing complex data structures with ease. Some specific Terms and registers roles for addressing are as follows:

Base register (BX, BP): These registers are used for pointing to base of an array, stack etc.

Index register (SI, DI): These registers are used as index registers in data and/or extra segments.

Displacement: It represents offset from the segment address.

The following table represents the Addressing modes of 8086.

Mode	Description	Example
Direct	Effective address is the displacement of memory variable.	
Register Indirect	Effective address is the contents of a register.	[BX]
		[SI]
		[DI]
		[BP]
Based	Effective address is the sum of a base register and a displacement.	LIST[BX] (OFFSET LIST + BX)
		[BP + 1]

Indexed	Effective address is the sum of an index register and a displacement.	LIST[SI] [LIST + DI] [DI + 2]
Based Indexed	Effective address is the sum of a base an index register	[BX + SI] [BX] [DI] [BP + DI]
Based Indexed with displacement	Effective address is the sum of a base register, an index register, and a displacement	[BX + SI + 2]

8.9.1 Register Addressing Mode

In this mode Operand can be a 16-bit register.

The following table represents the complete description of Register Addressing mode:

Addressing Mode	Description	Example
AX, BX, CX, DX, SI, DI, BP, IP, CS, DS, ES, SS Or it may be AH, AL, BH, BL, CH, CL, DH, DL	In general, the register addressing mode is the most efficient because registers are within the CPU and do not require memory access.	MOV AL, CH MOV AX, CX

8.9.2 Immediate Addressing Mode

An immediate operand can be a constant expression, such as a number, a character, or an arithmetic expression. The only constraint is that the assembler must be able to determine the value of an immediate operand at assembly time. The value is directly inserted into the machine instruction.

MOV AL, 05

The following table represents the complete description of Immediate Addressing mode:

Mode	Description	Example

Immediate	Please note in the last examples the expression (2 + 3)/5, is evaluated at assembly time.	MOV AL, 10
		MOV AL, 'A'
		MOV AX, 'AB'
		MOV AX, 64000
		MOV AL, (2 + 3)/5

8.9.3 Direct Addressing Mode

A direct operand refers to the contents of memory at an address implied by the name of the variable.

The following table represents the complete description of Direct Addressing mode:

Mode	Description	Example
DIRECT	The direct operands are also called as relocatable operands as they represent the offset of a Label from the beginning of a segment. On reloading a program even in a different	MOV COUNT, CL : move CL to COUNT (a ; byte variable)

	segment will not cause change in the offset that is why we call them relocatable. Please note that a variable is considered in Data segment (DS) and code label in code segment (SS) by default. Thus, in the example, COUNT, by default will be assumed to be in data segment, while LABEL 1, will be assumed to be in code segment. If we specify, as a direct operand then the address is non-relocatable. Please note the value of segment register will be known only at the run time.	MOV AL. COUNT
		: move COUNT to AL
		JMP LABEL 1
		: jump to LABEL 1
		MOV AX, DS:5
		; segment register and
		: offset
		MOV BX, CSEG:2Ch
		; segment name and offset
		MOV AX, ES:COUNT
		; segment register and
		; variable.
		; The offsets of these
		; variables are calculated
		; with respect to the
		; segment name (register)
		; specified in the
		; instruction.

8.9.4 Indirect Addressing Mode

In indirect addressing modes, operands use registers to point to locations in memory. So it is actually a register indirect addressing mode. This is a useful mode for handling strings/ arrays etc. For this mode two types of registers are used. These are:

- Base register BX, BP
- Index register SI, DI

BX contains offset/ pointer in Data Segment.
BP contains offset/ pointer in Stack segment.
SI contains offset/pointer in Data segment.
DI contains offset /pointer in extra data segment.
There are five different types of indirect addressing modes:

1. Register indirect
2. Based indirect
3. Indexed indirect
4. Based indexed
5. Based indexed with displacement

The following table represents the complete description of the above Addressing modes:-

Mode	Description	Example

Register indirect	Indirect operands are particularly powerful when processing list of arrays, because a base or an index register may be modified at runtime.	MOV BX. OFFSET ARRAY
		; point to start of array
		MOV AL, [BX]
		; get first element
		INC BX
		; point to next
		MOV DL, [BX]
		; get second element

		The brackets around BX signify that we are referring to the contents of memory location, using the address stored in BX. In the following example, three byres in an array are added together:
		MOV SI OFFSET ARRAY
		; address of first byte
		MOV AL, [SI]
		; move the first byte to AL
		INC SI
		; point to next byte
		ADD AL, [SI]
		; add second byte
		ESC SI
		; point to the third byte
		ADD AL, [SI]
		; add the third byte

Based Indirect and Indexed Indirect	Based and indirect addressing modes are used in the same manner. The contents of a register are added to a displacement to generate an effective address. The register must be one of the following: SI, DI, BX or BP If the registers used for displacement are base registers, BX or BP, it is said to be base addressing or else it is called indexed addressing. A displacement is either a number or a label whose offset is known at assembly time. The notation may take several equivalent forms. If BX, SI or DI is used, the effective address is usually an offset from the DS register; BP on the other hand, usually contains an offset from the SS register.	; Register added to an ; offset MOV DX, ARRAY[BX] MOV DX, [DI + ARRAY] MOV DX, [ARRAY – SI] ; Register added to a ; constant MOV AX, [BP – 2] MOV DL, [DI – 2]; DI + (–2) MOV DX, 2 [SI]
Based Indexed	In this type of addressing the operand's effective address is formed by combining a base register with an index register.	MOV AL, [BP] [SI] MOV DX, [BX – SI] ADD CX, [DI] [BX] ; Two base registers or ; two index registers ; cannot be combined so ; the following would be ; incorrect MOV DL, [BP + BX] ; error : two base registers MOV AX, [SI + DI] ; error: two index registers
Based Indexed with Displacement	The operand's effective address is formed by combining a base register, an index register, and a displacement.	MOV DX ARRAY [BX][SI] MOV AX, [BX + SI - ARRAY]

		ADD DL, [BX – SI + 3] SUB CX, ARRAY [BP +SI] Two base registers or two index registers cannot be combined, so the following would be in-correct: MOV AX, [BP + BX + 2] MOV DX, ARRAY [SI – DI]

8.10 THE 8086 INTERRUPT MECHANISM

The original 8088/8086 PCs used an Intel 8259A PIC (*Programmable Interrupt Controller*) to manage its eight hardware interrupts (also called *IRQs*, which is short for *Interrupt Requests*). This separate chip communicates with the processor and tells it when an interrupt needs to be serviced and which ISR (Interrupt Service Routine) to call. Therefore, any device that generates interrupts is connected to one the interrupt pins on the PIC, not to the CPU. The interrupt (or IRQ) pins on the PIC are numbered 0 to 7 where IRQ 0 is the highest priority interrupt and IRQ 7 is the lowest priority. The PIC connects to the processor's single maskable interrupt pin. If an interrupt occurs, the PIC lets the processor know by asserting this interrupt pin.

8.10.1 The Interrupt Flag

One of the flags in the 8086 processor is the interrupt flag, usually referred to as IF. This flag can be set or cleared with the sti and cli instructions respectively. When set (1), the processor monitors its interrupt pin. When it is cleared (0), the processor ignores the signals on its interrupt pin. Therefore, interrupts can be enabled or disabled by setting and resetting the IF flag. However, this essentially disables communication between the PIC and the processor, disabling all external interrupts. To disable individual interrupts, a mask value can be written to the PIC instead. However, this will not be necessary for this class.

8.10.2 The 8259A's Registers

The 8259A has three 8-bit registers that determine its behavior: the IMR (*Interrupt Mask Register*), the ISR (*In-Service Register*), and the IRR (*Interrupt Request Register*). The bits of these registers are numbered 0 through 7, where 0 is the least significant and 7 is the most significant bit. Each bit of each of these registers corresponds to the respective interrupt pin on the PIC. That is, bit 7 corresponds to IRQ 7, bit 6 corresponds to IRQ 6, and so on.

The IMR. This register allows the programmer to disable or "mask" individual interrupts so that the PIC doesn't interrupt the processor when the corresponding interrupt is signaled. For an interrupt to be disabled, its corresponding bit in the IMR must be 1. To be enabled, its bit must be 0. Interrupts can be enabled or disabled by the programmer by reading the IMR, setting or clearing the appropriate bits, then writing the new value back to the IMR.

The IRR. This register indicates when an interrupt has been signaled by a device. As soon as a device signals an interrupt, the corresponding bit in the IRR is set to a 1. This register can only be modified by the PIC and its contents are usually not important to the programmer. However, it can be used to tell which interrupts are waiting for service.

The ISR. This register indicates which interrupts are currently being serviced (*i.e.*, which ISRs have begun execution and have not yet finished). A 1 bit indicates that the corresponding ISR is currently in-service. Several interrupts can be in-service at the same time because of interrupt nesting. The PIC uses this register to determine the highest priority of the interrupts currently being serviced. With this information, the PIC will only interrupt the processor if the highest priority set bit in the IRR has a higher priority than the highest priority set bit in the ISR. In other words, the PIC will never interrupt an in-service interrupt in order to service another interrupt of the same or lower priority. Before an ISR finishes executing, it must send to the PIC the *end of interrupt* command (*EOI*) so that the PIC knows that it can safely clear the highest priority bit in the ISR and signal any other pending interrupts. Be careful not to confuse "In-Service Register" with "Interrupt Service Routine", both of which use the "ISR" acronym.

8.10.3 Processor/PIC Actions

When an interrupt occurs, if the interrupt is enabled in the IMR (*i.e.*, it's IMR bit is 0) and if the processor's IF flag is set (IF=1), then hardware does the following:

1. The PIC informs the processor that an interrupt has occurred by asserting the processor's interrupt pin.
2. The processor finishes the currently executing instruction.
3. The processor sends an acknowledgement signal to the PIC.
4. The PIC then passes to the processor the vector number for the interrupt that occurred.
5. The processor uses this vector number to determine the address where the ISR (interrupt service routine) is stored. The vector number is used as an index into the *interrupt vector table* (or interrupt descriptor table), which starts at address 0:0. The corresponding entry in the interrupt vector table contains the address (segment and offset) for the ISR.
6. The processor pushes the flags, CS, and IP onto the stack (in that order).
7. The processor clears IF, disabling interrupts.
8. The processor then sets CS and IP to the segment and offset of the ISR that was read from the vector table and begins execution.

The vector number that is passed to the CPU by the PIC corresponds to the interrupt that occurs. IRQ 0 corresponds to vector number 8, IRQ 1 corresponds to vector number 9, and so on.

8.10.4 ISR Actions

Once execution begins in the ISR, your code that makes up the ISR should do the following (in the order given):

1. Save the context of whatever was running by pushing all registers onto the stack, except SP, SS, CS, IP, and the flags.
2. Enable interrupts to allow higher priority IRQs to interrupt.
3. Run the interrupt handler, which is usually a C function that does the work for this interrupt.
4. Disable interrupts.
5. Send the EOI command to the PIC, informing it that the handler is finished.
6. Restore the context of what was running before the interrupt occurred by popping the registers saved in step 1 off the stack.
7. Execute the iret instruction. This restores in one atomic operation the values for IP, CS, and the flags, which were automatically saved on the stack by the processor when the interrupt occurred. This effectively restores execution to the point where the interrupt occurred.

When using an RTOS (Real-Time Operating System) it is also usually necessary to call an OS function after step 1 to inform the OS that an ISR has begun execution. Then another OS function must be called after step 5 to inform the RTOS that the ISR is finished. In the RTOS you will build, these functions correspond to YKEnterISR() and YKExitISR(). This will allow the RTOS to know when execution is supposed to return to task code and allows the RTOS to return control to the highest prority ready task rather than the interrupted task.

Remember that the handling of the IRR and ISR registers in the PIC is done for you automatically as interrupts occur. Additionally, you do not need to read or set the IMR since the PIC will automatically ignore interrupts of the same or lower priority while an interrupt is being serviced. For the labs in this class, if you do not wish to use an interrupt, you should simply create an ISR that does not call an interrupt handler.

Do not push or pop the registers SP or SS when saving and restoring context! This is not necessary since SS and SP must already be correct before executing "pop ss" or "pop sp". Also, the "push sp" instruction behaves differently on some 8086 clones. In an RTOS it will be necessary to save and restore SS and SP as part of a task's context, but this should be handled in a different way (e.g., the mov instruction).

8.10.5 The EOI Command

In order for interrupts to work correctly, the PIC must be told when the end of an interrupt has occurred. To tell the PIC that the end of an interrupt has occurred we send it what is called a nonspecific EOI command (or unspecific EOI). To do this we simply write the value 0x20 to port 0x20, which is one of the PIC's ports.

```
mov   al, 0x20  ; Load nonspecific EOI value (0x20) into register al
out   0x20, al  ; Write EOI to PIC (port 0x20)
```

This code should be executed by every ISR right before it restores the context and executes the iret instruction.

EXERCISE

1. State and explain the instruction formats of 8086.
2. What is addressing mode? Explain different addressing schemes used in 8086 with examples of each.
3. Which instruction of 8086 can be used for look up table manipulations? Justify your answer with an example.
4. Examine the overall result of the following 8086 ALP blocks and give single 8086 instructions that can do the same job as the blocks.

   (*i*) JNC DOWN (*ii*) PUSH AX

   (*iii*) PUSH AX

   INC AX PUSH BX PUSH BX

   DOWN: NOP POP AX POP BX

   POP BX POP CX
5. Explain the functions of the following instructions in 8086:

   (*i*) XLAT (*ii*) DAA

   (*iii*) XCHG AX, BX (*iv*) MOVSB
6. Consider the following pairs of partial programs:

   (*i*) MOV AX, 4000H (*ii*) MOV AX, 4000H

ADD AX, AX ADD AX, AX
ADC AX, AX RCL AX, 1
JZ DOWN JZ DOWN

For each of these cases (*i*) and (*ii*), what will be the data in AX after executing instruction 3 and from where will the instruction be fetched after executing instruction 4?

7. How 8086 does identifies between 8-bit and 16-bit operations?
8. With suitable example, explain the following instructions by stating the addressing mode they belong to:
   (*i*) LDS (*ii*) XLAT (*iii*) CWD (*iv*) SCAS
9. Find out the machine code for the following instructions:
   (*i*) LEA SI, [BX = 500H] (*ii*) CALL [5000h]
   (*iii*) NEG 50 [BP] (*iv*) IN DX, AX
10. Write the flow chart and the 8086 ALP to convert the BCD numbers 0 to 9 to their equivalent seven segment codes using the look-up table technique. Assume the CODELIST at the relative addresses from 0 to 9. The BCD number (CHAR) is taken in AL.
11. Explain briefly the instruction used to handle look up tables in 8086.
12. Write en assembly language program in 8086 to arrange the ten 8-bit numbers stored in a memory starting with address ARRAY in ascending order.
13. Write an 8086 to find the GCD and the LCM of two 16-bit numbers. Explain your program ONLY through proper comments.
14. Describe the interrupt sequence of 8086.
15. What do mean by interrupt priorities? List out interrupt priorities in 8086.
16. What is LOCK prefix? What is its use?
17. What is the difference between Jump and loop instruction?
18. How will you enter the single step mode of 8086?
19. Bring out the development in 8086 instruction set over 8085 instruction set in detail.
20. What is REP prefix? What is its use?

9

Assembly Language Programming Based on 8085

A program is a sequence of instructions written to tell a computer to perform a specific function. The instructions are selected from the instruction set of the microprocessor. The first step in writing a program is to think very carefully about the problem that you want the program to solve. In other words, ask yourself many times, "What do I really want this program to do?" If you don't do this, you may write a program that works great but does not do what you need it to do. As you think about the problem, it is a good idea to write down exactly what you want the program to do and the order in which you want the program to do it. At this point, you do not write down the instructions, you just divide the given problem in small steps in terms of the operations the 8085 can perform. This sequence of operations used to solve a programming problem is often called the *algorithm*. For more complex problems, however, we develop a more extensive outline before translating the algorithm into instructions. The most common way of representing the algorithm for a program is flowchart. The steps necessary to write the program can be represented in a pictorial format, called a *flowchart*. Generally, a flowchart is used for two purposes: to assist and clarify the thinking process and to communicate the programmer's thoughts and logic to others. The below figure shows some of the common flowchart symbols:

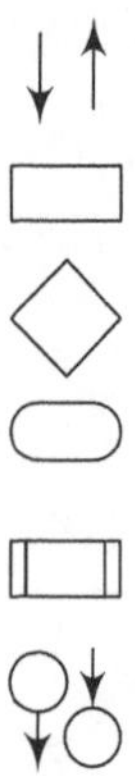

Arrow	indicates the direction of the program execution
Rectangle	represents a process or an operation
Diamond	represents a decision making block
Oval	indicates the beginning or the end of the operation
Double-sided Rectangle	represents a predefined process such as a sub-routine.
Circle with an Arrow	represents a communication (an entry or an exit) to a different page

9.1 WRITING AND EXECUTING A PROGRAM IN A SINGLE BOARD MICROCOMPUTER

Writing a simple program of adding two hexadecimal numbers 23H and 84H and saving the result in a register in assembly language is illustrated as follows:

Algorithm

1. Load the 1st number 23H in one register.
2. Load the 2nd number 84H in another register.
3. Add the contents of the two registers.
4. Save the result in any register.
5. End the program.

Flow Chart

The above steps can be represented in a pictorial format with the help of flowchart.

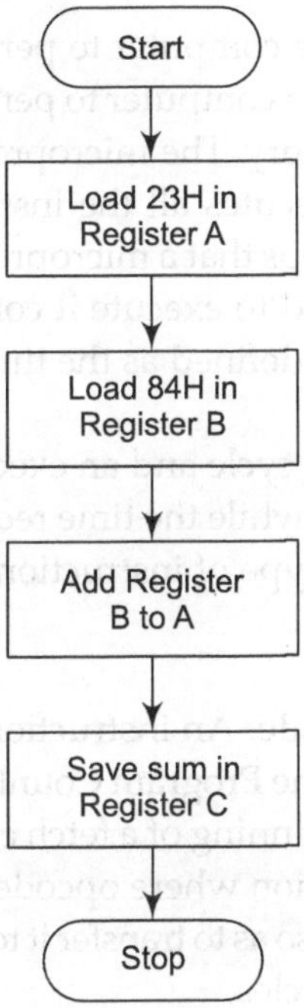

Assembly Language

The steps described in the flowchart are translated into an Assembly Language as follows:

	Opcode	Operand
1.	MVI	A, 23H
2.	MVI	B, 84H
3.	ADD	B
4.	MOV	C, A
5.	HLT	

Machine Language

Now, the above mnemonics should be converted into machine language. By looking up the machine code for each instruction in the instruction set, we can translate the program into machine language as follows:

	Mnemonics	Machine Code (Hex)
1.	MVI A, 23H	2E}
2.	MVI	23}
3.	ADD	
4.	MOV	
5.	HLT	

9.2 INSTRUCTION CYCLE

An instruction is a command given to the computer to perform a specified operation on the given data. Sequence of instructions written for a computer to perform a particular task is called program. Program and data are stored in the memory. The microprocessor fetches one instruction from the memory at a time and executes it. It executes all the instructions of the program one by one to produce the final result. The necessary steps that a microprocessor carries out to fetch an instruction and necessary data from the memory and to execute it constitute an instruction cycle.

In other words, an instruction cycle is defined as the time required completing the execution of an instruction.

An instruction cycle consists of a fetch cycle and an execute cycle. The time required to fetch an opcode (fetch cycle) is a fixed slot of time while the time required to execute an instruction (execute cycle) is variable which depends on the type of instruction to be executed.

9.2.1 Fetch Operation

The first byte of an instruction is its opcode. An instruction may be more than one byte long. The other bytes are data of operand address. The Program Counter (PC) keeps the memory address of the next instruction to be executed. In the beginning of a fetch cycle the content of the program counter, which is the address of the memory location where opcode is available, is sent to the memory. The memory places the opcode on the data bus so as to transfer it to the microprocessor. The entire operation of fetching an opcode takes three clock cycles.

9.2.2 Execute Operation

The opcode fetched from the memory goes to the Instruction Register (IR). From the instruction register it goes to the decoder circuitry which decodes the instruction. After the instruction is decoded, execution begins. If the operand is in general purpose registers execution is immediately performed.

The time taken for decoding and execution is one clock cycle. If an instruction contains data or operand and address which are still in the memory, the microprocessor has to perform some read operations to get the desired data. After receiving the data it performs execute operation. A read cycle is similar to a fetch cycle. In case of a read cycle the quantity received from the memory are data or operand address instead of an opcode. In some instructions write operation is performed. In write cycle data are sent from the microprocessor to the memory or an output device. Thus, we see that in some cases an execute cycle may involve one or more read or write cycles or both.

9.3 MACHINE CYCLE

Machine cycle is defined as the time required completing the operation of accessing either memory or I/O. In the 8085, the machine cycle may consist of three to six T states.

9.3.1 T-State

T-State is defined as one sub-division of the operation performed in one clock period. These sub-divisions are internal states synchronized with the system clock.

9.4 PROGRAMMING TECHNIQUES

Microprocessor is very fast and accurate in processing the data. It is more efficient than human beings when it is required to perform the repeated tasks. To perform repetitions tasks programmer must use different programming techniques such as looping, indexing etc.

9.4.1 Looping

Looping is the programming technique which is used to tell the processor to repeat the task. A loop can be constructed by using jump instructions. Continuous loops can be constructed by using unconditional jump instructions. Unless you reset the system continuous loop does not stop repeating the tasks. For example: modulo ten counter which will be discussed later.

Conditional loops can be constructed by using conditional jump instructions. The specified tasks will be repeated only when the conditions are met. For example: the delay loop, which will be discussed in the next section.

9.4.2 Indexing

Indexing is the programming technique in which objects will be sorted in a sequential manner. In this data bytes are stores in memory locations sequentially and those data bytes are referred to by their memory address.

9.5 PROGRAMS BASED ON 8085

Now, we will discuss various programs based on Assembly Language Programming in 8085 environment.

Program 1. *Write an Assembly Language Program to add 2–16 bit numbers.*

Solution:

```
LHLD 9501H
XCHG
LHLD 9503H
MVI C, 00
DADD
JNC LOOP1
INR C
LOOP1 SHLD 9505H
MOV A, C
STA 9507H
HLT
```

Program 2. *Write an ALP to add n- 8 bit numbers.*

Solution:

```
MVI D, 00
MVI C, 05
```

```
LXI H, 8030
MOV A, M
DCR C
AHEAD INX H
ADD M
JNC LOOP-1
INR D
LOOP-1 DCR C
JNZ AHEAD
STA 8056
MOV A, D
STA 8051
HLT
```

Program 3. *Write an ALP to perform 32-bit addition.*

Solution:

```
LXI H, 8500
MOV C, M
INX H
LXI D, 8600
XRA A
LOOP-1 LDAX D
ADC M
MOV M, A
INX H
INX D
DCR C
JNZ LOOP-1
MVI A, 00
RAL
MOV M, A
RST 5
```

Program 4. *Write an ALP to subtract 2–8 bit numbers.*

Solution:

```
LXI H, 8501
MOV A, M
INX H
SUB M
JNC LOOP-1
INR C
LOOP-1 INX H
MOV M, A
MOV A, C
INX H
MOV M, C
RST5
```

Program 5. *Write an ALP to ADD 2-BCD numbers.*

Solution:

```
LXI H, 8A00
MVI D, 00
MOV A, M
INX H
ADD M
DAA
JNC LOOP-1
INR D
LOOP-1 STA 8A03
MOV A, D
STA 8A04
RST5
```

Program 6. *Write an ALP to SUBTRACT 2–16 BIT NUMBERS.*

Solution:

```
LHLD 8100
XCHG
LHLD 8102
MOV A, E
SUB L
STA 8104
JNC LOOP
DCR D
LOOP MOV A, D
SUB H
STA 8105
RST5
```

Program 7. *Write an ALP to multiply 2-8bit numbers.*

Solution:

```
LXI H, 8A00
MOV B, M
XRA A
MOV C, A
INX H
AHEAD ADD M
JNC LOOP-1
INR C
LOOP-1 DCR C
JNZ AHEAD
INX H
MOV M, A
INX H
MOV M, C
RST5
```

Program 8. *Write an ALP to perform division of 1-8 bit number by another 8-bit number.*

Solution:

```
LXI H, 8900H
MOV A, M
INX H
MOV B, M
MVI C, FF
LOOP INR C
SUB B
JNC LOOP
ADD B
STA 8902H
MOV A, C
STA 8903H
HLT
```

9.6 TIME DELAYS

Counters are constructed using software instructions to keep track of the events. Since the counting is performed at such high speed, only the last count can be observed. To observe all the counts, there must be an appropriate time delay between counts.

Designing a delay is very simple. A register is loaded with a number, depending on the delay required, and then the register is decremented until it reaches zero by setting up a loop with a conditional JUMP instruction. The loop causes the delay, depending upon the clock period of the system.

9.6.1 Single Register Delay

A count is loaded in a register and the loop is executed until the count reaches zero. The set of instructions necessary to set up a delay loop are:

```
LOOPMVI B, FF
DCR B
JNZ LOOP
```

To calculate the time delay we must consider the T-states required for each instruction, and for the number of times the instruction are executed in the loop. The clock frequency of 8085 is **3MHZ.**

$$\text{Clock period } T = 1/f = 1/3*106 = 0.33*10^{-6}$$

Register B is loaded with FFH (25510) therefore the loop is repeated 255 times.

The time delay can be calculated as follows:

$$T_L = (T*\text{Loop T states}*N_{10})$$

Where:

T_L = Time Delay in loop

T = System Clock Period.

N_{10} = Equivalent decimal number of Hexadecimal count loaded in the delay register.

DCR & JNZ forms a 0 loop with a total of 14 (4+10) T- states.

Therefore:

$$T_L = (0.33*10–6*14*255)$$

$$T_L = 1.1781 *10^{-3}$$

Example: *Write an ALP to display FF and 00 alternatively with a delay.*

Solution:

```
MVI A, FF
BACK PUSH PSW
STA 8FF1
CALL UPDDT
CALL DELAY
POP PSW
CMA
JMP BACK
HLT
```

EXERCISE

1. What are the assembly language Programming techniques explain with examples?
2. Write a short note on flowchart.
3. Write a program to sort given 10 numbers from memory location 2200H in descending order.
4. What do you mean by looping, counting and indexing? Explain with the help of examples.
5. Write a program to find the smallest number from a given block of data.
6. Write a program to find even numbers among ten number series.
7. What is subroutine?
8. Explain stack operation in detail.
9. Write an Assembly language program for 8085 to multiply two 8-bit numbers. Explain your program with only comments.
10. Explain the working of CALL address and RET instructions, in terms of machine cycles.
11. What is a stack? On what principle it works? Is it necessary to initialize stack pointer while writing program? Explain with example.
12. Using JMP, can you call a subroutine? Explain with an example and a corresponding ALP.
13. Explain the various steps involved while executing CALL instruction with an example.
14. What is a stack? On what principle it works? Is it necessary to intialize stack pointer while writing program? Explain with example.

10

Assembly Language Programming based on 8086

In the previous Chapters, we have discussed the 8086 microprocessor. We have discussed the register set, instruction set and addressing modes for this microprocessor. In this and two later units we will discuss the assembly language for 8086/8088 microprocessor. In this Chapter, we will discuss the importance of assembly language, basic components of an assembly program followed by discussions on the program developmental tools available. We will then discuss what are COM programs and EXE programs. Finally we will present a complete example. For all our discussions, we have used Microsoft Assembler (MASM). However, for different assemblers the assembly language directives may change. Therefore, before running an assembly program you must consult the reference manuals of the assembler you are using.

10.1 THE NEED AND USE OF THE ASSEMBLY LANGUAGE

Machine language code consists of the 0-1 combinations that the computer decodes directly. However, the machine language has the following problems:

- It greatly depends on machine and is difficult for most people to write in 0-1 forms.
- DEBUGGING is difficult.
- Deciphering the machine code is very difficult. Thus program logic will be difficult to understand.

To overcome these difficulties computer manufacturers have devised English-like words to represent the binary instruction of a machine. This symbolic code for each instruction is called a mnemonic. The mnemonic for a particular instruction consists of letters that suggest the operation to be performed by that instruction. For example, ADD mnemonic is used for adding two numbers. Using these mnemonics machine language instructions can be written in symbolic form with each machine instruction represented by one equivalent symbolic instruction. This is called an assembly language.

10.1.1 Pros and Cons of Assembly Language

The following are some of the advantages/disadvantages of using assembly language:

- Assembly Language provides more control over handling particular hardware and software, as it allows you to study the instructions set, addressing modes, interrupts etc.
- Assembly Programming generates smaller, more compact executable modules: as the programs are closer to machine, you may be able to write highly optimized programs. This results in faster execution of programs.

Assembly language programs are at least 30% denser than the same programs written in high-level language. The reason for this is that as of today the compilers produce a long list of code for every instruction as compared to assembly language, which produces single line of code for a single instruction. This will be true especially in case of string related programs. On the other hand assembly language is machine dependent. Each microprocessor has its own set of instructions. Thus, assembly programs are not portable. Assembly language has very few restrictions or rules; nearly everything is left to the discretion of the programmer. This gives lots of freedom to programmers in construction of their system.

10.1.2 Uses of Assembly Language

Assembly language is used primarily for writing short, specific, efficient interfacing modules/ subroutines. The basic idea of using assembly is to support the HLL with some highly efficient but non-portable routines. It will be worth mentioning here that UNIX mostly is written in C but has about 5-10% machine dependent assembly code. Similarly in telecommunication application assembly routine exists for enhancing efficiency.

10.2 ASSEMBLY PROGRAM EXECUTION

An assembly program is written according to a strict set of rules. An editor or word processor is used for keying an assembly program into the computer as a file, and then the assembler is used to translate the program into machine code.

There are 2 ways of converting an assembly language program into machine language:

1. Manual assembly
2. By using an assembler.

10.2.1 Manual Assembly

It was an old method that required the programmer to translate each opcode into its numerical machine language representation by looking up a table of the microprocessor instructions set, which contains both assembly and machine language instructions. Manual assembly is acceptable for short programs but becomes very inconvenient for large programs. The Intel SDK-85 and most of the earlier university kits were programmed using manual assembly.

10.2.2 By Using an Assembler

The symbolic instructions that you code in assembly language is known as—Source program. An assembler program translates the source program into machine code, which is known as object program.

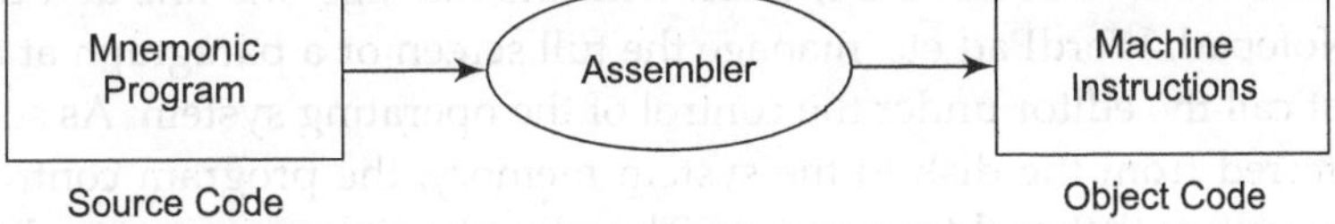

The steps required to assemble, link and execute a program are:

Step 1: The assembly step involves translating the source code into object code and generating an intermediate .OBJ (object file) or module. The assembler also creates a header immediately in front of the generated .OBJ module; part of the header contains information about incomplete addresses. The .OBJ module is not quite in executable form.

Step 2: The link step involves converting the .OBJ module to an .EXE machine code module. The linker's tasks include completing any address left open by the assembler and combining separately assembled programs into one executable module.

The linker:

- combines assembled module into one executable program
- generates an .EXE module and initializes with special instructions to facilitate its subsequent loading for execution.

Step 3: The last step is to load the program for execution. Because the loader knows where the program is going to load in memory, it is now able to resolve any remaining address still left incomplete in the header. The loader drops the header and creates a Program Segment Prefix (PSP) immediately before the program is loaded in memory.

All this conversion and execution of Assembly language performed by two-pass assembler.

10.2.2.1 Two-pass Assembler

Assemblers typically make two or more passes through a source program in order to resolve forward references in a program. A forward reference is defined as a type of instruction in the code segment that is referencing the label of an instruction, but the assembler has not yet encountered the definition of that Instruction.

Pass 1: Assembler reads the entire source program and constructs a symbol table of names and labels used in the program, that is, name of data fields and programs labels and their relative location (offset) within the segment. Pass 1 determines the amount of code to be generated for each instruction.

Pass 2: The assembler uses the symbol table that it constructed in Pass 1. Now it knows the length and relative position of each data field and instruction, it can complete the object code for each instruction. It produces .OBJ (Object file), .LST (list file) and cross reference (.CRF) files.

10.3 TOOLS REQUIRED FOR ASSEMBLY LANGUAGE PROGRAMMING

The tools of the assembly process described below may vary in details.

10.3.1 Editor

The editor is a program that allows the user to enter, modify, and store a group of instructions or text under a file name. The editor programs can be classified in 2 groups.

- Line editors
- Full screen editors.

Line editors, such as EDIT in MS DOS, work with the manage one line at a time. Full screen editors, such as Notepad, WordPad etc. manage the full screen or a paragraph at a time. To write text, the user must call the editor under the control of the operating system. As soon as the editor program is transferred from the disk to the system memory, the program control is transferred from the operating system to the editor program. The editor has its own command and the user can enter and modify text by using those commands. Some editor programs such as WordPerfect are very easy to use. At the completion of writing a program, the exit command of the editor program will save the program on the disk under the file name and will transfer the control to the operating system. If the source file is intended to be a program in the 8086 assembly language the user should follow the syntax of the assembly language and the rules of the assembler.

10.3.2 Assembler

An assembly program is used to transfer assembly language mnemonics to the binary code for each instruction, after the complete program has been written, with the help of an editor it is then assembled with the help of an assembler. An assembler works in 2 phases, *i.e.*, it reads your source code two times. In the first pass the assembler collects all the symbols defined in the program, along with their offsets in symbol table. On the second pass through the source program, it produces binary code for each instruction of the program, and give all the symbols an offset with respect to the segment from the symbol table. The assembler generates three files. The object file, the list file and cross reference file. The object file contains the binary code for each instruction in the program. It is created only when your program has been successfully assembled with no errors. The errors that are detected by the assembler are called the symbol errors.

For example,

MOVE AX1, ZX1;

In the statement, it reads the word MOVE, it tries to match with the mnemonic sets, as there is no mnemonic with this spelling, it assumes it to be an identifier and looks for its entry in the symbol table. It does not even find it there therefore gives an error as undeclared identifier.

List file is optional and contains the source code, the binary equivalent of each instruction, and the offsets of the symbols in the program. This file is for purely documentation purposes. Some of the assemblers available on PC are MASM, TURBO etc.

10.3.3 Linker

For modularity of your programs, it is better to break your program into several sub routines. It is even better to put the common routine, like reading a hexadecimal number, writing hexadecimal number, etc., which could be used by a lot of your other programs into a separate file. These files are assembled separately. After each file has been successfully assembled, they can be linked together to form a large file, which constitutes your complete program. The file containing the common routines can be linked to your other program also. The program that links your program is called the linker. The linker produces a link file, which contains the binary code for all compound modules. The linker also produces link maps, which contains the address information about the linked files. The linker however does not assign absolute addresses to your program. It only assigns continuous relative addresses to all the modules linked starting from the zero. This form a program is said to be reloadable because it can be put anywhere in memory to be run.

10.3.4 Loader

Loader is a program which assigns absolute addresses to the program. These addresses are generated by adding the address from where the program is loaded into the memory to all the offsets. Loader comes into action when you want to execute your program. This program is brought from the secondary memory like disk. The file name extension for loading is .exe or .com, which after loading can be executed by the CPU.

10.3.5 Debugger

The debugger is a program that allows the user to test and debug the object file. The user can employ this program to perform the following functions.

- Make changes in the object code.
- Examine and modify the contents of memory.
- Set breakpoints, execute a segment of the program and display register contents after the execution.

- Trace the execution of the specified segment of the program and display the register and memory contents after the execution of each instruction.
- Disassemble a section of the program, *i.e.*, convert the object code into the source code or mnemonics.

In summary, to run an assembly program you may require your computer:

- A word processor like notepad
- MASM, TASM or Emulator
- LINK.EXE, it may be included in the assembler
- DEBUG.COM for debugging if the need so be.

10.4 ERRORS

Two possible kinds of errors can occur in assembly programs:

(*a*) **Programming errors:** They are the familiar errors you can encounter in the course of executing a program written in any language.

(*b*) **System errors:** These are unique to assembly language that permits low-level operations. A system error is one that corrupts or destroys the system under which the program is running. In assembly language there is no supervising interpreter or compiler to prevent a program from erasing itself or even from erasing the computer operating system.

10.5 AN ASSEMBLY PROGRAM AND ITS COMPONENTS

Sample Program

In this program we just display:

```
Offset
        ———————— Source Code ————————
        DATA SEGMENT
0000            MESSAGE DB "HAVE A NICE DAY!$"
        DATA ENDS
        STACK SEGMENT
                STACK 0400H
        STACK ENDS
        CODE SEGMENT
                ASSUME CS: CODE, DS: DATA SS: Stack
Offset  Machine Code
0000    B8XXXX      MOV AX, DATA
0003    8ED8        MOV DS. AX
0005    BAXXXX      MOV DX, OFFSET MESSAGE
0008    B409        MOV AH, 09H
000A    CD21        INT 21H
000C    BB004C      MOV AX, 4C00H
O00F    CD21        INT 21H
                    CODE ENDS
                    END
```

The details of this program are as follows:

10.5.1 The Program Annotation

The program annotation consists of 3 columns of data: line numbers, offset and machine code.

- The assembler assigns line numbers to the statements in the source file sequentially. If the assembler issues an error message; the message will contain a reference to one of these line numbers.
- The second column from the left contains offsets. Each offset indicates the address of an instruction or a datum as an offset from the base of its logical segment, *e.g.*, the statement at line 0010 produces machine language at offset 0000H of the CODE SEGMENT and the statement at line number 0002 produces machine language at offset 0000H of the DATA SEGMENT.
- The third column in the annotation displays the machine language produce by code instruction in the program.

10.5.1.1 Segment Numbers

There is a good reason for not leaving the determination of segment numbers up to the assembler. It allows programs written in 8086 assembly language to be almost entirely reloadable. They can be loaded practically anywhere in memory and run just as well. Program1 has to store the message "Have a nice day$" somewhere in memory. It is located in the DATA SEGMENT. Since the characters are stored in ASCII, therefore it will occupy 15 bytes (please note each blank is also a character) in the DATA SEGMENT.

10.5.1.2 Missing Offset

The xxxx in the machine language for the instruction at line 0010 is there because the assembler does not know the DATA segment location that will be determined at loading time. The loader must supply that value.

10.5.1.3 Program Source Code

Each assembly language statement appears as:

{identifier} Keyword {{parameter},} {;comment}.

The element of a statement must appear in the appropriate order, but significance is attached to the column in which an element begins. Each statement must end with a carriage return, a line feed.

(*a*) **Keyword:** A keyword is a statement that defines the nature of that statement. If the statement is a directive then the keyword will be the title of that directive; if the statement is a data-allocation statement the keyword will be a data definition type. Some examples of the keywords are: SEGMENT (directive), MOV (statement) etc.

(*b*) **Identifiers:** An identifier is a name that you apply to an item in your program that you expect to reference. The two types of identifiers are name and label.

1. Name refers to the address of a data item such as counter, arr etc.
2. Label refers to the address of our instruction, process or segment. For example MAIN is the label for a process as:

MAIN PROC FAR

A20: BL, 45; defines a label A20.

Identifier can use alphabet, digit or special character but it always starts with an alphabet.

(*c*) **Parameters:** A parameter extends and refines the meaning that the assembler attributes to the keyword in a statement. The number of parameters is dependent on the Statement.

(*d*) **Comments:** A comment is a string of a text that serves only as internal document action for a program. A semicolon identifies all subsequent text in a statement as a comment.

10.5.2 Directives

Assembly languages support a number of statements. This enables you to control the way in which a source program assembles and list. These statements, called directives, act only when the assembly is in progress and generate no machine executable code. Let us discuss some common directives.

1. **List:** A list directive causes the assembler to produce an annotated listing on the printer, the video screen, a disk drive or some combination of the three. An annotated listing shows the text of the assembly language programs, numbers of each statement in the program and the offset associated with each instruction and each datum. The advantage of list directive is that it produces much more informative output.
2. **HEX:** The HEX directive facilitates the coding of hexadecimal values in the body of the program. That statement directs the assembler to treat tokens in the source file that begins with a dollar sign as numeric constants in hexadecimal notation.
3. **PROC Directive:** The code segment contains the executable code for a program, which consists of one or more procedures defined initially with the PROC directive and ended with the ENDP directive.
   Procedure-name PROC FAR; Beginning of Procedure
   Procedure-name ENDP FAR; End Procedure
4. **END DIRECTIVE:** ENDS directive ends a segment, ENDP directive ends a procedure and END directive ends the entire program that appears as the last statement.
5. **ASSUME Directive:** An .EXE program uses the SS register to address the base of stack, DS to address the base of data segment, CS to address base of the code segment and ES register to address the base of Extra segment. This directive tells the assembler to correlate segment register with a segment name. For example,
   ASSUME SS: stack_seg_name, DS: data_seg_name, CS: code_seg_name.
6. **SEGMENT Directive:** The segment directive defines the logical segment to which subsequent instructions or data allocations statement belong. It also gives a segment name to the base of that segment. The address of every element in a 8086 assembly program must be represented in segment - relative format. That means that every address must be expressed in terms of a segment register and an offset from the base of the segmented addressed by that register. By defining the base of a logical segment, a segment directive makes it possible to set a segment register to address that base and also makes it possible to calculate the offset of each element in that segment from a common base.

An 8086 assembly language program consists of logical segments that can be a code segment, a stack segment, a data segment, and an extra segment. A segment directive indicates to assemble all statements following it in a single source file until an ENDS directive.

10.5.3 Code Segment

The logical program segment is named code segment. When the linker links a program it makes a note in the header section of the program's executable file describing the location of the code segment when the DOS invokes the loader to load an executable file into memory, the loader reads that note. As it loads the program into memory, the loader also makes notes to itself of exactly where in memory it actually places each of the program's other logical segments. As the loader hands execution over to the program it has just loaded, it sets the CS register to address the base of the segment identified by the linker as the code segment. This renders every instruction in the code segment addressable in segment relative terms in the form CS: xxxx. The linker also assumes by default that the first instruction in the code segment is intended to be the first instruction to be executed. That instruction will appear in memory at an offset of 0000H from the base of the code segment, so the linker passes that value on to the loader by leaving an another note in the header of the program's executable file. The loader sets the IP (Instruction Pointer) register to that value. This sets CS:IP to the segment relative address of the first instruction in the program.

10.5.4 Stack Segment

8086 Microprocessor supports the **Word stack**. The stack segment parameters tell the assembler to alert the linker that this segment statement defines the program stack area. A program must have a stack area in that the computer is continuously carrying on several background operations that are completely transparent, even to an assembly language programmer, for example, a real time clock. Every 55 milliseconds the real time clock interrupts. Every 55 ms the CPU is interrupted. The CPU records the state of its registers and then goes about updating the system clock. When it finishes servicing the system clock, it has to restore the registers and go back to doing whatever it was doing when the interruption occurred. All such information gets recorded in the stack. If your program has no stack and if the real time clock were to pulse while the CPU is running your program, there would be no way for the CPU to find the way back to your program when it was through updating the clock. 0400H byte is the default size of allocation of stack. Please note if you have not specified the stack segment it is automatically created.

10.5.5 Data Segment

It contains the data allocation statements for a program. This segment is very useful as it shows the data organization.

(a) Defining Types of Data

The following format is used for defining data definition:

Format for data definition:

{Name} <Directive> <expression>

Name - a program references the data item through the name although it is optional.

Directive: Specifying the data type of assembly.

Expression: Represent a value or evaluated to value.

The lists of directives are given below:

Directive	Description	Number of Bytes
DB	Define byte	1
DW	Define word	2
DD	Define double word	4
DQ	Define Quad word	8
DT	Define 10 bytes	10

DUP Directive is used to duplicate the basic data definition to '*n*' number of times

ARRAY DB 10 DUP (0)

In the above statement ARRAY is the name of the data item, which is of byte type (DB). This array contains 10 duplicate zero values; that is 10 zero values.

EQU directive is used to define a name to a constant

CONST EQU 20

Type of number used in data statements can be octal, binary, hexadecimal, decimal and ASCII. The above statement defines a name CONST to a value 20.

Some other examples of using these directives are as follows:

```
TEMP     DB   0111001B   ; Binary value in byte operand
                         ; named temp
VALI     DW   7341Q      ; Qctal value assigned to word
                         ; variable
Decimal  DB   49         ; Decimal value 49 contained in
                         ; byte variable
HEX      DW   03B2AH     ; Hex decimal value in word
                         ; operand
ASCH     DB   EXAMPLE    ; ASCII array of values
```

10.6 INPUT OUTPUT IN ASSEMBLY PROGRAM

A software interrupt is a call to an Interrupt servicing program located in the operating system. Usually the input-output routine in 8086 is constructed using these interrupts.

10.6.1 Interrupts

An **interrupt** causes interruption of an ongoing program. Some of the common interrupts are: keyboard, printer, monitor, an error condition, trap etc.

8086 recognizes two kinds of interrupts: **Hardware** interrupts and **Software** interrupts.

Hardware interrupts are generated when a peripheral Interrupt servicing program requests for some service. A software interrupt causes a call to the operating system. It usually is the **input-output** routine.

Let us discuss the software interrupts in more detail. A software interrupt is initiated using the following statements: INT number

In 8086, this interrupt instruction is processing using the **interrupt vector table (IVT).** The IVT is located in the first 1K bytes of memory, and has a total of 256 entities, each of 4 bytes. An entry in the interrupt vector table is identified by the number given in the interrupt instruction. The entry stores the address of the operating system subroutine that is used to process the interrupt. This address may be different for different machines. Figure 1 shows the processing of an interrupt.

10.6.1.1 Processing of an Interrupt

The following are the steps that we have to follow to process an interrupt:

Step 1: The number field in INT instruction is multiplied by 4 to find its entry in the interrupt vector table. For example, the IVT entry for instruction INT 10h will be found at IVT at an address 40h. Similarly the entry of INT 3h will be placed at 0Ch.

Step 2: The CPU locates the interrupt servicing routine (ISR) whose address is stored at IVT entry of the interrupt. For example, in the figure above the ISR of INT 10h is stored at location at a segment address F000h and an offset F065h.

Step 3: The CPU loads the CS register and the IP register, with this new address in the IVT, and transfers the control to that address, just like a far CALL.

Step 4: IRET (interrupt return) causes the program to resume execution at the next instruction in the calling program.

10.6.1.2 Keyboard Input and Video Output

A Keystroke read from the keyboard is called a console input and a character displayed on the video screen is called a console output. In assembly language, reading and displaying character is most tedious to program. However, these tasks were greatly simplified by the convenient architecture of the 8086/8088. That architecture provides for a pack of software interrupt vectors beginning at address 0000:0000.

The advantage of this type of call is that it appears static to a programmer but flexible to a system design engineer. For example, INT 00H is a special system level vector that points to the "recovery from division by zero" subroutine. If new designer come and want to move interrupt location in memory, it adjusts the entry in the IVT vector of interrupt 00H to a new location. Thus from the system programmer point of view, it is relatively easy to change the vectors under program control. One of the most commonly used Interrupts for Input /Output is called DOS function call. Let us discuss more about it in the next subsection:

10.6.2 DOS Function Calls (Using INT 21H)

INT 21H supports about 100 different functions. A function is identified by putting the function number in the AH register. For example, if we want to call function number 01, then we place this value in AH register first by using MOV instruction and then call INT 21H:

Some important DOS function calls are as follows:

DOS Function Call	Purpose	Example

AH = 01H	For reading a single character from key-board and echo it on monitor. The input value is put in AL register.	To get one character input in a variable in data segment you may include the following in the code segment:
		MOV AH, 01
		LNT 21H
		MOV X, AL
		(Please note that interrupt call will return value in AL which is being transferred to variable of data segment X. X must be byte type).

AH = 02H	This function prints 8 bit data (normally ASCII) that is stored in DL register on the screen	To print a character let say '?' on the screen we may have to use following set of commands:
		MOV AH, 02H;
		MOV DL, '?'
		INT 21H

AH = O8H	This is an input function for inputting one character. This is same as AH = 01H functions with the only difference that value does not get displayed on the screen.	Same example as 01 can be used only diff-erence in this case would be that the input character wouldn't get displayed
		MOV AH, 08H
		INT 21H
		MOV X, AL

AH = 09H	This program outputs a string whose offset is stored in DX register and that is terminated using a S character. One can print newline, tab character also.	To print a string "hello world'" followed by a carriage return (control character) we may have to use the following assembly program segment

Example of AH = 09H	CR EQU ODH ; ASCII code of carriage return. DATA SEGMENT STRING DB 'HELLO WORLD', CR, '$' DATA ENDS CODE SEGMENT	

	:
	MOV AX, DATA
	MOV DS, AX
	MOV AH, 09H
	MOV DX, OFFSET STRING
	: Store the offset of string in DX register.
	INT 21H	

AH = 0AH	For input of string up to 255 characters. The string is stored in a buffer.	Look in the examples given.
AH = 4CH	Return to DOS	

Problems of Input:

Problem 1. *Input a single ASCII character into BL register without echo on screen.*

Solution:

```
CODE SEGMENT
MOV AH, 08H ; Function 08H
INT 21H ; The character input in AL is
MOV BL, AL; transfer to BL
:
CODE ENDS
```

Problem 2. *Input a Single Digit for example (0,1, 2, 3, 4, 5, 6, 7, 8, 9).*

Solution:

```
CODE SEGMENT
...
; Read a single digit in BL register with echo. No error check in the Program
        MOV AH, 01H
        INT 21H
; Assuming that the value entered is digit, then its ASCII will be stored in AL.
; Suppose the key pressed is 1 then ASCII '31' is stored in the AL. To get the
; digit 1 in AL subtract the ASCII value '0' from the AL register.
; Here it store 0 as ASCII 30,
; 1 as 31, 2 as 32.......9 as 39
; to store 1 in memory subtract 30 to get 31 – 30 = 1
        SUB BL, ' 0' ; ' 0' is digit 0 ASCII
        ; OR
        SUB BL, 30H
; Now BL contain the single digit 0 to 9
; The only code missing here is to check whether the input is in the specific
; range.
...
CODE ENDS.
```

Problem 3. *Input numbers like (10, 11.............99).*

Solution:

```
; If we want to store 39, it is actually 30 + 9
; and it is 3 × 10 + 9
; to input this value through keyboard, first we input the tenth digit e.g., 3 and ; then type 9
        MOV AH, 08H
        INT 21H
        MOV BL, AL ; If we have input 39 then, BL will first have character
; 3, we can convert it to 3 using previous logic that is 33 – 30 = 3.
        SUB BL, '0'
        MUL BL, AH ; To get 30 Multiply it by 10.
        ; Now BL Store 30
        ; Input another digit from keyboard
        MOV AH, 08H
        INT 21H;
        MOV DL, AL ; Store AL in DL
        SUB DL, '0' ; (39 – 30) = 9.
; Now BL contains the value: 30 and DL has the value 9 add them and get the ; required numbers.
        ADD BL, DL
        ; Now BL store 39. We have 2 digit value in BL.
```

Let us try to summarize these segments as:

```
CODE SEGMENT
; Set DS register
  MOV AX, DATA     ; }  boiler plate code to set the DS register so that the
  MOV DS, AX       ; }  program can access the data segment.
; read first digit from keyboard
        MOV     AH, 08
        INT     21H
        MOV     BL, AL
        SUB     BL, '0'
        MUL     BL, 10H
; read second digit from keyboard
        MOV     AH, 08H
        INT     21H
        MOV     DL, AL
        SUB     DL, '0'
; DL = 9 AND BL = 30
        SUM     BL, DL
; now BL store 39
CODE ENDS.
```

Note: Boilerplate code is the code that is present more or less in the same form in every assembly language program.

Problem 4. *Write a code segment to input a string?*

Solution:

```
CODE SEGMENT
...
MOV AH, 0AH ; Move 04 to AH register
MOV DX, BUFF ; BUFF must be defined in data segment.
INT 21H
...
CODE ENDS
DATA SEGMENT
BUFF DB 50 ; max length of string,
; including CR, 50 characters
DB ? ; actual length of string not known at present
DB 50 DUP(0) ; buffer having 0 values
DATA ENDS.
```

Explanation:

The above DATA segment creates an input buffer BUFF of maximum 50 characters. On input of data 'JAIN' followed by enter data would be stored as:

50	4	J	A	I	N	#

Examples of Display on Video Monitor

Example 1. *Write a code segment to display a single character.*

Solution:

```
; display contents of BL register (assume that it has a single character)
MOV AH, 02H
MOV DL, BL.
INT 21 H.
```

Here data from BL is moved to DL and then data display on monitor function is called which displays the contents of DL register.

Example 2. *Write a code segment to display a single digit (0-9).*

Solution:

Assume that a value 5 is stored in BL register, then to output BL as ASCII value add character '0' to it

```
ADD BL, '0'
MOV AH, 02H
MOV DL, BL
INT 21H
```

Example 3. *Write a code segment to display a number from 10 to 99.*

Solution:

Assuming that the two digit number 59 is stored as number 5 in BH and number 9 in BL, to convert them to equivalent ASCII we will add '0' to each of them.

```
ADD BH, '0'
ADD BL, '0'
MOV AH, 02H
MOV DL, BH
INT 21H
```

```
MOV DL, BL
INT 21H
```

Example 4. *Write a code segment to display a String.*

Solution:

```
MOV AH, 09H
MOV DX, OFFSET BUFF
INT 21H
```

Here data in input buffer stored in data segment is going to be displayed on the monitor.

The Complete Program

Write a program to input a letter from keyboard and respond. "The letter you typed is ___".

```
CODE SEGMENT
; set the DS register
MOV AX, DATA
MOV DS, AX
; Read Keyboard
MOV AH, 08H
INT 21H
; Save input
MOV BL, AL
; Display first part of Message
MOV AH, 09H
MOV DX, OFFSET MESSAGE
INT 21 H
; Display character of BL register
MOV AH, 02H
MOV DL, BL
INT 21 H
; Exit to DOS
MOV AX, 4C00H
INT 21H
CODE ENDS
DATA SEGMENT
MESSAGE DB "The letter you typed is $"
DATA ENDS
END.
```

10.7 THE TYPES OF ASSEMBLY PROGRAMS

Assembly language programs can be written in two ways:

- COM Program: Having all the segments as part of one segment
- EXE Program: Which have more than one segment.

10.7.1 COM Programs

A COM (Command) program is the binary image of a machine language program. It is loaded in the memory at the lowest available segment address. The program code begins at an offset 100h, the first 1K locations being occupied by the IVT. A COM program keeps its code, data, and stack segments within the same segment. Since the offsets in a physical segment can be of 16 bits, therefore the size of COM program is limited to 216 = 64K which includes code, data and stack. The following program shows a COM program:

```
; Title add two numbers and store the result and carry in memory variables.
; name of the segment in this program is chosen to be CSEG
CSEG SEGMENT
        ASSUME CS: CSEG, DS: CSEG, SS:CSEG
        ORG 100h
START: MOV AX, CSEG ; Initialize data segment
        MOV DS, AX ; register using AX
        MOV AL, NUM1 ; Take the first number in AL
ADD AL, NUM2; Add the 2nd number to it
MOV RESULT, AL ; Store the result in location RESULT
RCL AL, 01 ; Rotate carry into LSB
AND AL, 00000001B; Mask out all but LSB
MOV CARRY, AL ; Store the carry result
MOV AX,4C00h
INT 21h
NUM1 DB 15h ; First number stored here
NUM2 DB 20h ; Second number stored here
RESULT DB ? ; Put sum here
CARRY DB ? ; Put any carry here
CSEG ENDS
END START
```

These programs are stored on a disk with an extension .com. A COM program requires less space on disk rather than equivalent EXE program. At run-time the COM program places the stack automatically at the end of the segment, so they use at least one complete segment

10.7.2 EXE Programs

An EXE program is stored on disk with extension .exe. EXE programs are longer than the COM programs, as each EXE program is associated with an EXE header of 256 bytes followed by a load module containing the program itself. The EXE header contains information for the operating system to calculate the addresses of segments and other components. We will not go into such details in this Chapter.

The load module of EXE program consists of up to 64K segments, although at the most only four segments may be active at any time. The segments may be of variable size, with maximum size being 64K.

We will write only EXE programs for the following reasons:

- EXE programs are better suited for debugging.
- EXE-format assembler programs are more easily converted into subroutines for high-level languages.
- EXE programs are more easily reloadable. Because, there is no ORG statement, forcing the program to be loaded from a specific address.
- To fully use multitasking operating system, programs must be able to share computer memory and resources. An EXE program is easily able to do this.

An example of equivalent EXE program for the COM program is:

```
;ABSTRACT       this program adds 2 8-bit numbers in the memory locations
;               NUM1 andNUM2. The result 15 stored in the
;               memory location RESULT. If there was a carry
;               from the addition it will be stored as 0000 0001 in
;               the location CARRY
;REGISTERS      Uses CS, DS, AX
DATA SEGMENT
    NUM1   DB   15h   ; First number
    NUM2   DB   20h   ; Second number
    RESULT DB   ?     ; Put sum here
    CARRY  DB   ?     ; Put any carry here
DATA ENDS
CODE SEGMENT
    ASSUME CS:CODE, DS:DATA
START:MOV AX, DATA          ; Initialise data segment
    MOV DS, AX              ; register using AX
    MOV AL, NUM1            ; Bring the first number in AL
    ADD AL, NUM2            ; Add the 2nd number to AL
    MOV RESULT, AL          ; Store the result
    RCL AL, 01              ; Rotate cany into Least Significant Bit (LSB)
    AND AL, 00000001B       ; Mask out all but LSB
    MOV CARRY, AL           ; Store the carry
    MOV AX, 4C00h           ; Terminate to DOS
    INT 21h
CODE ENDS
    END START
```

10.8 HOW TO WRITE GOOD ASSEMBLY PROGRAMS ?

Now that we have seen all the details of assembly language programming, let us discuss the art of writing assembly programs in brief.

10.8.1 Preparation of Writing the Program

The following are the important points which we have to follow for the preparation of writing the program:

1. Write an algorithm for your program closer to assembly language. For example, the algorithm for preceding program would be:

```
        get NUM1
        add NUM2
                put sum into memory at RESULT
                position carry bit in LSB of byte
                mask off upper seven bits
                store the result in the CARRY location.
```

2. Specify the input and output required.
   Input required — two 8-bit numbers
   Output required — an 8-bit result and a 8-bit carry in memory.

3. Study the instruction set carefully. This step helps in specifying the available instructions and their format and constraints. For example, the segment registers cannot be directly initialized by a memory variable. Instead we have to first move the offset for segment into a register, and then move the contents of register to the segment register.

You can exit to DOS, by using interrupt routine 21h, with function 4Ch, placed in AH register.

It is a nice practice to first code your program on paper, and use comments liberally. This makes programming easier, and also helps you understand your program later. Please note that the number of comments do not affect the size of the program. After the program development, you may assemble it using an assembler and correct it for errors, finally creating exe file for execution.

10.9 SIMPLE ASSEMBLY PROGRAMS

As part of this chapter, we will discuss writing assembly language programs. We shall start with very simple programs, and later graduate to more complex ones.

10.9.1 Data Transfer

Two most basic data transfer instructions in the 8086 microprocessor are MOV and XCHG. Let us give examples of the use of these instructions.

Program 1. *The following program shows the difference of MOV and XCHG instructions:*

```
DATA SEGMENT
        VAL     DB      5678H               ; initialize variable VAL
DATA ENDS
CODE SEGMENT
                ASSUME          CS:     CODE, DS: DATA
MAINP:          MOV             AX, 1234H       ; AH=12 & AL=34
                XCHG            AH, AL          ; AH=34 & AL=12
                MOV             AX, 1234H       ; AH=12 & AL=34
                MOV             BX, VAL         ; BH=56 & BL=78
                XCHG            AX, BX          ; AX=5678 & BX=1234
                XCHG            AH, BL          ; AH=34, AL=78, BH=12, & BL=56
                MOV             AX, 4C00H       ; Halt using INT 21h
                INT             21H
CODE ENDS
END MAINP
```

Program 2. *The following Program is for interchanging the values of two Memory locations.*

```
DATA SEGMENT
        VALUE 1 DB      OAh         Variables
        VALUE 2 DB      14h
DATA ENDS
CODE SEGMENT
        ASSUME CS: CODE: DS:DATA
        MOV AX, DATA                    ; Initialise dam segments
        MOV DS, AX                      ; using AX
        MOV AL, VALUE 1                 ; Load Value1 into AL
        XCHG VALU 3, 2, AL              ; exchange AL with Value 2.
```

```
        MOV VALUE1, AL              ; Store A1 in Value 1
        INT 21h                     ; Return to Operating system
        CODE ENDS END
END
```

The explanation of the above code segment is as follows:

The question is why cannot we simply use XCHG instruction with two memory variables as operand? To answer the question let us look into some of constraints for the MOV & XCHG instructions:

The MOV instruction has the following constraints and operands:

- CS and IP may never be destination operands in MOV;
- Immediate data value and memory variables may not be moved to segment registers;
- The source and the destination operands should be of the same size;
- Both the operands cannot be memory locations;
- If the source is immediate data, it must not exceed 255 (FFh) for an 8-bit destination or 65,535 (FFFFh) for a 16-bit destination.

The statement MOV AL, VALUE1, copies the VALUE1 that is 0Ah in the AL register:

```
AX : 00  0A  <----------  0A    (VALUE 1)
     AH  AL               14    (VALUE 2)
```

The instruction, XCHG AL: VALUE2 ; exchanges the value of AL with VALUE2

Now AL and VALUE2 contains and values as under:

```
AX: 00  14 <----------+   OA    (VALUE 1)
                      +-> OA    (VALUE 2)
```

The statement, MOV VALUE1 AL :. now puts the value of AL to VALUE 1.

Thus the desired exchange is complete

```
AX: 00  14  ---------->   14    (VALUE 1)
                          OA    (VALUE 2)
```

10.9.2 Simple Arithmetic Application

Let us discuss an example that uses simple arithmetic:

Program 1. *Find the average of two values stored in memory locations named FIRST and SECOND and puts the result in the memory location AVGE.*

Solution:

Input: Two memory variables stored in memory locations FIRST and SECOND

```
; REGISTERS                 ; Uses DS, CS, AX, BL
; PORTS                     ; None used
DATA    SEGMENT
        FIRST   DB 90h      ; FIRST number,     90h is a sample value
        SECOND  DB 78h      ; SECOND number,    78h is a sample value
        AVGE    DB ?        ; Store average here
DATA    ENDS
CODE SEGMENT
        ASSUME CS:CODE, DS:DATA
START:  MOV  AX, DATA       ; Initialise data segment, i.e., set
        MOV  DS, AX         ; Register DS to point to Data Segment
        MOV  AL, FIRST      ; Get first number
        ADD  AL, SECOND     ; Add second to it
```

```
        MOV  AH, 00h        ; Clear all of AH register
        ADC  AH, 00h        ; Put carry in LSB of AH
        MOV  BL, 02h        ; Load divisor in BL register
        DIV  BL             ; Divide AX by BL. Quotient in AL.
                            ; and remainder in AH
        MOV  AVGE, AL       ; Copy result to memory
CODE    ENDS
     END  START
```

The explanation of the above code segment is as follows:

An add instruction cannot add two memory locations directly, so we moved a single value in AL first and added the second value to it.

Please note, on adding the two values, there is a possibility of carry bit. (The values here are being treated as unsigned binary numbers). Now the problem is how to put the carry bit into the AH register such that the AX (AH: AL) reflects the added value. This is done using ADC instruction. The ADC AH,00h instruction will add the immediate number 00h to the contents of the carry flag and the contents of the AH register. The result will be left in the AH register. Since we had cleared AH to all zeros, before the add, we really are adding 00h + 00h + CF. The result of all this is that the carry flag bit is put in the AH register, which was desired by us. Finally, to get the average, we divide the sum given in AX by 2. A more general program would require positive and negative numbers. After the division, the 8-bit quotient will be left in the AL register, which can then be copied into the memory location named AVGE.

10.10 PROGRAMMING WITH LOOPS AND COMPARISONS

Let us now discuss a few examples which are slightly more advanced than what we have been doing till now. This section deals with more practical examples using loops, comparison and shift instructions.

10.10.1 Simple Program Loops

The loops in assembly can be implemented using:

- Unconditional jump instructions such as JMP, or
- Conditional jump instructions such as JC, JNC, JZ, JNZ etc. and
- Loop instructions.

Let us consider some examples, explaining the use of conditional jumps.

Example 1. CMP AX, BX; compare instruction: sets flags
JE THERE; if equal then skip the ADD instruction
ADD AX, 02; add 02 to AX
THERE: MOV CL, 07; load 07 to CL

In the example above the control of the program will directly transfer to the label THERE if the value stores in AX register are equal to that of the register BX. The same example can be rewritten in the following manner, using different jumps.

Example 2. CMP AX, BX; compare instruction: sets flags
JNE FIX; if not equal do addition
JMP THERE; if equal skips next instruction
FIX: ADD AX, 02; add 02 to AX
THERE: MOV CL, 07

The above code is not efficient, but suggests that there are many ways through which a conditional jump can be implemented. Select the most optimum way.

10.10.2 Programs Based on Looping

Program 1. *Assume a constant inflation factor that is added to a series of prices stored in the memory. The program copies the new price over the old price. It is assumed that price data is available in BCD form.*

Solution:

```
; The algorithm:

; Repeat
;           Read a price from the array
;           Add inflation factor
;           Adjust result to correct BCD
;           Put result back in array
;           Until all prices are inflated

; REGISTERS : Uses DS. CS. AX. BX. CX
; PORTS     : Not used
ARRAYS      SEGMENT
            PRICE        DB    36h, 55h, 27h. 42h. 38h, 4lh, 29h. 39h
ARRAYS      ENDS
CODE        SEGMENT
            ASSUME CS:CODE. DS:ARRAYS
START:      MOV  AX, ARRAY    ; Initialize data segment
            MOV  DS, AX       ; register using AX
            LEA  BX, PRICES   ; initialize pointer to base of array
            MOV  CX, 0008h    ; Initialise counter to 8 as array have 8
                              ; values.
DO_NEXT:    MOV  AL, [BX]     ; Copy a price to AL. BX is addressed in
                              ; indirect mode.
            ADD  AL, 0Ah      ; Add inflation factor
            DAA               ; Make sure that result is BCD
            MOV  [BX], AL     ; Copy result back to the memory
            INC  BX           ; increment BX to make it point to next price
            DEC  CX           ; Decrement counter register
            JNZ  DO_NEXT      ; If not last, (last would be when CX will
                              ; become 0) Loop back to DO_NEXT
            MOV  AH, 4CH      ; Return to DOS
            INT  21H
CODE        ENDS
            END  START
```

The explanation of the above code segment is as follows:

Please note the use of instruction: LEA BX, PRICES: It will load the BX register with the offset of the array PRICES in the data segment. [BX] is an indirection through BX and contains the value stored at that element of array. PRICES. BX is incremented to point to the next element of the array. CX register acts as a loop counter and is decremented by one to keep a check of the bounds of the array. Once the CX register becomes zero, zero flag is set to 1. The JNZ instruction keeps track of the value of CX, and the loop terminates when zero flag is 1 because JNZ does not loop back. The same program can be written using the LOOP instruction, in such case, DEC CX and JNZ DO_NEXT instructions are replaced by LOOP DO_NEXT instruction. LOOP decrements the value of CX and jumps to the given label, only if CX is not equal to zero.

Program 2. *This following program prints the alphabets (A-Z).*

Solution:

```
; Register used : AX CX, DX

CODE  SEGMENT
        ASSUME : CS:CODE.
MAINP:  MOV   CX, 1AH     ; 26 in decimal = 1A in hexadecimal Counter.
        MOV   DL, 41H     ; Loading DL with ASCII hexadecimal of A.
NEXTC:  MOV   AH, 02H     ; display result character in DL
        INT   21H         ; DOS interrupt
        INC   DL          ; Increment DL for next char
        LOOP  NEXTC       ; Repeat until CX=0.(loop automatically decrements
                          ; CS and checks whether it is zero or not)
        MOV   AX 4C00H    ; Exit DOS
        INT   21H;        ; DOS Call
CODE ENDS
END MAINP
```

Program 3. *This program compares a pair of characters entered through keyboard.*

Solution:

```
; Registers used: AX BX, CX, DX

DATA SEGMENT
        XX DB ?
        YY DB ?
DATA ENDS

CODE SEGMENT
        ASSUME   CS: CODE,  DS: DATA
MAINP:       MOV AX DATA          ; initialize data
             MOV DS, AX           ; segment using AX
             MOV CX, 03H          ; set counter to 3.
NEXTP:       MOV AH. 01H          ; Waitings for user to enter a char.
             INT 21H
             MOV XX, AL           ; store the 1st input character in XX
             MOV AH, 01H          ; waiting for user to enter second
             INT 21H              ; character.
             MOV YY, AL           ; store the character to YY
             MOV BH, XX           ; load first character in BH
             MOV BL, YY           ; load second character in BL
             CMP BH. BL           ; compare the characters
             JNE NOT_EQUAL        ;
EQUAL:       MOV AH. 02H          ; if characters are equal then control
```

```
            MOV DL, 'Y'          ; will execute this block and
            INT 21H              ; display 'Y'
            JMP CONTINUE         ; Jump to continue loop.

NOT_EQUAL:  MOV AH. 02H          ; if characters are not equal then control
            MOV DL, 'N'          ; will execute this block and
            INT 21 H             ; display "N"

CONTINUE:   LOOP NEXT P          ; Get the next character
            MOV AH, 4CH          ; Exit to DOS
            INT 21 H
CODE ENDS
END MAINP
```

EXERCISE

1. What is assembly language and what are its advantages?
2. What are the different types of assemblers available in the market? What is their respective linker program?
3. What are the differences between the .EXE and .COM files?
4. What is the use of segment directive? Explain the difference SEGMENT and PROC directives.
5. Explain the two types of identifiers used in assembly language program. What are their differences?
6. What are the automatically generated statements that are assembler produces when we use simplified segment directive?
7. What happens when reserved words are used incorrectly?
8. List the two instructions required to initialize the DS register in an assembly language program. Assume the name of the data segment is DATA_SEG.
9. What are the uses and reasons for comments in an assembly language program?
10. List the different steps required to:
    - Assemble a program
    - Link a program.

11

MICROPROCESSOR PERIPHERAL INTERFACING

Peripheral Interfacing is considered to be a main part of Microprocessor, as it is the only way to interact with the external world. The interfacing happens with the ports of the Microprocessor.

- The main IC's which are to be interfaced with 8085 are:
    1. 8255 PPI
    2. 8259 PIC
    3. 8251 USART
    4. 8279 Key board display controller
    5. 8253 Timer/Counter
    6. A/D and D/A converter interfacing.

11.1 8237A DMA CONTROLLER

The 8237A Multimode Direct Memory Access (DMA) Controller is a peripheral interface circuit for microprocessor systems. It is designed to improve system performance by allowing external devices to directly transfer information from the system memory. Memory-to-memory transfer capability is also provided. The 8237A offers a wide variety of programmable control features to enhance data throughput and system optimization and to allow dynamic reconfiguration under program control. The 8237A is designed to be used in conjunction with an external 8-bit address latch. It contains four independent channels and may be expanded to any number of channels by cascading additional controller chips. The three basic transfer modes allow programmability of the types of DMA service by the user. Each channel can be individually programmed to Auto initialize to its original condition following an End of Process (EOP). Each channel has a full 64K address and word count capability.

The block diagram of 8237A is represented in Fig. 11.1.

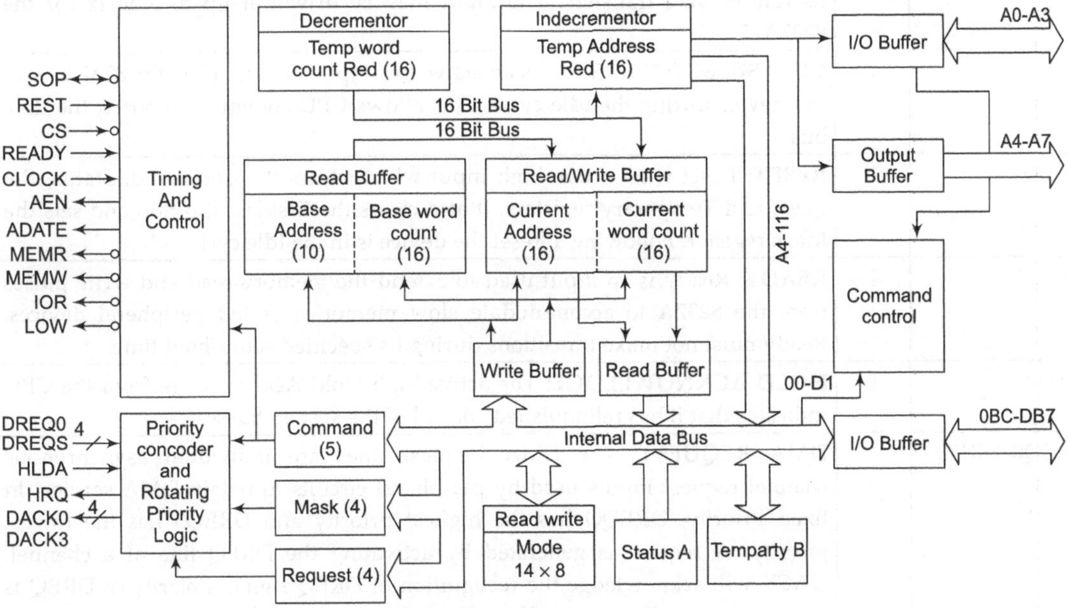

Fig. 11.1 Block Diagram

The pin description is represented in Fig 11.2.

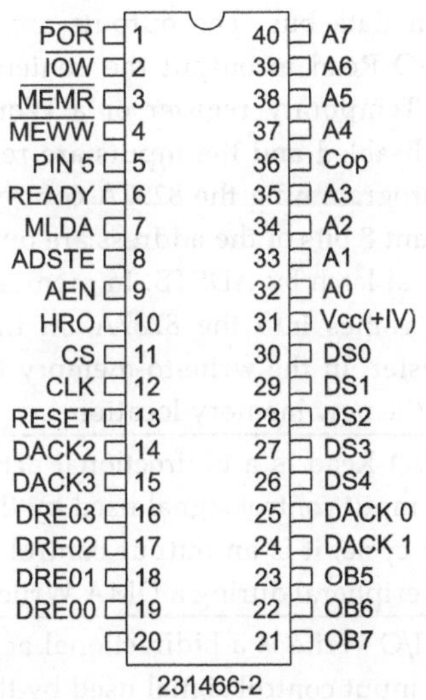

Fig. 11.2 Pin Configuration

11.1.1 Pin Description

The complete pin description is represented in Table 11.1.

Table 11.1 Pin Description

Symbol	Type	Name and Function
V_{CC}		**POWER:** +5V supply.
V_{SS}		**GROUND:** Ground.

CLK	I	**CLOCK INPUT:** Clock Input controls the internal operations of the 8237A and its rate of data transfers. The input may be driven at up to 5 MHz for the 8237A-5.
$\overline{CS}$	I	**CHIP SELECT:** Chip Select is an active low input used to select the 8237A as an I/O device during the Idle cycle. This allows CPU communication on the data bus.
RESET	I	**RESET:** Reset is an active high input which clears the Command, Status, Request and Temporary registers. It also clears the first/last flip/flop and sets the Mask register. Following a Reset the device is in the Idle cycle.
READY	I	**READY:** Ready is an input used to extend the memory read and write pulses from the 8237A to accommodate slow memories or I/O peripheral devices. Ready must not make transitions during its specified setup/hold time.
HLDA	I	**HOLD ACKNOWLEDGE:** The active high Hold Acknowledge from the CPU indicates that it has relinquished control of the system busses.
DREQ0–DREQ3	I	**DMA REQUEST:** The DMA Request lines are individual asynchronous channel request inputs used by peripheral circuits to obtain DMA service. In fixed Priority, DREQO has the highest priority and DREQ3 has the lowest priority. A request is generated by activating the DREQ line of a channel. DACK will acknowledge the recognition of DREQ signal. Polarity of DREQ is programmable. Reset initializes these lines to active high. DREQ must be maintained until the corresponding DACK goes active.
DB0-DB7	I/O	**DATA BUS:** The Data Bus lines are bidirectional three-state signals connected to the system data bus. The outputs are enabled in the Program condition during the I/O Read to output the contents of an Address register, a Status register, the Temporary register or a Word Count register to the CPU. The outputs are disabled and the inputs are read during an I/O Write cycle when the CPU is programming the 8237A control registers. During DMA cycles the most significant 8 bits of the address are output onto the data bus to be strobed into an external latch by ADSTB. In memory-to-memory operations, data from the memory comes into the 8237A on the data bus during the read-from-memory transfer. In the write-to-memory transfer, the data bus outputs place the data into the new memory location.
$\overline{IOR}$	I/O	**I/O READ:** I/O Read is a bidirectional active low three-state line. In the Idle cycle, it is an input control signal used by the CPU to read the control registers. In the Active cycle, it is an output control signal used by the 8237A to access data from a peripheral during a DMA Write transfer.
$\overline{IOW}$	I/O	**I/O WRITE:** I/O Write is a bidirectional active low three-state line. In the Idle cycle, it is an input control signal used by the CPU to load information into the 8237A. In the Active cycle, it is an output control signal used by the 8237A to load data to the peripheral during a DMA Read transfer.
$\overline{EOP}$	I/O	**END OF PROCESS:** End of Process is an active low bidirectional signal. Information concerning the completion of DMA services is available at the bidirectional $\overline{EOP}$ pin. The 8237A allows an external signal to terminate an active DMA service. This is accomplished by pulling the $\overline{EOP}$ input low with an external EOP signal. The 8237A also generates a pulse when the terminal

		count (TC) for any charnel is reached. This generates an $\overline{\text{EOP}}$ signal which is output through the $\overline{\text{EOP}}$ line. The reception of $\overline{\text{EOP}}$, either internal or external, will cause the 8237A to terminate the service, reset the request, and, if Autoinitialize is enabled, to write the base registers to the current registers of that channel. The mask bit and TC bit in the status word will be set for the currently active channel by $\overline{\text{EOP}}$ unless the channel is programmed for Autoinitialize. In that case, the mask hit remains unchanged. During memory-to-memory transfers, $\overline{\text{EOP}}$ will be output when the TC for channel 1 occurs. EOP should be tied high with a pull-up resistor if it is not used to prevent erroneous end of process inputs.
A0-A3	I/O	**ADDRESS:** The four least significant address lines are bidirectional three-state signals. In the Idle cycle they are inputs and are used by the CPU to address the register to be loaded or read. In the Active cycle they are outputs and provide the lower 4 bits of the output address.
A4-A7	O	**ADDRESS:** The four most significant address lines are three-state outputs and provide 4 bits of address. These lines are enabled only during the DMA service.
HRQ	O	**HOLD REQUEST:** This is the Hold Request to the CPU and is used to request control of the system bus. If the corresponding mask bit is clear, the presence of any valid DREQ causes 8237Ato issue the HRQ.
DACK0-DACK3	O	**DMA ACKNOWLEDGE:** DMA Acknowledge is used to notify the individual peripherals when one has been granted a DMA cycle. The sense of these lines is programmable. Reset initializes them to active low.
AEN	O	**ADDRESS ENABLE:** Address Enable enables the 8-bit latch containing the upper 8 address bits onto the system address bus. AEN can also be used to disable other system bus drivers during DMA transfers. AEN is active HIGH.
ADSTB	O	**ADDRESS STROBE:** The active high, Address Strobe is used to strobe the upper address byte into an external latch.
$\overline{\text{MEMR}}$	O	**MEMORY READ:** The Memory Read signal is an active low three-state output used to access data from the selected memory location during a DMA Read or a memory-to-memory transfer.
$\overline{\text{MEMW}}$	O	**MEMORY WRITE:** The Memory Write is an active low three-state output used to write data to the selected memory location during a DMA Write or a memory-to-memory transfer.
PIN5	I	**PIN5:** This pin should always be at a logic HIGH level. An internal pull-up resistor will establish a logic high when the pin is left floating. It is recommended however, that PIN5 be connected to V_{CC}.

11.1.2 Functional Description

The 8237A block diagram includes the major logic blocks and all of the internal registers. The data interconnection paths are also shown. Not shown are the various control signals between the blocks. The 8237A contains 344 bits of internal memory in the form of registers. Following figure lists these registers by name and shows the size of each. A detailed description of the registers and their functions can be found under Register Description.

The following table represents the information about 8237A internal registers.

Name	Size	Number
Base Address Registers	16 bits	4
Base Word Count Registers	16 bits	4
Current Address Registers	16 bits	4
Current Word Count Registers	16 bits	4
Temporary Address Register	16 bits	1
Temporary Word Count Register	16 bits	1
Status Register	8 bits	1
Command Register	8 bits	1
Temporary Register	8 bits	1
Mode Registers	6 bits	4
Mask Register	4 bits	1
Request Register	4 bits	1

The 8237A contains three basic blocks of control logic. The Timing Control block generates internal timing and external control signals for the 8237A. The Program Command Control block decodes the various commands given to the 8237A by the microprocessor prior to servicing a DMA Request. It also decodes the Mode Control word used to select the type of DMA during the servicing. The Priority Encoder block resolves priority contention between DMA channels requesting service simultaneously. The Timing Control block derives internal timing from the clock input. In 8237A systems, this input will usually be the w2 TTL clock from an 8224 or CLK from an 8085AH or 8284A. 33% duty cycle clock generators, however, may not meet the clock high time requirement of the 8237A of the same frequency. For example, 82C84A-5 CLK output violates the clock high time requirement of 8237A-5. In this case 82C84A CLK can simply be inverted to meet 8237A-5 clock high and low time requirements. For 8085AH-2 systems above 3.9 MHz, the 8085 CLK(OUT) does not satisfy 8237A-5 clock LOW and HIGH time requirements. In this case, an external clock should be used to drive the 8237A-5.

11.1.3 DMA Operation

The 8237A is designed to operate in two major cycles. These are called Idle and Active cycles. Each device cycle is made up of a number of states. The 8237A can assume seven separate states, each composed of one full clock period. State I (SI) is the inactive state. It is entered when the 8237A has no valid DMA requests pending. While in SI, the DMA controller is inactive but may be in the Program Condition, being programmed by the processor. State S0 (S0) is the first state of a DMA service. The 8237A has requested a hold but the processor has not yet returned an acknowledge. The 8237A may still be programmed until it receives HLDA from the CPU. An acknowledge from the CPU will signal that DMA transfers may begin. S1, S2, S3 and S4 are the working states of the DMA service. If more time is needed to complete a transfer than is available with normal timing, wait states (SW) can be inserted between S2 or S3 and S4 by the use of the Ready line on the 8237A. Note that the data is transferred directly from the I/O device to memory (or vice versa) with IOR and MEMW (or MEMR and IOW) being active at the same time. The data is not read into or driven out of the 8237A in I/O-to-memory or memory- to-I/O DMA transfers.

Memory-to-memory transfers require a read-from and a write-to-memory to complete each transfer. The states, which resemble the normal working states, use two digit numbers for identification. Eight states are required for a single transfer. The first four states (S11, S12, S13, S14) are used for the read from- memory half and the last four states (S21, S22, S23, S24) for the write-to-memory half of the transfer.

11.1.3.1 IDLE Cycle

When no channel is requesting service, the 8237A will enter the Idle cycle and perform "SI" states. In this cycle the 8237A will sample the DREQ lines every clock cycle to determine if any channel is requesting a DMA service. The device will also sample CS, looking for an attempt by the microprocessor to write or read the internal registers of the 8237A. When CS is low and HLDA is low, the 8237A enters the Program Condition. The CPU can now establish, change or inspect the internal definition of the part by reading from or writing to the internal registers. Address lines A0±A3 are inputs to the device and select which registers will be read or written. The IOR and IOW lines are used to select and time reads or writes. Due to the number and size of the internal registers, an internal flip-flop is used to generate an additional bit of address. This bit is used to determine the upper or lower byte of the 16-bit Address and Word Count registers. The flip-flop is reset by Master Clear or Reset. A separate software command can also reset this flip-flop. Special software commands can be executed by the 8237A in the Program Condition. These commands are decoded as sets of addresses with the CS and IOW. The commands do not make use of the data bus. Instructions include Clear First/Last Flip-Flop and Master Clear.

11.1.3.2 Active Cycle

When the 8237A is in the Idle cycle and a no masked channel requests a DMA service, the device will output an HRQ to the microprocessor and enter the Active cycle. It is in this cycle that the DMA service will take place, in one of four modes.

(*a*) **Single Transfer Mode:** In Single Transfer mode the device is programmed to make one transfer only. The word count will be decremented and the address decremented or incremented following each transfer. When the word count "rolls over" from zero to FFFFH, a Terminal Count (TC) will cause an Auto initialize if the channel has been programmed to do so.
DREQ must be held active until DACK becomes active in order to be recognized. If DREQ is held active throughout the single transfer, HRQ will go inactive and release the bus to the system. It will again go active and, upon receipt of a new HLDA, another single transfer will be performed. In 8080A, 8085AH, 8088, or 8086 system, this will ensure one full machine cycle execution between DMA transfers. Details of timing between the 8237A and other bus control protocols will depend upon the characteristics of the microprocessor involved.

(*b*) **Block Transfer Mode:** In Block Transfer mode the device is activated by DREQ to continue making transfers during the service until a TC, caused by word count going to FFFFH, or an external End of Process (EOP) is encountered. DREQ need only be held active until DACK becomes active. Again, an Auto initialization will occur at the end of the service if the channel has been programmed for it.

(*c*) **Demand Transfer Mode:** In Demand Transfer mode the device is programmed to continue making transfers until a TC or external EOP is encountered or until DREQ goes inactive. Thus transfers may continue until the I/O device has exhausted its data capacity. After the I/O device has had a chance to catch up, the DMA service is re-established by means of a DREQ.

During the time between services when the microprocessor is allowed to operate the intermediate values of address and word count are stored in the 8237A Current Address and Current Word Count registers. Only an EOP can cause an Auto initialize at the end of the service. EOP is generated either by TC or by an external signal. DREQ has to be low before S4 to prevent another Transfer.

(*d*) **Cascade Mode:** This mode is used to cascade more than one 8237A together for simple system expansion. The HRQ and HLDA signals from the additional 8237A are connected to the DREQ a DACK signals of a channel of the initial 8237A. This allows the DMA requests of the additional device to propagate through the priority network circuitry of the preceding device. The priority chain is preserved and the new device must wait for its turn to acknowledge requests. Since the cascade channel of the initial 8237A is used only for prioritizing the additional device, it does not output any address or control signals of its own. These could conflict with the outputs of the active channel in the added device. The 8237A will respond to DREQ and DACK but all other outputs except HRQ will be disabled. The ready input is ignored. The following figure shows two additional devices cascaded into an initial device using two of the previous channels. This forms a two level DMA system. More 8237As could be added at the second level by using the remaining channels of the first level. Additional devices can also be added by cascading into the channels of the second level device, forming a third level.

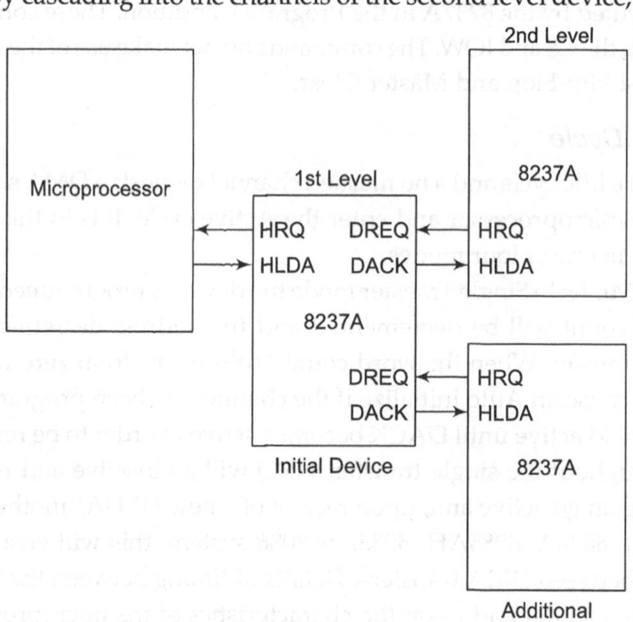

Fig. 11.3 Cascaded 8237 As

11.1.4 Transfer Types

Each of the three active transfer modes can perform three different types of transfers. These are Read, Write and Verify. Write transfers move data from an I/O device to the memory by activating MEMW and IOR. Read transfers move data from memory to an I/O device by activating MEMR and IOW. Verify transfers are pseudo transfers. The 8237A operates as in Read or Write transfers generating addresses, and responding to EOP, etc. However, the memory and I/O control lines all remain inactive. The ready input is ignored in verify mode.

11.1.4.1 Memory-to-Memory

To perform block moves of data from one memory address space to another with a minimum of program effort and time, the 8237A includes a memory-to-memory transfer feature. Programming a bit in the Command register selects channels 0 and 1 to operate as memory-to memory transfer channels. The transfer is initiated by setting the software DREQ for channel 0. The 8237A requests a DMA service in the normal manner. After HLDA is true, the device, using four state transfers in Block Transfer mode, reads data from the memory. The channel 0 Current Address register is the source for the address used and is decremented or incremented in the normal manner. The data byte read from the memory is stored in the 8237A internal Temporary register. Channel 1 then performs a four-state transfer of the data from the Temporary register to memory using the address in its Current Address register and incrementing or decrementing it in the normal manner. The channel 1 current Word Count is decremented. When the word count of channel 1 goes to FFFFH, a TC is generated causing an EOP output terminating the service. Channel 0 may be programmed to retain the same address for all transfers. This allows a single word to be written to a block of memory.

The 8237A will respond to external EOP signals during memory-to-memory transfers. Data comparators in block search schemes may use this input to terminate the service when a match is found. The timing of memory-to-memory transfers is found in the following figure. Memory-to-memory operations can be detected as an active AEN with no DACK outputs.

MEMORY-TO-MEMORY TRANSFER TIMING

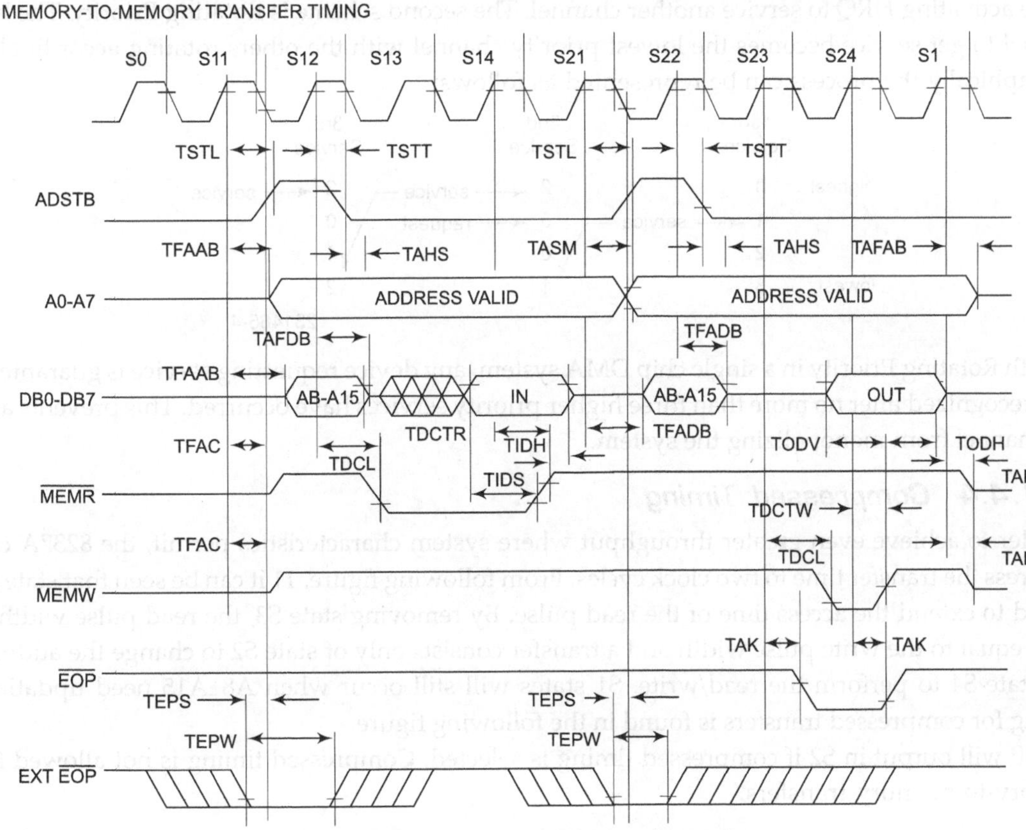

Fig. 11.4 Memory-to-Memory Transfer

11.1.4.2 Auto Initialize

By programming a bit in the Mode register, a channel may be set up as an Auto initialize channel. During Auto initialize initialization, the original values of the Current Address and Current Word Count registers are automatically restored from the Base Address and Base Word count registers of that channel following EOP. The base registers are loaded simultaneously with the current registers by the microprocessor and remain unchanged throughout the DMA service. The mask bit is not altered when the channel is in Auto initialize. Following Auto initialize the channel is ready to perform another DMA service, without CPU intervention, as soon as a valid DREQ is detected. In order to Auto initialize both channels in a memory-to-memory transfer, both word counts should be programmed identically. If interrupted externally, EOP pulses should be applied in both bus cycles.

11.1.4.3 Priority

The 8237A has two types of priority encoding available as software selectable options. The first is Fixed Priority which fixes the channels in priority order based upon the descending value of their number. The channel with the lowest priority is 3 followed by 2, 1 and the highest priority channel, 0. After the recognition of any one channel for service, the other channels are prevented from interfering with that service until it is completed.

After completion of a service, HRQ will go inactive and the 8237A will wait for HLDA to go low before activating HRQ to service another channel. The second scheme is Rotating Priority. The last channel to get service becomes the lowest priority channel with the others rotating accordingly.

Graphically the process can be represented as follows:

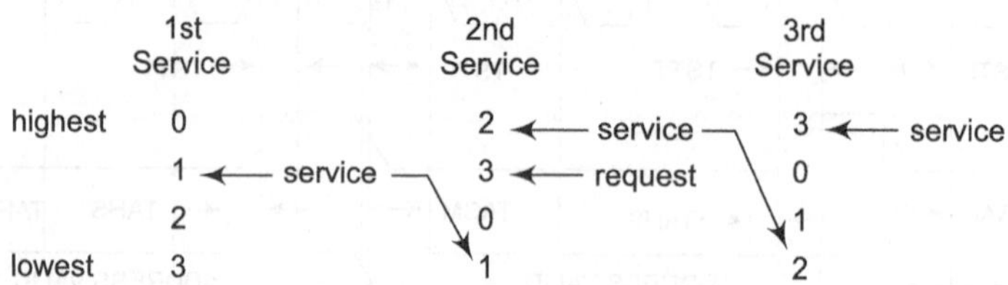

231466-4

With Rotating Priority in a single chip DMA system, any device requesting service is guaranteed to be recognized after no more than three higher priority services have occurred. This prevents any one channel from monopolizing the system.

11.1.4.4 Compressed Timing

In order to achieve even greater throughput where system characteristics permit, the 8237A can compress the transfer time to two clock cycles. From following figure, 11 it can be seen that state S3 is used to extend the access time of the read pulse. By removing state S3, the read pulse width is made equal to the write pulse width and a transfer consists only of state S2 to change the address and state S4 to perform the read/write. S1 states will still occur when A8±A15 need updating. Timing for compressed transfers is found in the following figure.

EOP will output in S2 if compressed timing is selected. Compressed timing is not allowed for memory-to-memory transfers.

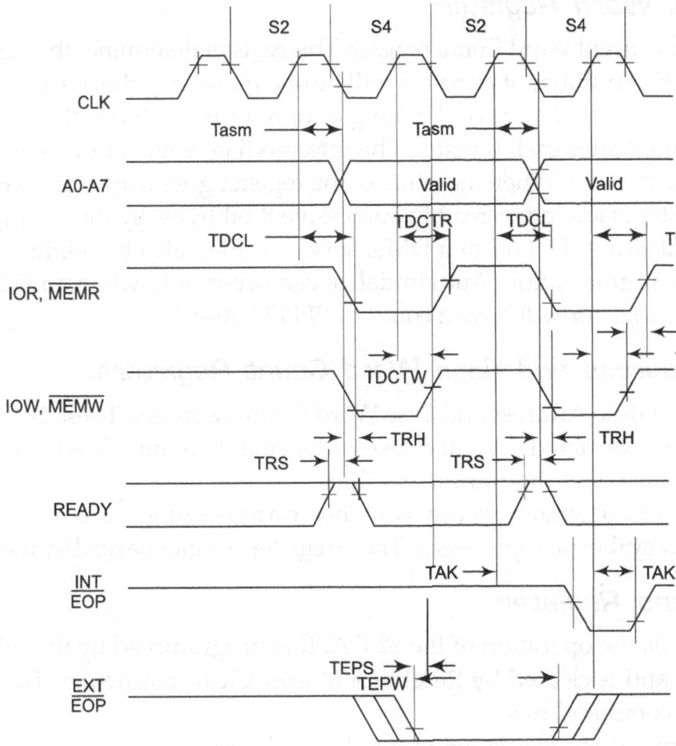

231466-18

11.1.4.5 Address Generation

In order to reduce pin count, the 8237A multiplexes the eight higher order address bits on the data lines. State S1 is used to output the higher order address bits to an external latch from which they may be placed on the address bus. The falling edge of Address Strobe (ADSTB) is used to load these bits from the data lines to the latch. Address Enable (AEN) is used to enable the bits onto the address bus through a three-state enable. The lower order address bits are output by the 8237A directly. Lines A0±A7 should be connected to the address bus. Figure 11 shows the time relationships between CLK, AEN, ADSTB, DB0±DB7 and A0±A7. During Block and Demand Transfer mode services, which include multiple transfers, the addresses generated will be sequential. For many transfers the data held in the external address latch will remain the same. This data need only change when a carry or borrow from A7 to A8 takes place in the normal sequence of addresses. To save time and speed transfers, the 8237A executes S1 states only when updating of A8±A15 in the latch is necessary. This means for long services, S1 states and Address Strobes may occur only once every 256 transfers, a savings of 255 clock cycles for each 256 transfers.

11.1.5 Register Description

11.1.5.1 Current Address Register

Each channel has a 16-bit Current Address register. This register holds the value of the address used during DMA transfers. The address is automatically incremented or decremented after each transfer and the intermediate values of the address are stored in the Current Address register during the transfer. This register is written or read by the microprocessor in successive 8-bit bytes. It may also be reinitialized by an Auto initialize back to its original value. Auto initialize takes place only after an EOP.

11.1.5.2 Current Word Register

Each channel has a 16- bit Current Word Count register. This register determines the number of transfers to be performed. The actual number of transfers will be one more than the number programmed in the Current Word Count register (*i.e.,* programming a count of 100 will result in 101 transfers). The word count is decremented after each transfer. The intermediate value of the word count is stored in the register during the transfer. When the value in the register goes from zero to FFFFH, a TC will be generated. This register is loaded or read in successive 8-bit bytes by the microprocessor in the Program Condition. Following the end of a DMA service it may also be reinitialized by an Auto initialization back to its original value. Auto initialize can occur only when an EOP occurs. If it is not Auto initialized, this register will have a count of FFFFH after TC.

11.1.5.3 Base Address and Base Word Count Registers

Each channel has a pair of Base Address and Base Word Count registers. These 16-bit registers store the original value of their associated current registers. During Auto initialize these values are used to restore the current registers to their original values.

The base registers are written simultaneously with their corresponding current register in 8-bit bytes in the Program Condition by the microprocessor. These registers cannot be read by the microprocessor.

11.1.5.4 Command Register

This 8-bit register controls the operation of the 8237A. It is programmed by the microprocessor in the Program Condition and is cleared by Reset or a Master Clear instruction. The following table lists the function of the command bits.

Graphically the command register is represented as follows:

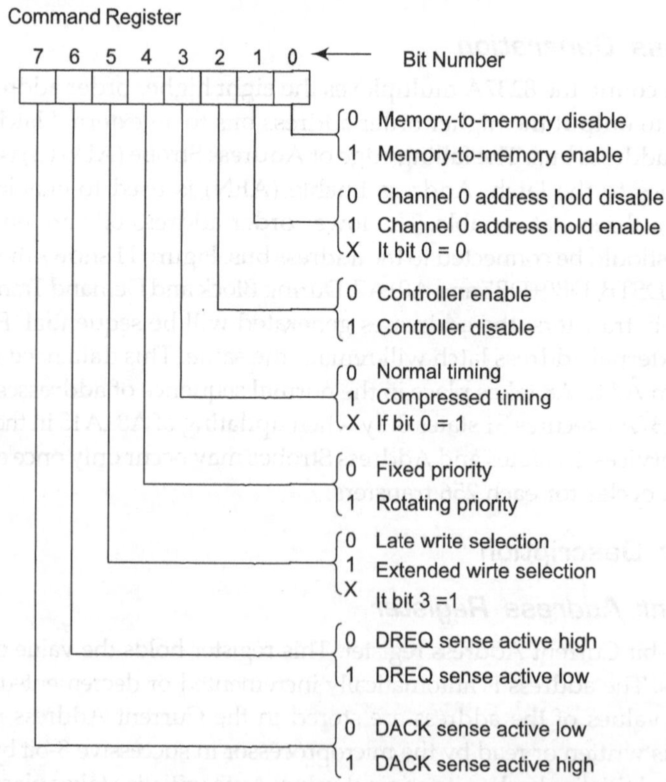

11.1.5.5 Mode Register

Each channel has a 6-bit Mode register associated with it. When the register is being written to by the microprocessor in the Program Condition, bits 0 and 1 determine which channel Mode register is to be written.

Graphically the mode register is represented as follows:

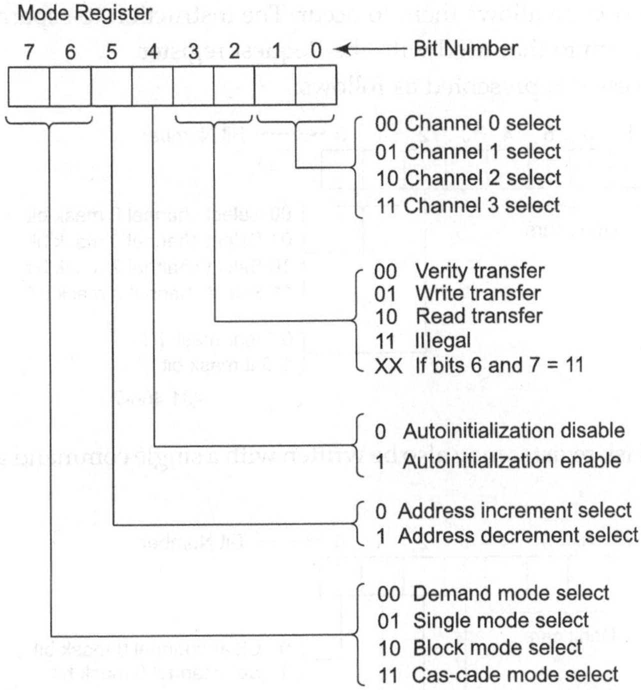

231 466-6

11.1.5.6 Request Register

The 8237A can respond to requests for DMA service which are initiated by software as well as by a DREQ. Each channel has a request bit associated with it in the 4-bit Request register. These are non-maskable and subject to prioritization by the Priority Encoder network. Each register bit is set or reset separately under software control or is cleared upon generation of a TC or external EOP. The entire register is cleared by a Reset. To set or reset a bit, the software loads the proper form of the data word. See the following figure to register address coding. In order to make a software request, the channel must be in Block Mode.

Graphically the request register is represented as follows:

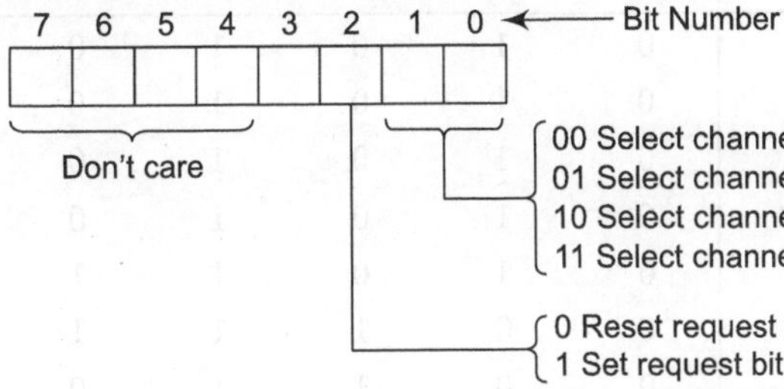

231 466-7

11.1.5.7 Mask Register

Each channel has associated with it a mask bit which can be set to disable the incoming DREQ. Each mask bit is set when its associated channel produces an EOP if the channel is not programmed for Auto initialize. Each bit of the 4-bit Mask register may also be set or cleared separately under software control. The entire register is also set by a Reset. This disables all DMA requests until a clear Mask register instruction allows them to occur. The instruction to separately set or clear the mask bits is similar in form to that used with the Request register.

Graphically the process is represented as follows:

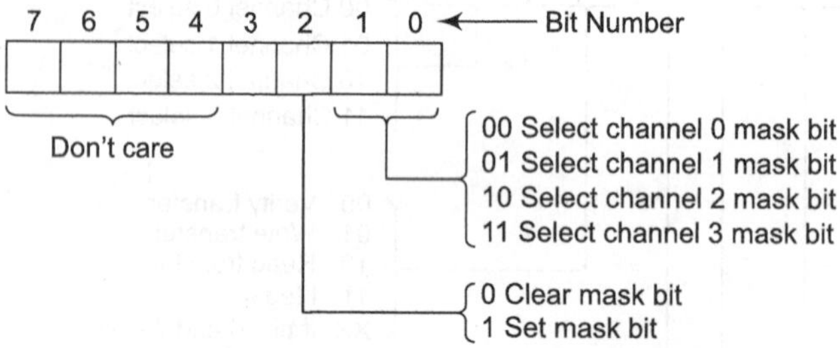

231 466-7

All four bits of the Mask register may also be written with a single command and it is represented as follows:

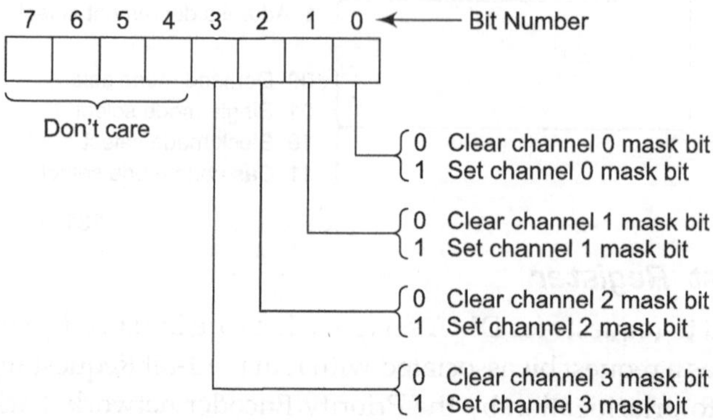

231 466-9

Definition of Register Codes is represented as follows:

Register	Operation	Signals						
		CS	IOR	IOW	A3	A2	A1	AO
Command	Write	0	1	0	1	0	0	0
Mode	Write	0	1	0	1	0	1	1
Request	Write	0	1	0	1	0	0	1
Mask	Set/Reset	0	1	0	1	0	1	0
Mask	Write	0	1	0	1	1	1	1
Temporary	Read	0	0	1	1	1	0	1
Status	Read	0	0	1	1	0	0	0

11.1.5.8 Status Register

The Status register is available to be read out of the 8237A by the microprocessor. It contains information about the status of the devices at this point. This information includes which channels have reached a terminal count and which channels have pending DMA requests. Bits 0 ± 3 are set every time a TC is reached by that channel or an external EOP is applied. These bits are cleared upon Reset and on each Status Read. Bits 4 ± 7 are set whenever their corresponding channel is requesting service.

11.1.5.9 Temporary Register

The Temporary register is used to hold data during memory-to-memory transfers. Following the completion of the transfers, the last word moved can be read by the microprocessor in the Program Condition. The Temporary register always contains the last byte transferred in the previous memory-to-memory operation, unless cleared by a Reset.

11.2 8255A PROGRAMMABLE PERIPHERAL INTERFACE

The 8255A is a Programmable Peripheral Interface (PPI) device designed for use in Intel microcomputer systems. Its function is that of a general purposes I/O component to Interface peripheral equipment to the microcomputer system bush. The functional configuration of the 8255A is programmed by the systems software so that normally no external logic is necessary to interface peripheral devices or structures.

11.2.1 Block Diagram

The following is the block diagram of 8255A Programmable Peripheral Interface:

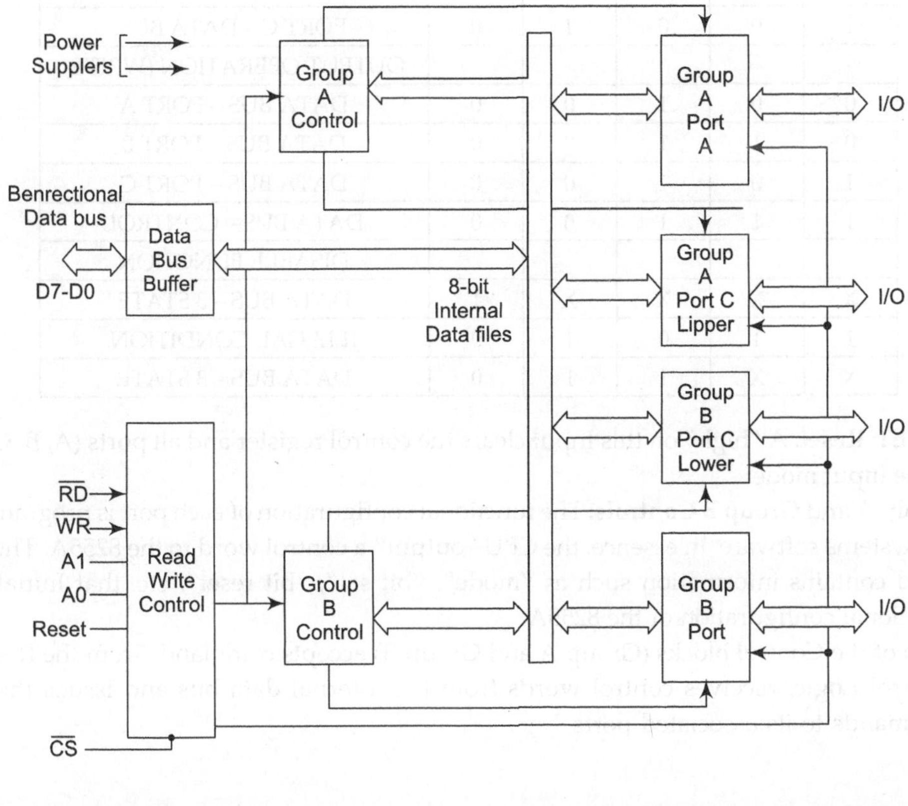

The following is the explanation of the above diagram:

(*a*) **Data Bus Buffer:** This 3-stable bi-directional 8-bit buffer is used to interface the 8255A to the systems data bus. Data is transmitted or received by the buffer upon execution of input or output instructions by the CPU. Control words and status information are also transferred through the data bus buffer.

(*b*) **Read/Write and Control Logic:** The function of this block is to manage all of the Internal and External transfers of both Data and Control or Status words. It accepts inputs from the CPU Address and Control business and in turn, issues commands to both of the Control Groups.

(*c*) **CS:** Chip Select. A "low" on this input pin enables the communication between the 8255A, and the CPU.

(*d*) **RD:** Read. A "low" on this Input pin enables the 8255A to send the data or status information to the CPU on the data bus. In essence, it allows the CPU to "read from the 8255A".

(*e*) **WR:** Write. A. "low" on the input pin enables the CPU to write data or control words into the 8255A.

(*f*) **A0 and A1:** Port Select 0 and Port Select 1. The Input signals, in conjunction with the RD and WR Inputs, controls the selection of one of the three ports or the control word registers. They are normally connected to the least significant bits of the address bus (A0 and A1).

(*g*) **8255A Basic Operation:** The following table represents the basic operation of 8255A.

A_1	A_0	$\overline{RD}$	$\overline{WR}$	$\overline{CS}$	INPUT OPERATION (READ)
0	0	0	1	0	PORT A – DATA BUS
0	1	0	1	0	PORT B – DATA BUS
1	0	0	1	0	PORT C – DATA BUS
					OUTPUT OPERATION (WRITE)
0	0	1	0	0	DATA BUS – PORT A
0	1	1	0	0	DATA BUS – PORT B
1	0	1	0	0	DATA BUS – PORT C
1	1	1	0	0	DATA BUS – CONTROL
					DISABLE FUNCTION
X	X	X	X	1	DATA BUS – 3 STATE
1	1	0	1	0	ILLEGAL CONDITION
X	X	1	1	0	DATA BUS – 3 STATE

(*h*) **RESET: Reset.** A "high" on this Input clears the control register and all ports (A, B, C) are set to the Input mode.

(*i*) **Group A and Group B Controls:** The functional configuration of each port is programmed by the systems software. In essence, the CPU "output" a control word to the 8255A. The control word contains information such as "mode", "bit set", "bit reset", etc. that Initializes the functional configuration of the 8255A.

Each of the Control blocks (Group A and Group B) accepts commands from the Read/Write Control Logic, receives control words from the internal data bus and issues the proper commands to its associated ports.

Control Group A – Port A and Port C upper (C7 C4)
Control Group B – Port B and Port C lower (C3 C0)
The Control Word Register can only be written into. No.
Read operation of the Control Word Register is allowed.

(*j*) **Ports A, B, and C:** The 8255A contains three 8-bit ports (A , B, and C). All can be configured in a wide variety of functional characteristics by the system software but each has its own special features or personally to further enhance the power and flexibility of the 8255A.

Port A. One 8 bit data output latch/buffer and one 8-bit data input latch.

Port B. One 8-bit data output latch/buffer and one 8-bit data input buffer.

Port C. One 8-bit data output latch/buffer and one 8-bit data input buffer (no latch for input). This port can be divided into two 4-bit ports under the mode control. Each 4-bit port contains a 4-bit latch and it can be used for the controls signal outputs and status signal inputs in conjunction with ports A and B.

11.2.2 Pin Configuration

The Pin Configuration of 8255A is as follows:

PIN CONFIGURATION

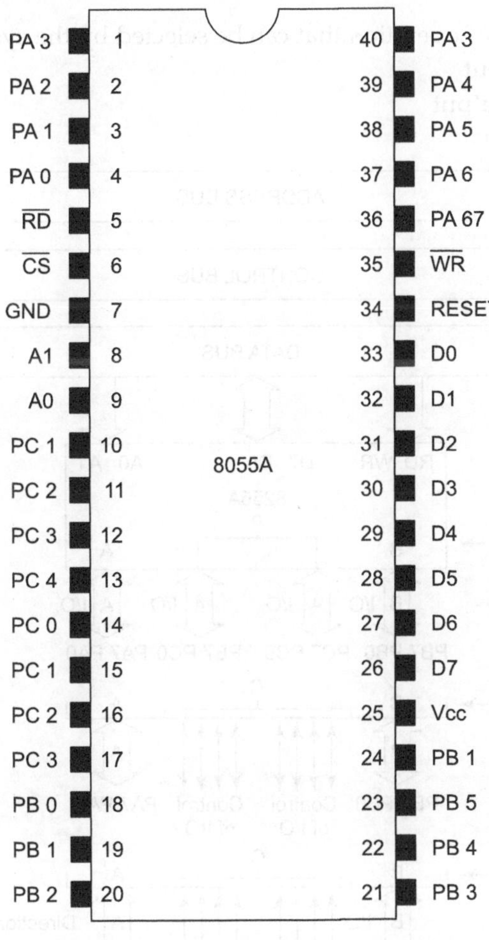

The explanation table of the above diagram is as in the following table.

$D_7 - D_0$	DATA BUS DIRECTIONAL
RESET	RESET INPUT
CS	CHIP SELECT
RD	READ INPUT
WR	WRITE INPUT
A0 – A1	PORT ADDRESS
PA 7 PA 0	PORT A (BIT)
PB 7 PB 0	PORT B (BIT)
PC 7 PC 0	PORT C (BIT)
V_{CC}	5 VOLTS
GND	0 VOLTS

11.2.2.1 8255A Operational Description

Mode Selection

There are three basic modes of operation that can be selected by the systems software:

Mode 0 – Basic Input/output
Mode 1 – Strobed Input/output
Mode 2 – Bi-Directional Bus

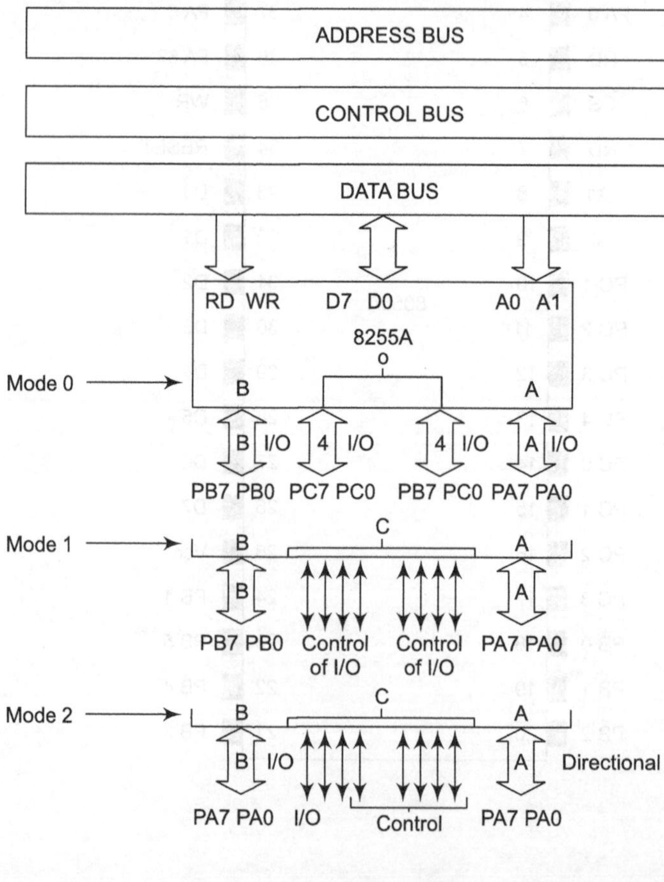

When the reset Input goes "high" all ports will be set to the Input mode (*i.e.*, all 24 lines will be in the high Impedance state). After the reset is removed the 8255A can remain in the input mode with no additional Initialization required. During the execution of the systems program any of the other modes may be selected using a single output Instruction. This allows a single 8255A to service a variety of peripheral devices with a simple software maintenance routine.

The modes for Ports A and Port B can be separately defined, while Port C is divided into two portions as required by the Port A and Port B definitions. All of the output registers, including the status flip-flops, will be reset whenever the mode is changed. Modes may be combined so that their functional definition can be "tailored" to almost any I/O stricture. For instance; Group B can be programmed in Mode 0 to monitor simple switch closing or display computational results, Group A could be programmed in Mode 1 to monitor a keyboard or tape reader on an interrupt-driven basis.

Graphically the Mode Definition format is represented as follows:

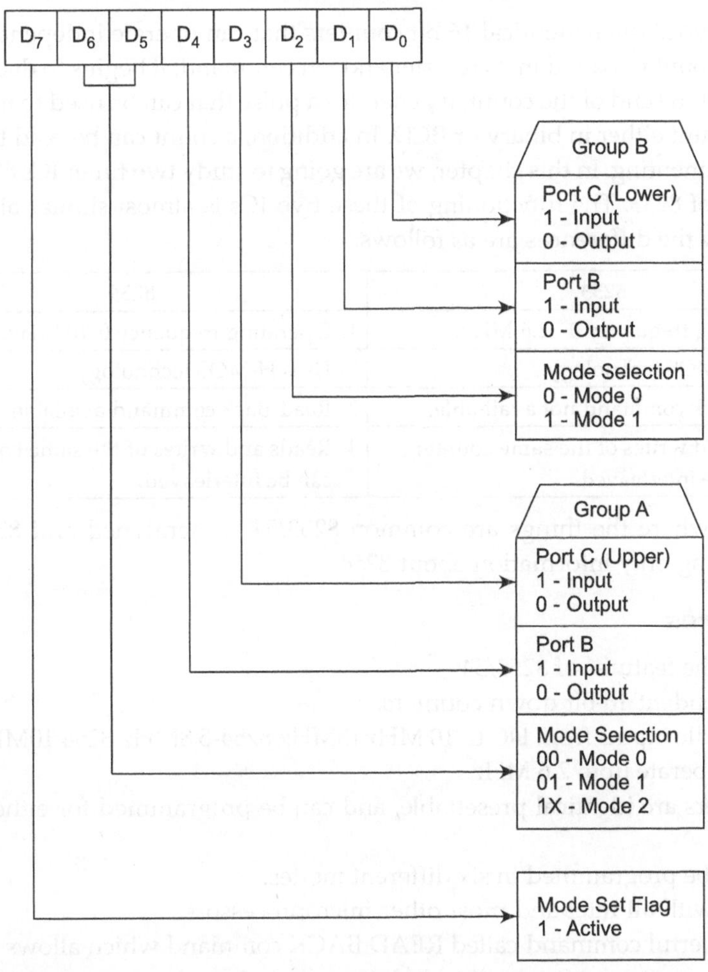

The Mode definitions and possible mode combinations may seem confusing at first but after a cursory review of the complete device operation a simple , logical I/O approach will surface. The design of the 8255A has taken into account things such as efficient PC board layout, control signal definition vs PC layout and complete functional flexibility to support almost any peripheral device with no use of the available pints.

11.3 8253/8254 PROGRAMMABLE TIMER/COUNTER

The Intel **8253** and **8254** are Programmable Interval Timers (PITs), which perform timing and counting functions. They are found in all x86 PCs.

11.3.1 Necessity and Introduction of 8253/54

The 8253/54 solves one of most common problem in any microcomputer system, the generation of accurate time delays under software control. Instead of setting up timing loops in system software, the programmer configures the 8253/54 to match his requirements, initializes one of the counters of the 8253/54 with the desired quantity, then upon command the 8253/54 will count out the delay and interrupt the CPU when it has completed its tasks. It is easy to see that the software overhead is minimum and that multiple delays can be easily be maintained by assignment of priority levels.

The 8253/54 includes three identical 16 bit counters that can operate independently. To operate a counter, a 16-bit count is loaded in its register and, on command; it begins to decrement the count until it reaches 0. At the end of the count, it generates a pulse that can be used to interrupt the CPU. The counter can count either in binary or BCD. In addition, a count can be read by the CPU while the counter is decrementing. In this chapter, we are going to study two timer ICs 8253 and 8254. The 8254 is a superset of 8253. The functioning of these two ICs is almost similar along with the pin configuration. Only the differences are as follows:

8253	8254
1. Operating frequency 0 - 2.6 MHz.	1. Operating frequency 0-10 MHz.
2. Uses N-MOS technology.	2. Uses H-MOS technology.
3. Read-Back command not avail-able.	3. Read-Back command available.
4. Reads and writes of the same counter cannot be interleaved.	4. Reads and writes of the same counter can be interleaved.

In this section, where the things are common 8253/54 is mentioned and 8254 is mentioned specifically for giving only information about 8254.

11.3.2 Features

The following are the features of 8253/54:

1. Three independent 16-bit down counters.
2. 8254 can handle inputs from DC to 10 MHz (5MHz 8254-5 8MHz 8254 10MHz 8254-2) where as 8253 can operate upto 2.6 MHz.
3. Three counters are identical presettable, and can be programmed for either binary or BCD count.
4. Counter can be programmed in six different modes.
5. Compatible with all Intel and most other microprocessors.
6. 8254 has powerful command called READ BACK command which allows the user to check the count value, programmed mode and current mode and current status of the counter.

11.3.3 Pin Diagram of 8254

The Pin diagram of 8254 is represented in the following table.

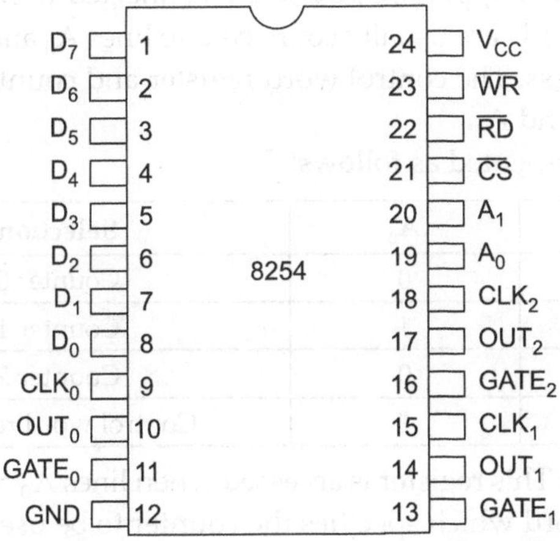

11.3.4 Block Diagram

The block diagram of 8254 is represented as follows:

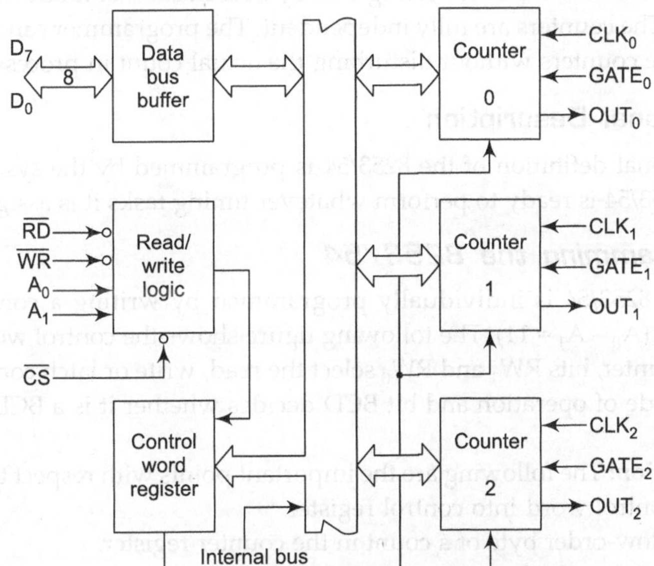

The above figure shows the block diagram of 8253/54. It includes three counters, a data bus buffer, Read/Write control logic, and a control register. Each counter has two input signals CLOCK and GATE and one output signal OUT.

The explanation of the above figure is as follows:

(*a*) **Data Bus Buffer:** This tri-state, bi-directional, 8-bit buffer is used to interface the 8253/54 to the system data bus. The Data bus buffer has three basic functions.

1. Programming the modes of 8253/54.
2. Loading the count registers.
3. Reading the count values.

(*b*) **Read/Write Logic:** The Read/Write logic has five signals: RD, WR, CS and the address lines A_0 and A_1. In the peripheral I/O mode, the RD, and WR signals are connected to IOR and IOW, respectively. In memory-mapped I/O, these are connected to MEMR and MEMW. Address lines A0 and A1 of the CPU are usually connected to lines A_0 and A_1 of the 8253/54, and CS is tied to a decoded address. The control word register and counters are selected according to the signals on lines A_0 and A_1.
The process can be represented as follows:

A_1	A_0	Selection
0	0	Counter 0
0	1	Counter 1
1	0	Counter 2
1	1	Control word register

(*c*) **Control Word Register:** This register is accessed when lines A_0 and A_1 are at logic 1. It is used to write a command word which specifies the counter to be used (binary or BCD), its mode, and either a read or write operation.

(*d*) **Counters:** These three functional blocks are identical in operation. Each counter consists of a single, 16-bit, pre-settable, down counter. The counter can operate in either binary or BCD and its input, gate and output are configured by the selection of modes stored in the control word register. The counters are fully independent. The programmer can read the contents of any of the three counters without disturbing the actual count in process.

11.3.5 Operational Description

The complete functional definition of the 8253/54 is programmed by the system software. Once programmed, the 8253/54 is ready to perform whatever timing tasks it is assigned to accomplish.

11.3.5.1 Programming the 8253/54

Each counter of the 8253/54 is individually programmed by writing a control word into the control word register ($A_0 - A_1 = 11$). The following figure shows the control word format. Bits SC_1 and SC_0 select the counter, bits RW_1 and RW_0 select the read, write or latch command, bits M_2, M_1 and M_0 select the mode of operation and bit BCD decides whether it is a BCD counter or binary counter.

(*a*) **WRITE Operation:** The following are the important points with respect to Write Operation:-

1. Write a control word into control register.
2. Load the low-order byte of a count in the counter register.
3. Load the high-order byte of count in the counter register.

Graphically the control word format is represented as follows:

(*b*) **READ Operation:** In some applications, especially in event counters, it is necessary to read the value of the count in process. This can be done by three possible methods:

1. **Simple Read:** It involves reading a count after inhibiting the counter by controlling the gate input or the clock input of the selected counter, and two I/O read operations are performed by the CPU. The first I/O operation reads the low-order byte, and the second I/O operation reads the high order byte.
2. **Counter Latch Command:** In the second method, an appropriate control word is written into the control register to latch a count in the output latch, and two I/O read operations

are performed by the CPU. The first I/O operation reads the low-order byte, and the second I/O operation reads the high order byte.

D_7	D_6	D_5	D_4	D_3	D_2	D_1	D_0
SC_1	SC_0	RW_1	RW_0	M_2	M_1	M_0	BCD

SC - Select counter

SC_1	SC_0	
0	0	Select counter 0
0	1	Select counter 1
1	0	Select counter 2
1	1	Illegal for 8253 Read-Back command for 8254 (See Read operations)

M - Mode

M_2	M_1	M_0	
0	0	0	Mode 0
0	0	1	Mode 1
x	1	0	Mode 2
x	1	1	Mode 3
1	0	0	Mode 4
1	0	1	Mode 5

RW - Read/Write

RW_1	RW_0	
0	0	Counter latch command (See read operations)
0	1	Read/Write least significant byte only
1	0	Read/Write most significant byte only
1	1	Read/Write most significant byte first, then most significant byte

BCD:

0	Binary counter 16-bits
1	Binary coded decimal (BCD) Counter (4 Decades)

Note: Don't care bits (x) should be 0 to ensure compatibility with future Intel products

3. **Read-Back Command (Available only for 8254):** The third method uses the Read-Back command. This command allows the user to check the count value, programmed Mode, and current status of the OUT pin and Null count flag of the selected counter(s). The following figure shows the format of the control word register for Read-Back command.

$A_0, A_1 = 11$ $\overline{CS} = 0$ $\overline{RD} = 1$ $\overline{WR} = 0$

D_7	D_6	D_5	D_4	D_3	D_2	D_1	D_0
1	1	COUNT	STATUS	CNT_2	CNT_1	CNT_0	0

D_5 : 0 = Latch count of selected counter(s)
D_4 : 0 = Latch status of selected countcr(s)
D_3 : 1 = Select counter 2
D_2 : 1 = Select counter 1
D_1 : 1 = Select counter 0
D_0 : Reserved for future expansion : must be 0.

The Read-Back command may be used to latch multiple counter output latches by setting the COUNT bit $D_5 = 0$ and selecting the desired counter (s). Each counter's latch count in held until it is read (or the counter is reprogrammed). That counter is automatically unlatched when read.

Other Features of Read-Back Command (Available only for 8254)

The Read-Back command may also be used to latch status information of selected counter(s) by setting STATUS bit $D_4 = 0$. The contents of the counter must be latched before reading. The status of a counter is then accessed by a read from that counter. The following figure shows the counter status format.

D_7	D_6	D_5	D_4	D_3	D_2	D_1	D_0
OUTPUT	NULL COUNT	RW_1	RW_0	M_2	M_1	M_0	BCD

D_7 1 = OUT pin in 1.
0 = OUT pin in 0.

D_6 1 = NULL count.
0 = Count available for reading.

D_5 - D_0 = Counter programmed mode.

Bit $D_5 - D_0$ contains the counter's programmed mode exactly as written in the last mode control word. Bit D_7 contains the current status of the output pin. In 8254, it is not possible to read count from the counter, if the count is not loaded into the counting element (CE). The Bit D_6 indicates whether the counting element has count or not. If $D_6 = 0$, counting element has count otherwise null count.

Interleaved Read and Write

Another feature of the 8254 is that reads and writes of the same counter may be interleaved. For example, if the counter is programmed for the two byte counts, the following sequence is valid.

1. Read least significant byte.
2. Write new least significant byte.
3. Read most significant byte.
4. Write new most significant byte.

11.3.6 Mode Definition

(I) **Mode 0:** The following figure represents the Mode 0 interrupt on terminal count.
The following is the explanation of the above figure:

(*a*) **Normal Operation:**
- The output will be initially low after the mode set operation.
- After the count is loaded into the selected count Register the output will remain low and the counter will count.
- When the terminal count is reached the output will go high and remain high until the selected count is reloaded.

(*b*) **Gate Disable:** The following are the information related to Gate:
- Gate = 1 enables counting.
- Gate = 0 disables counting.

Note: Gate has no effect on OUT.

(*c*) **New Count:** If a new count is written to the counter, it will be loaded on the next CLK pulse and counting will continue from the new count.

In case of two byte count we have to follow the following steps:

(*i*) Writing the first byte disables counting.

(*ii*) Writing the second byte loads the new count on the next CLK pulse and counting will continue from the new count.

(II) **MODE 1: Hardware Retriggerable One-shot :** The following figure represents the Mode 1 hardware retriggerable one-shot:

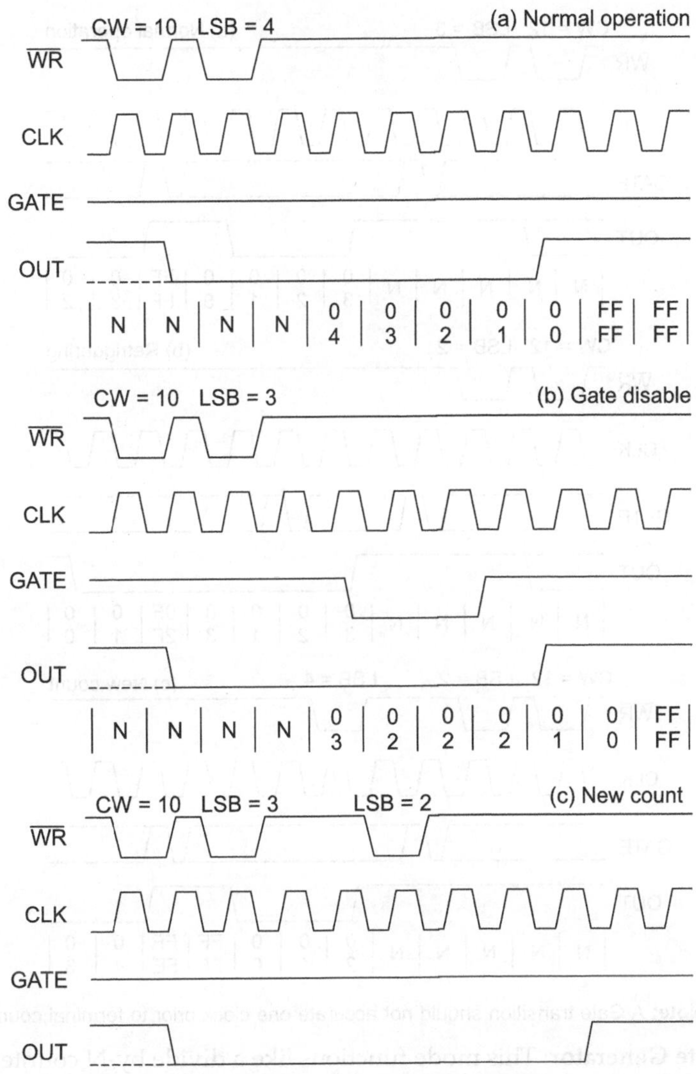

Mode 0 interrupts on terminal count

The explanation of the above figure is as follows:

(*a*) **Normal Operation:** The following are the important points related to Normal operation:-

1. The output will be initially high
2. The output will go low on the CLK pulse following the rising edge at the gate input.
3. The output will go high on the terminal count and remain high until the next rising edge at the gate input.

(*b*) **Retriggering:** The one shot is retriggerable; hence the output will remain low for the full count after any rising edge of the gate input.

(*c*) **New Count:** If the counter is loaded during one shot pulse, the current one shot is not affected unless the counter is retriggered. If retriggered, the counter is loaded with the new count and the one-shot pulse continues until the new count expires.

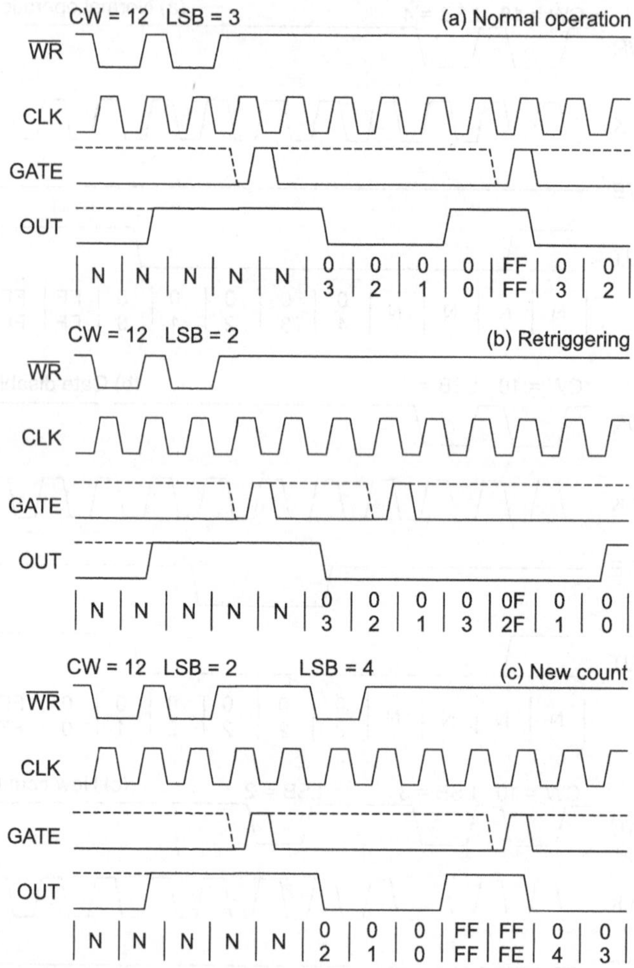

Note: A Gate transition should not accurate one clock prior to terminal count.

(III) **MODE 2 : Rate Generator:** This mode functions like a divide by-N counter.
Graphically **Mode 2 rate generator** can be represented as follows:
The explanation of the above figure is as follows:

(*a*) **Normal Operation:** The following are the important points related to Normal Operation:
1. The output will be initially high.
2. The output will go low for one clock pulse before the terminal count.
3. The output then goes high, the counter reloads the initial count and the process is repeated.
4. The period from one output pulse to the next equals the number of input counts in the count register.

(*b*) **Gate Disable:** The following are the important points related to Gate Disable:
1. If Gate = 1 it enables a counting otherwise it disables counting (Gate = 0).
2. If Gate goes low during an low output pulse, output is set immediately high. A trigger reloads the count and the normal sequence is repeated.

(*c*) **New Count:** The current counting sequence does not affect when the new count is written. If a trigger is received after writing a new count but before the end of the current period,

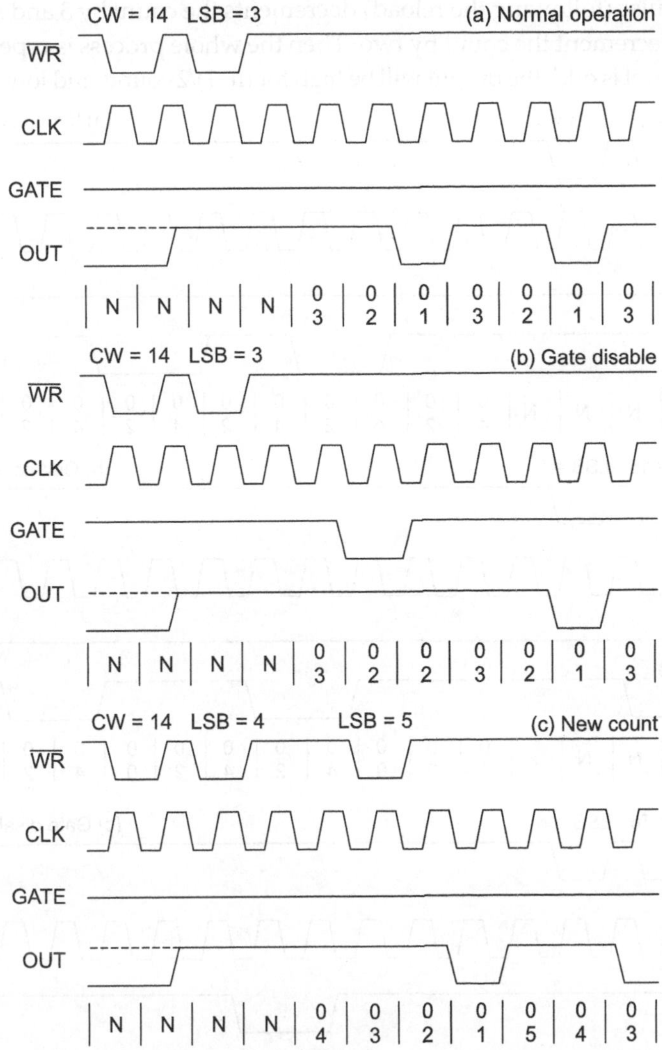

Note: A GATE transition should not occur one clock prior to terminal count.

the new count will be loaded with the new count on the next CLK pulse and counting will continue from the new count. Otherwise, the new count will be loaded at the end of the current counting cycle.

Note: In mode 2, a count of 1 is illegal.

(IV) **MODE 3: Square Wave Rate Generator:** Mode 3 square wave mode is represented as follows:

The explanation of the above figure is as follows:

(*a*) **Normal Operation:** The following are the important points related to Normal Operation:

1. Initially output is high.
2. For even count, counter is decremented by 2 on the falling edge of each clock pulse. When the counter reaches terminal count, the state of the output is changed and the counter is reloaded with the full count and the whole process is repeated.
3. If the count is odd and the output is high the first clock pulse (after the count is loaded) decrements the count by 1. Subsequent clock pulses decrement the clock by 2. After timeout, the output goes low and the full count is reloaded. The first

clock pulse (following the reload) decrements the count by 3 and subsequent clock pulse decrement the count by two. Then the whole process is repeated. In this way, if the count is odd, the output will be high for (n+1)/2 counts and low for (n-1)/2 counts.

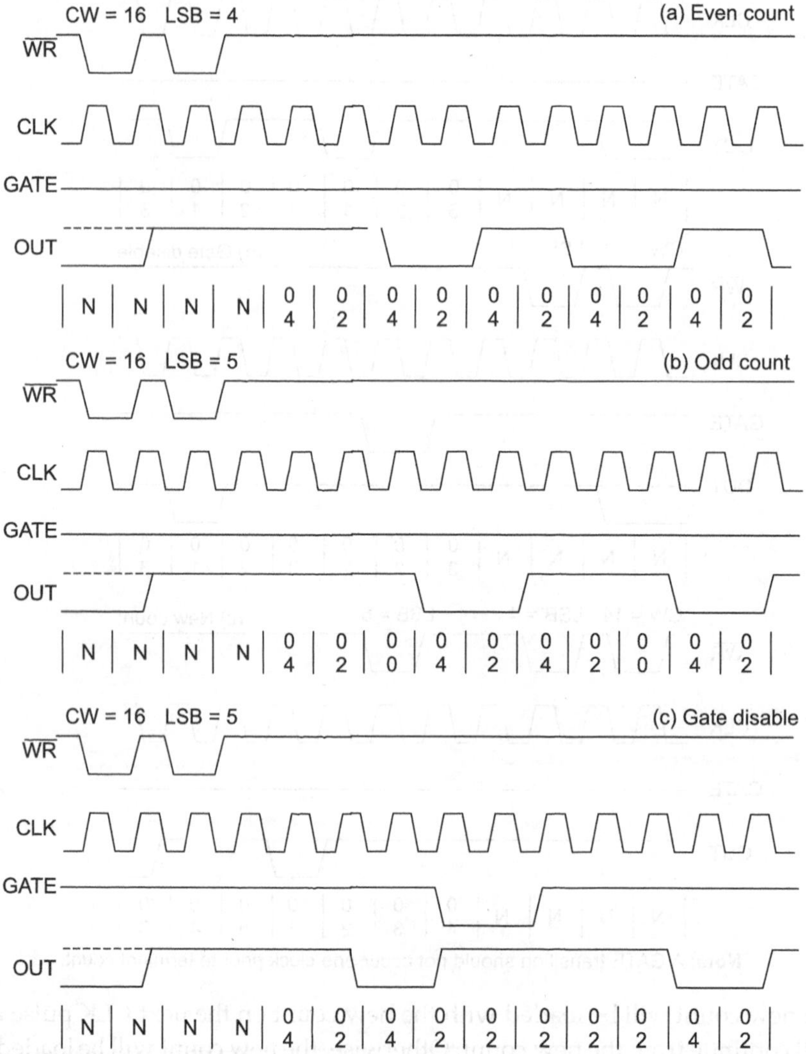

Note: A GATE transition should not occur one clock prior to terminal count.

(*b*) **Gate Disable:** If Gate is 1 counting is enabled otherwise it is disabled. If Gate goes low while output is low, output is set high immediately. After this, When Gate goes high, the counter is loaded with the initial count on the next clock pulse and the sequence is repeated.

(*c*) **New Count:** The current counting sequence does not affect when the new count is written. If a trigger is received after writing a new count but before the end of the current half-cycle of the square wave, the counter will be loaded with the new count on the next CLK pulse and counting will continue from the new count. Otherwise, the new count will be loaded at end of the current half-cycle.

(V) **MODE 4: Software Triggered Strobe:** Graphically it can be represented as follows:

Programming Examples

Example 1. *Write a program to initialize counter 2 in mode 0 with a count of C030H. Assume address for control register = 0BH, counter 0 = 08H, counter 1 = 09H and counter 2 = 0AH.*

Solution : Control word is represented as follows:

D_7	D_6	D_5	D_4	D_3	D_2	D_1	D_0	
SC_1	SC_2	RW_2	RW_0	M_2	M_1	M_0	BCD	
1	0	1	1	0	0	0	0	= B0H

Source Program

```
MOV AL, B0H
OUT 0BH, AL      ; Loads control word (B0H) in the control register.
MOV AL, 30H
OUT 0AH, AL      ; Loads lower byte of (30H)the count.
MOV AL, 0C0H
OUT 0AH, AL      ; Loads higher byte (C0H) of the count.
```

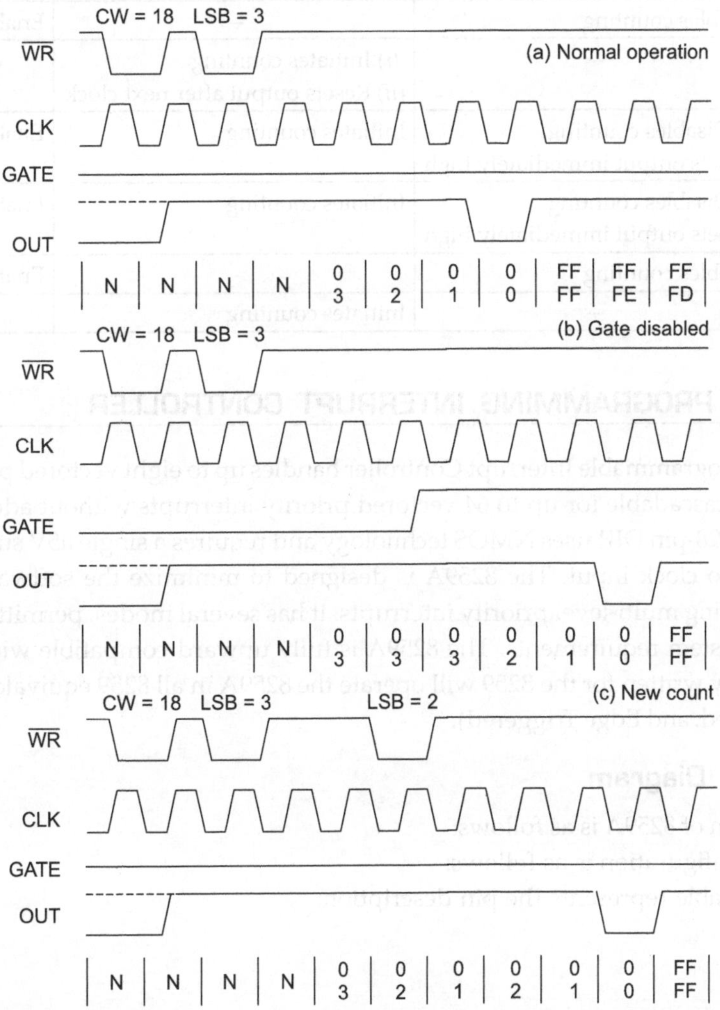

The explanation of the above figure is as follows:

(*a*) **Normal operation:** The following are the important points related to Normal Operation:

1. The output will be initially high.
2. The counting is triggered by the rising edge of the Gate.
3. The output will go low for one CLK pulse after the terminal count (TC).

(*b*) **Retriggering:** If the triggering occurs on the Gate input during the counting, the initial count is loaded on the next CLK pulse and the counting will be continued until the terminal count is reached.

(*c*) **New count:** If a new count is written during counting, the current counting sequence will not be affected. If the trigger occurs after the new count is written but before the terminal count, the counter will be loaded with the new count on the next CLK pulse and counting will continue from there.

The following table shows the gate pin operations:

Signal Status Modes	Low or Going Low	Rising	High
0	Disables counting	—	Enables counting
1	—	(*i*) Initiates counting (*ii*) Resets output after next clock	—
2	(*i*) Disables counting (*ii*) Sets output immediately high	Initiates counting	Enables counting
3	(*i*) Disables counting (*ii*) Sets output immediately high	Initiates counting	Enables counting
4	Disables counting	—	Enables counting
5	—	Initiates counting	—

11.4 8259A PROGRAMMING INTERRUPT CONTROLLER

The Intel 8259A Programmable Interrupt Controller handles up to eight vectored priority interrupts for the CPU. It is cascadable for up to 64 vectored priority interrupts without additional circuitry. It is packaged in a 28-pin DIP, uses NMOS technology and requires a single a5V supply. Circuitry is static, requiring no clock input. The 8259A is designed to minimize the software and real time overhead in handling multi-level priority interrupts. It has several modes, permitting optimization for a variety of system requirements. The 8259A is fully upward compatible with the Intel 8259. Software originally written for the 8259 will operate the 8259A in all 8259 equivalent modes (MCS-80/85, Non-Buffered, and Edge Triggered).

11.4.1 Block Diagram

The Block Diagram of 8259A is as follows:

And the pin configuration is as follows:

The following table represents the pin description:

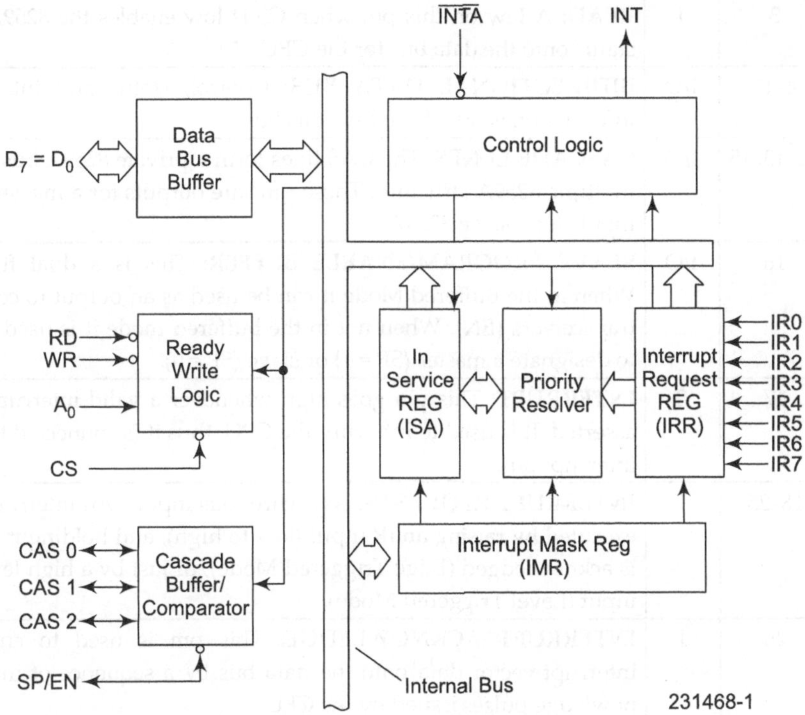

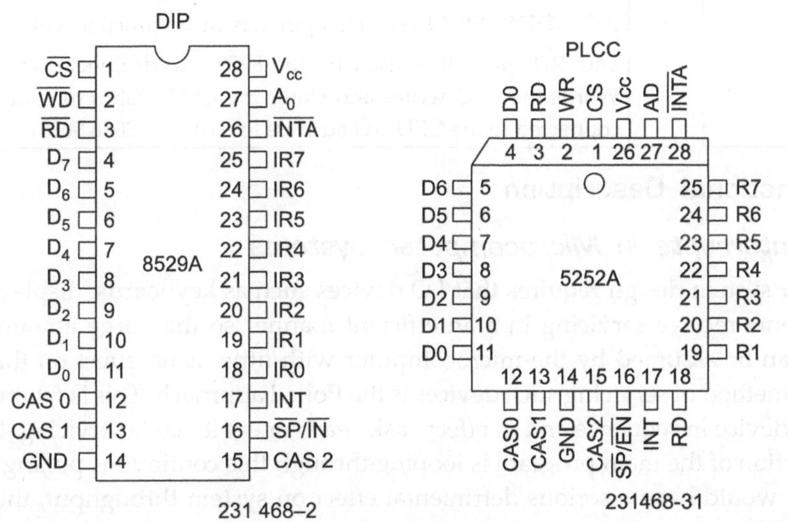

Pin Configurations

Symbol	Pin No.	Type	Name and Function
V_{CC}	28	I	**SUPPLY:** +5V Supply.
GND	14	I	**GROUND**
$\overline{CS}$	1	I	**CHIP SELECT:** A low on this pin enables RD and WR communication between the CPU and the 8259A. INTA functions are independent of CS.
WR	2	I	**WRITE:** A low on this pin when CS is low enables the 8259A to accept command words from the CPU.

$\overline{RD}$	3	I	**READ:** A low on this pin when CS is low enables the 8259A to release status onto the data bus for the CPU.
D_7–D_0	4–11	I/O	**BIDIRECTIONAL DATA BUS:** Control, status and interrupt-vector information is transferred via this bus.
CAS_0–CAS_2	12, 13, 15	I/O	**CASCADE LI NES:** The CAS lines form a private 8259A bus to control a multiple 8259A structure. These pins are outputs for a master 8259A and inputs for a slave 8259A.
$\overline{SP}/\overline{EN}$	16	I/O	**SLAVE PROGRAM/ENABLE BUFFER:** This is a dual function pin. When in the Buffered Mode it can be used as an output to control buffer transceivers (EN). When not in the buffered mode it is used as an input to designate a master (SP = 1} or slave (SP = 0).
INT	17	0	**INTERRUPT:** This pin goes high whenever a valid interrupt request is asserted. It is used to interrupt the CPU, thus it is connected to the CPU's interrupt pin.
IR_0–IR_7	18–25	I	**INTERRUPT REQUESTS:** Asynchronous inputs. An interrupt request is executed by raising an IR input (low to high), and holding it high until it is acknowledged (Edge Triggered Mode), or just by a high level on an IR input (Level Triggered Mode).
$\overline{INTA}$	26	I	**INTERRUPT ACKNOWLEDGE:** This pm is used to enable 8259A interrupt-vector data onto the data bus by a sequence of interrupt acknowledge pulses issued by the CPU.
A_0	27	I	**AO ADDRESS LI NE:** This pin acts in conjunction with the $\overline{CS}$, $\overline{WR}$, and $\overline{RD}$ pins. It is used by the 8259A to decipher various Command Words the CPU writes and status the CPU wishes to read. It is typically connected to the CPU AO address line (A1 for 8086.8088).

11.4.2 Functional Description

11.4.2.1 Interrupts in Microcomputer Systems

Microcomputer system design requires that I.O devices such as keyboards, displays, sensors and other components receive servicing in a an efficient manner so that large amounts of the total system tasks can be assumed by the microcomputer with little or no effect on throughput. The most common method of servicing such devices is the Polled approach. This is where the processor must test each device in sequence and in effect "ask" each one if it needs servicing. It is easy to see that a large portion of the main program is looping through this continuous polling cycle and that such a method would have a serious detrimental effect on system throughput, thus limiting the tasks that could be assumed by the microcomputer and reducing the cost effectiveness of using such devices. A more desirable method would be one that would allow the microprocessor to be executing its main program and only stop to service peripheral devices when it is told to do so by the device itself. In effect, the method would provide an external asynchronous input that would inform the processor that it should complete whatever instruction that is currently being executed and fetch a new routine that will service the requesting device. Once this servicing is complete, however, the processor would resume exactly where it left off.

This method is called Interrupt. It is easy to see that system throughput would drastically increase, and thus more tasks could be assumed by the microcomputer to further enhance its cost effectiveness. The Programmable Interrupt Controller (PIC) functions as an overall manager in an Interrupt-Driven system environment. It accepts requests from the peripheral equipment, determines which of the incoming requests is of the highest importance (priority), ascertains whether the incoming request has a higher priority value than the level currently being serviced, and issues an interrupt to the CPU based on this determination. Each peripheral device or structure usually has a special program or "routine" that is associated with its specific functional or operational requirements; this is referred to as a "service routine". The PIC, after issuing an Interrupt to the CPU, must somehow input information into the CPU that can "point" the Program Counter to the service routine associated with the requesting device. This "pointer" is an address in a vectoring table and will often be referred to, in this document, as vectoring data.

The 8259A is a device specifically designed for use in real time, interrupt driven microcomputer systems. It manages eight levels or requests and has built-in features for expandability to other 8259A's (up to 64 levels). It is programmed by the system's software as an I/O peripheral. A selection of priority modes is available to the programmer so that the manner in which the requests are processed by the 8259A can be configured to match his system requirements. The priority modes can be changed or reconfigured dynamically at any time during the main program. This means that the complete interrupt structure can be defined as required, based on the total system environment.

Graphically the polled method is represented as follows:

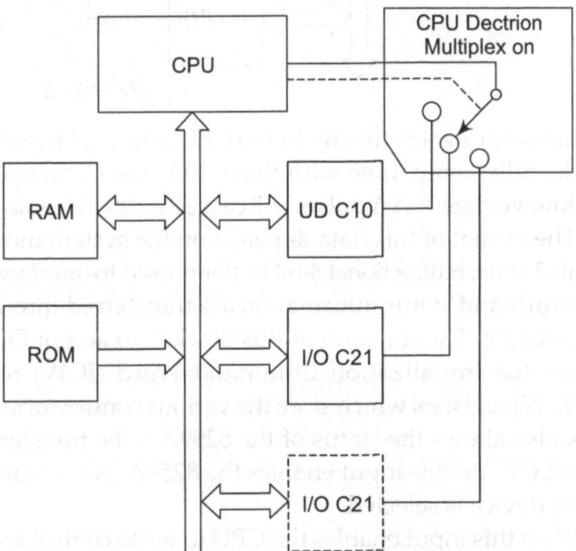

231 460–3

Graphically the interrupt method is represented as follows:

(*a*) **Interrupt Request Register (IRR) and In-Service Register (ISR):** The interrupts at the IR input lines are handled by two registers in cascade, the Interrupt Request Register (IRR) and the In-Service Register (ISR). The IRR is used to store all the interrupt levels which are requesting service; and the ISR is used to store all the interrupt levels which are being serviced.

(*b*) **Priority Resolver:** This logic block determines the priorities of the bits set in the IRR. The highest priority is selected and strobed into the corresponding bit of the ISR during INTA pulse.

(*c*) **Interrupt Mask Register (IMR):** The IMR stores the bits which mask the interrupt lines to be masked. The IMR operates on the IRR. Masking of a higher priority input will not affect the interrupt request lines of lower quality.

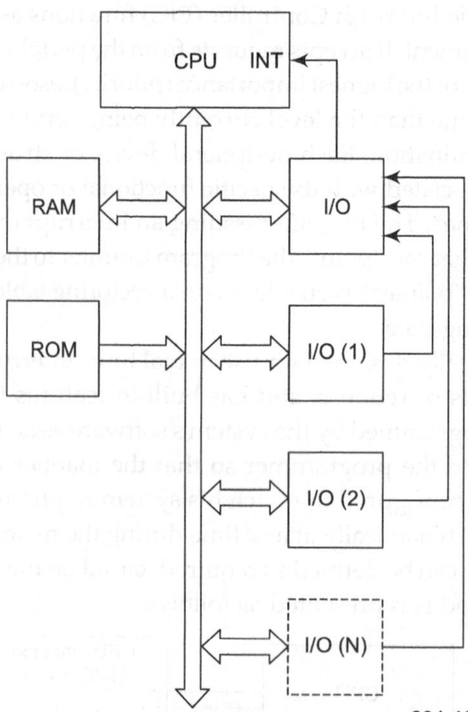

231 460–3

(*d*) **INT (Interrupt):** This output goes directly to the CPU interrupt input. The VOH level on this line is designed to be fully compatible with the 8080A, 8085A and 8086 input levels.

(*e*) **INTA (Interrupt Acknowledge):** INTA pulses will cause the 8259A to release vectoring information onto the data bus. The format of this data depends on the system mode (mPM) of the 8259A.

(*f*) **Data Bus Buffer:** This 3-state, bidirectional 8-bit buffer is used to interface the 8259A to the system Data Bus. Control words and status information are transferred through the Data Bus Buffer.

(*g*) **Read/Write Control Logic:** The function of this block is to accept OUTput commands from the CPU. It contains the Initialization Command Word (ICW) registers and Operation Command Word (OCW) registers which store the various control formats for device operation. This function block also allows the status of the 8259A to be transferred onto the Data Bus.

(*h*) **CS (Chip Select):** A LOW on this input enables the 8259A. No reading or writing of the chip will occur unless the device is selected.

(*i*) **WR (Write):** A LOW on this input enables the CPU to write control words (ICWs and OCWs) to the 8259A.

(*j*) **RD (Read):** A LOW on this input enables the 8259A to send the status of the Interrupt Request Register (IRR), In Service Register (ISR), the Interrupt Mask Register (IMR), or the Interrupt level onto the Data Bus.

(*k*) **A0:** This input signal is used in conjunction with WR and RD signals to write commands into the various command registers, as well as reading the various status registers of the chip. This line can be tied directly to one of the address lines.

11.5 8251 USART AND RS-232C

The 8251 is a USART (Universal Synchronous Asynchronous Receiver Transmitter) for serial data communication. As a peripheral device of a microcomputer system, the 8251 receives parallel data from the CPU and transmits serial data after conversion. This device also receives serial data from the outside and transmits parallel data to the CPU after conversion.

The Block diagram of the 8251 USART (Universal Synchronous Asynchronous Receiver Transmitter) is as follows:

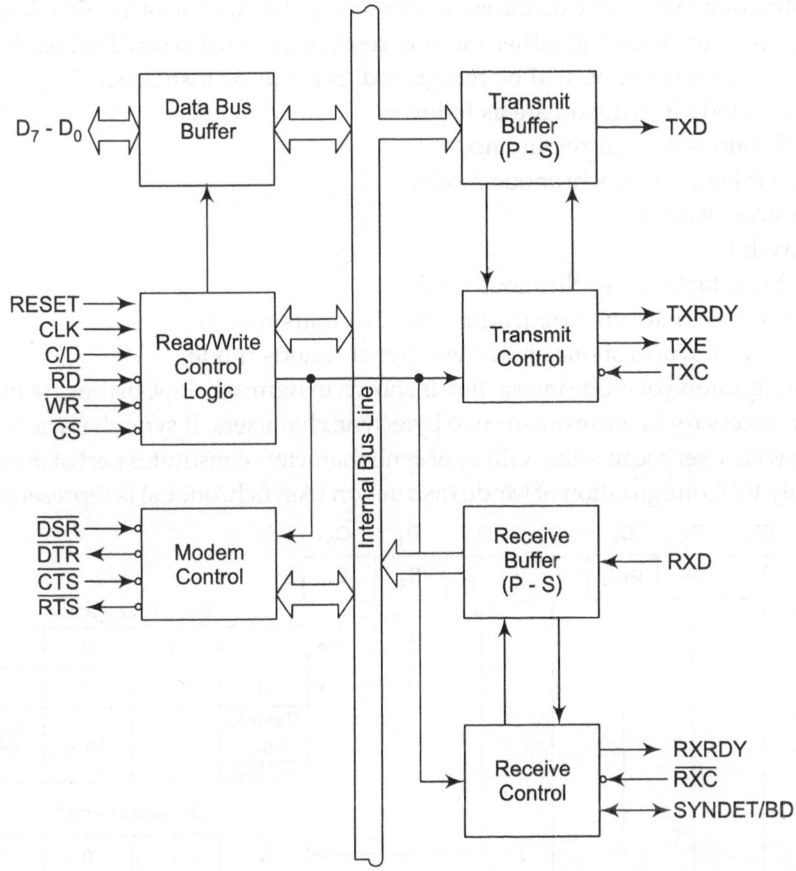

The 8251 functional configuration is programmed by software. Operation between the 8251 and a CPU are executed by program control. The following Table shows the operation between a CPU and the device.

$\overline{CS}$	$C/\overline{D}$	$\overline{RD}$	$\overline{WR}$	
1	X	X	X	Data Bus 3-State
0	X	1	1	Data Bus 3-State
0	1	0	1	Status → CPU
0	1	1	0	Control Word ← CPU
0	0	0	1	Data → CPU
0	0	1	0	Data ← CPU

11.5.1 Control Words

There are two types of control word:

- Mode instruction (setting of function)
- Command (setting of operation)

1. **Mode Instruction :** Mode instruction is used for setting the function of the 8251. Mode instruction will be in "wait for write" at either internal reset or external reset. That is, the writing of a control word after resetting will be recognized as a "mode instruction."
   Items set by mode instruction are as follows:
   - Synchronous/asynchronous mode
   - Stop bit length (asynchronous mode)
   - Character length
   - Parity bit
   - Baud rate factor (asynchronous mode)
   - Internal/external synchronization (synchronous mode)
   - Number of synchronous characters (Synchronous mode)

   The bit configuration of mode instruction is shown in figures below. In the case of synchronous mode, it is necessary to write one-or two byte sync characters. If sync characters were written, a function will be set because the writing of sync characters constitutes part of mode instruction. Graphically Bit Configuration of Mode Instruction (Asynchronous) is represented as follows:

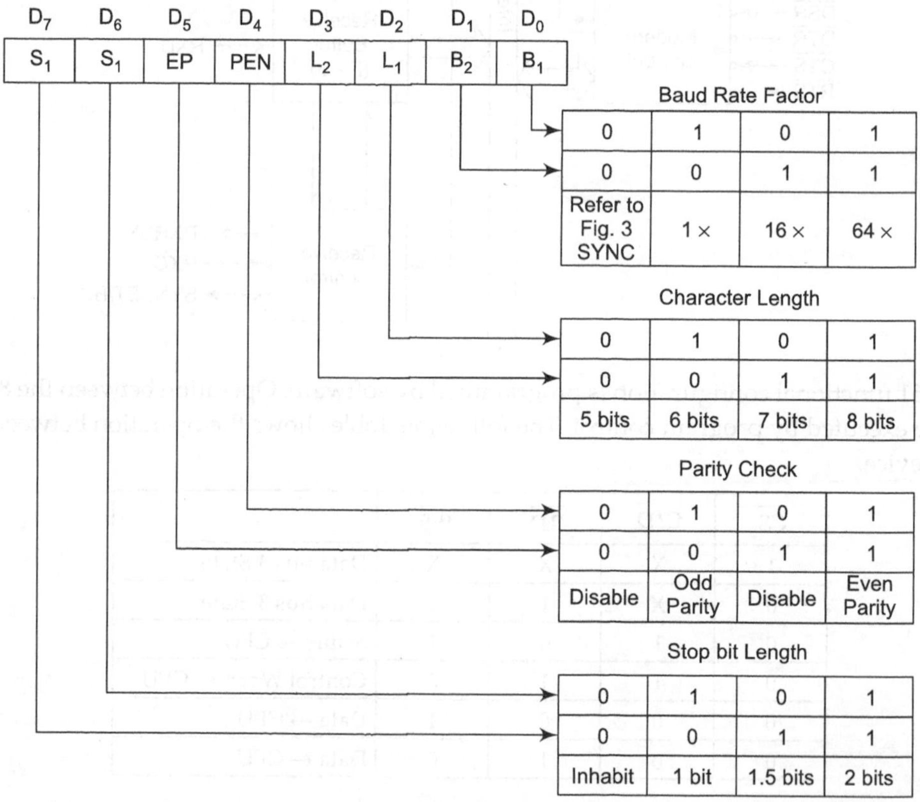

Graphically the bit configuration of mode instruction (Synchronous) is represented as follows:

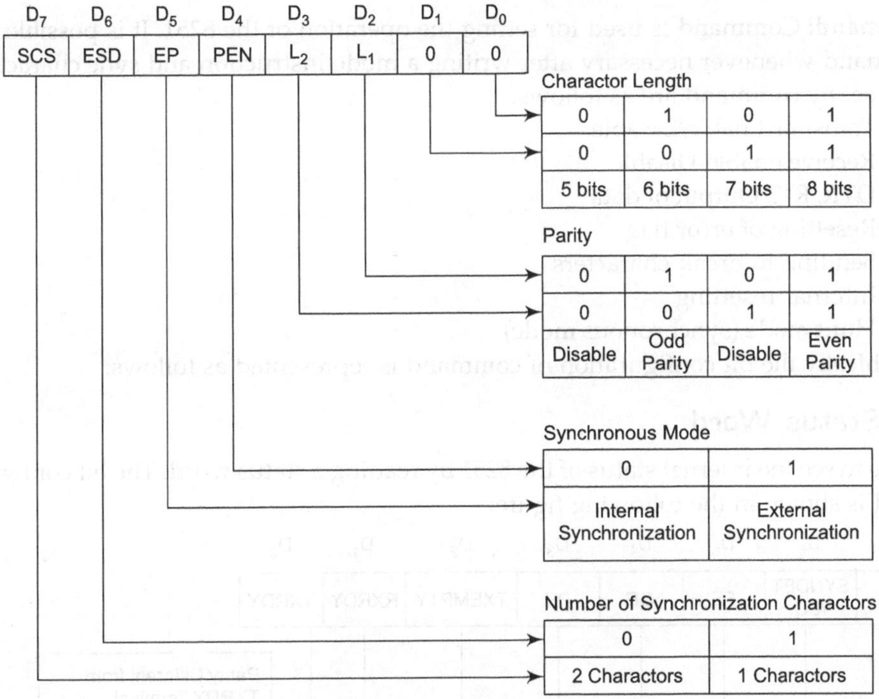

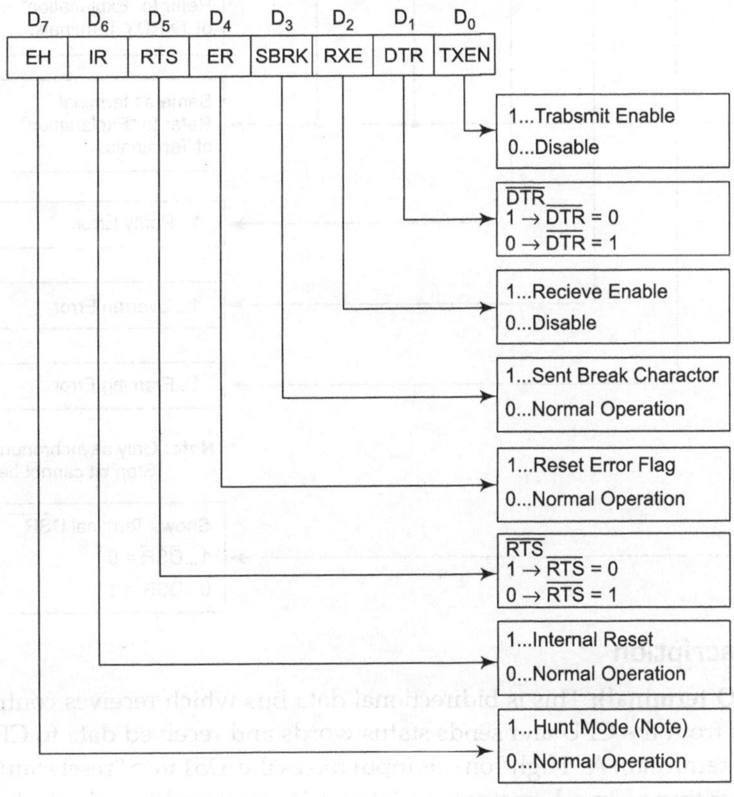

Note: Seach mode for synchronous charactors in synchronous mode.

2. **Command:** Command is used for setting the operation of the 8251. It is possible to write a command whenever necessary after writing a mode instruction and sync characters. Items to be set by command are as follows:
    - Transmit Enable/Disable
    - Receive Enable/Disable
    - DTR, RTS Output of data
    - Resetting of error flag
    - Sending to break characters
    - Internal resetting
    - Hunt mode (synchronous mode)

   Graphically the bit configuration of command is represented as follows:

11.5.2 Status Word

It is possible to see the internal status of the 8251 by reading a status word. The bit configuration of status word is shown in the following figure:

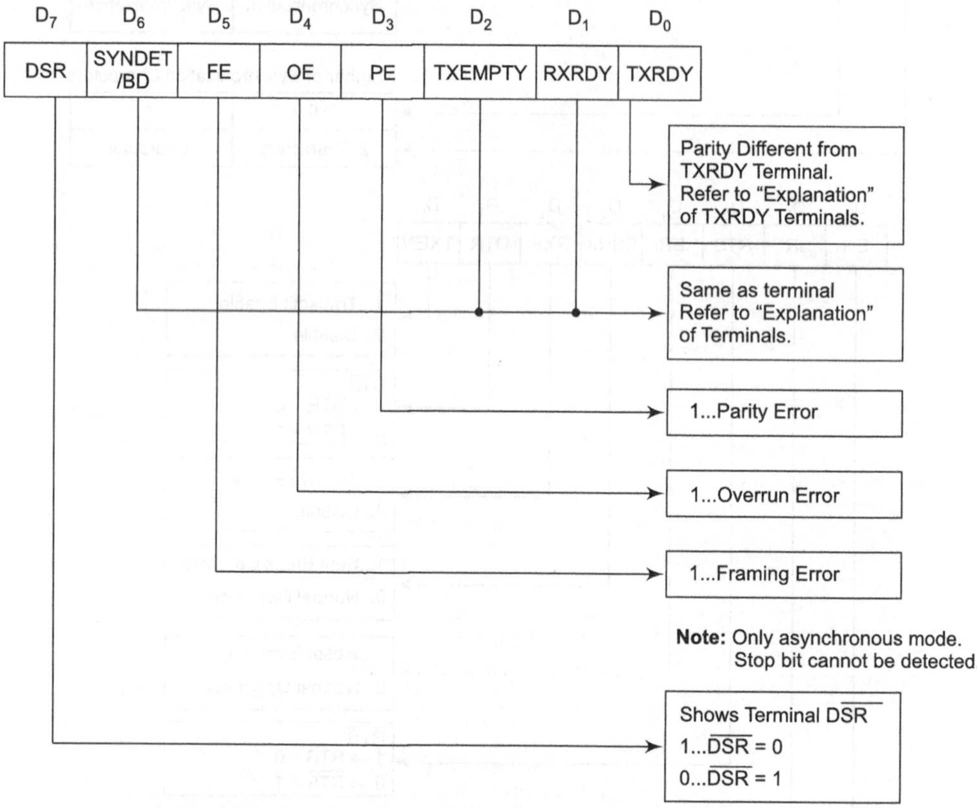

11.5.3 Pin Description

(*a*) **D 0 to D 7 (I/O terminal):** This is bidirectional data bus which receives control words and transmits data from the CPU and sends status words and received data to CPU.

(*b*) **RESET (Input terminal):** A "High" on this input forces the 8251 into "reset status." The device waits for the writing of "mode instruction." The min. reset width is six clock inputs during the operating status of CLK.

(c) **CLK (Input terminal):** CLK signal is used to generate internal device timing. CLK signal is independent of RXC or TXC. However, the frequency of CLK must be greater than 30 times the RXC and TXC at Synchronous mode and Asynchronous "x1" mode, and must be greater than 5 times at Asynchronous "x16" and "x64" mode.

(d) **WR (Input terminal):** This is the "active low" input terminal which receives a signal for writing transmit data and control words from the CPU into the 8251.

(e) **RD (Input terminal):** This is the "active low" input terminal which receives a signal for reading receive data and status words from the 8251.

(f) **C/D (Input terminal):** This is an input terminal which receives a signal for selecting data or command words and status words when the 8251 is accessed by the CPU. If C/D = low, data will be accessed. If C/D = high, command word or status word will be accessed.

(g) **CS (Input terminal):** This is the "active low" input terminal which selects the 8251 at low level when the CPU accesses. Note: The device won't be in "standby status"; only setting CS = High.

(h) **TXD (Output terminal):** This is an output terminal for transmitting data from which serial-converted data is sent out. The device is in "mark status" (high level) after resetting or during a status when transmit is disabled. It is also possible to set the device in "break status" (low level) by a command.

(i) **TXRDY (Output terminal):** This is an output terminal which indicates that the 8251is ready to accept a transmitted data character. But the terminal is always at low level if CTS = high or the device was set in "TX disable status" by a command. Note: TXRDY status word indicates that transmit data character is receivable, regardless of CTS or command. If the CPU writes a data character, TXRDY will be reset by the leading edge or WR signal.

(j) **TXEMPTY (Output terminal):** This is an output terminal which indicates that the 8251 has transmitted all the characters and had no data character. In "synchronous mode," the terminal is at high level, if transmit data characters are no longer remaining and sync characters are automatically transmitted. If the CPU writes a data character, TXEMPTY will be reset by the leading edge of WR signal.
Note: As the transmitter is disabled by setting CTS "High" or command, data written before disable will be sent out. Then TXD and TXEMPTY will be "High". Even if a data is written after disable, that data is not sent out and TXE will be "High". After the transmitter is enabled, it sent out. (Refer to Timing Chart of Transmitter Control and Flag Timing)

(k) **TXC (Input terminal):** This is a clock input signal which determines the transfer speed of transmitted data. In "synchronous mode," the baud rate will be the same as the frequency of TXC. In "asynchronous mode", it is possible to select the baud rate factor by mode instruction. It can be 1, 1/16 or 1/64 the TXC. The falling edge of TXC sifts the serial data out of the 8251.

(l) **RXD (Input terminal):** This is a terminal which receives serial data.

(m) **RXRDY (Output terminal):** This is a terminal which indicates that the 8251 contains a character that is ready to READ. If the CPU reads a data character, RXRDY will be reset by the leading edge of RD signal. Unless the CPU reads a data character before the next one is received completely, the preceding data will be lost. In such a case, an overrun error flag status word will be set.

(n) **RXC (Input terminal):** This is a clock input signal which determines the transfer speed of received data. In "synchronous mode," the baud rate is the same as the frequency of RXC. In "asynchronous mode," it is possible to select the baud rate factor by mode instruction. It can be 1, 1/16, 1/64 the RXC.

(*o*) **SYNDET/BD (Input or output terminal):** This is a terminal whose function changes according to mode. In "internal synchronous mode." this terminal is at high level, if sync characters are received and synchronized. If a status word is read, the terminal will be reset. In "external synchronous mode, "this is an input terminal. A "High" on this input forces the 8251 to start receiving data characters. In "asynchronous mode," this is an output terminal which generates "high level "output upon the detection of a "break" character if receiver data contains a "low-level" space between the stop bits of two continuous characters. The terminal will be reset, if RXD is at high level. After Reset is active, the terminal will be output at low level.

(*p*) **DSR (Input terminal):** This is an input port for MODEM interface. The input status of the terminal can be recognized by the CPU reading status words.

(*q*) **DTR (Output terminal):** This is an output port for MODEM interface. It is possible to set the status of DTR by a command.

(*r*) **CTS (Input terminal):** This is an input terminal for MODEM interface which is used for controlling a transmit circuit. The terminal controls data transmission if the device is set in "TX Enable" status by a command. Data is transmitable if the terminal is at low level.

(*s*) **RTS (Output terminal):** This is an output port for MODEM interface. It is possible to set the status RTS by a command.

EXERCISE

1. Draw and explain the block diagram of 8237 A.
2. What is the use of USART?
3. Draw and explain the internal block diagram of 8253.
4. Explain the data transfer types supported by 8237A.
5. Illustrates the different modes of operations of 8253\8254.
6. What is PPI?
7. Explain the DMA data transfer in the microprocessor.
8. Explain the different modes of operation of 8255. Describe the control word format of 8255 in I\O and BSR mode.
9. Explain the functional diagram of keyboard and display controller.
10. List the features of 8259.

12

INTRODUCTION TO DAC AND ADC

12.1 INTRODUCTION

Analog-to-digital converters (ADCs) are needed in all those applications which, interfacing with the analogue world, exploit the digital processing of data. As digital processing is more and more gaining ground over analogue signal processing, the importance of ADCs correspondingly increase. Digital signal processing systems are generally designed by considering analogue-to-digital converters (ADC's) as ideal components affected only by quantization and sampling errors. Vice versa, the effect of the ADC actual working conditions modifies the expected digital values and could compromise the effectiveness of the digital signal processing as a whole. ADC's are among the components that mostly influence metrological performance of digital measurement systems. ADC errors limit system dynamic, its frequency band and so on, by adding distortion and error effects to the output. Therefore, a deeper insight into the ADC characteristics is needed. Furthermore, the digital-to-analogue converters (DACs) are widely used in many fields, *i.e.*, in the automatic system to drive actuators, in the digital music, and so on. In particular, the possibility to integrate in one chip an ADC, a DAC and a DSP has improved the applicability of these components to new fields, as in mobile telecommunication systems. In order to improve the ADC and DAC performance many efforts are now carrying out in several Research Centres. The efforts are oriented to set up new models, testing methods and error correction techniques. As in other technical fields, also in the ADC error investigation the use of models has become more and more widespread. Modelling actual ADC allows the error characteristic to be analyzed and, hence, its correction to be carried out more easily. Then, in the diagnostic process, it allows a fault model to be set up in order to isolate a possible fault condition. In order to improve the ADCs and DACs performance, some ideas have been proposed, either linked to the improvement of the chip design or related to the correction of the ADC and DAC errors by means of correction techniques. This last field requires to set up opportune models and testing techniques to correctly estimate the metrological characteristics of the corrected component. These problems are of high relevance not only for producers but mainly for system integrators and end-users. Such customers build complex systems where the accuracy of ADC and DAC impacts in a decisive way the property of final product. Consequently user-friendly standards for ADC and DAC assessment will be highly welcome for these target users. It is worth to underlying that; now, there is a lack in the realization of a European standard for ADCs and DACs.

12.2 DAC

In electronics, a **digital-to-analog converter** (**DAC** or **D-to-A**) is a device that converts a digital (usually binary) code to an analog signal (current, voltage, or electric charge). An analog-to-digital converter (ADC) performs the reverse operation.

12.2.1 Basic Ideal Operation

A DAC converts an abstract finite-precision number (usually a fixed-point binary number) into a concrete physical quantity (*e.g.*, a voltage or a pressure). In particular, DACs are often used to convert finite-precision time series data to a continually varying physical signal.

A typical DAC converts the abstract numbers into a concrete sequence of impulses that are then processed by a reconstruction filter using some form of interpolation to fill in data between the impulses. Other DAC methods (*e.g.*, methods based on Delta-sigma modulation) produce a pulse-density modulated signal that can then be filtered in a similar way to produce a smoothly varying signal.

By the Nyquist-Shannon sampling theorem, sampled data can be reconstructed perfectly provided that its bandwidth meets certain requirements (*e.g.*, a baseband signal with bandwidth less than the Nyquist frequency). However, even with an ideal reconstruction filter, digital sampling introduces quantization error that makes perfect reconstruction practically impossible. Increasing the digital resolution (*i.e.*, increasing the number of bits used in each sample) or introducing sampling dither can reduce this error.

8-channel digital-to-analog converter Cirrus Logic CS4382 as used in a soundcard is represented as follows:

Ideally sampled signal. Signal of a typical interpolating DAC output is represented as follows:

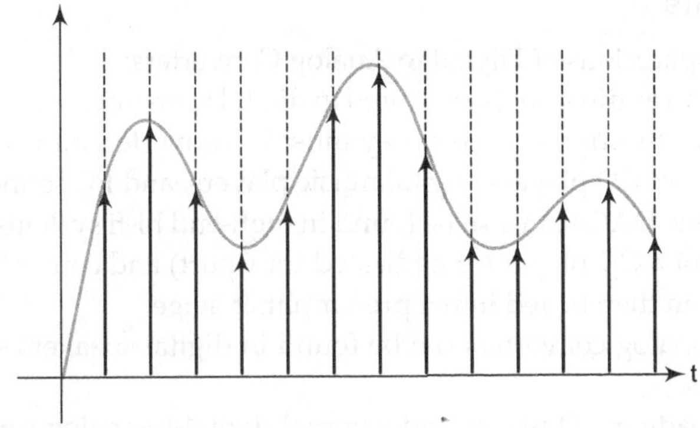

12.2.2 Practical Operation

Instead of impulses, usually the sequence of numbers updates the analogue voltage at uniform sampling intervals.

These numbers are written to the DAC, typically with a clock signal that causes each number to be latched in sequence, at which time the DAC output voltage changes rapidly from the previous value to the value represented by the currently latched number. The effect of this is that the output voltage is *held* in time at the current value until the next input number is latched resulting in a piecewise constant or 'staircase' shaped output. This is equivalent to a zero-order hold operation and has an effect on the frequency response of the reconstructed signal.

The fact that practical DACs output a sequence of piecewise constant values or rectangular pulses would cause multiple harmonics above the Nyquist frequency. These are typically removed with a low pass filter acting as a reconstruction filter.

However, this filter means that there is an inherent effect of the zero-order hold on the effective frequency response of the DAC resulting in a mild roll-off of gain at the higher frequencies (often a 3.9224 dB loss at the Nyquist frequency) and depending on the filter, phase distortion. Not all DACs have a zero order response however. This high-frequency roll-off is the output characteristic of the DAC, and is not an inherent property of the sampled data.

Piecewise constant signal typical of a zero-order (non-interpolating) DAC output is represented as follows:

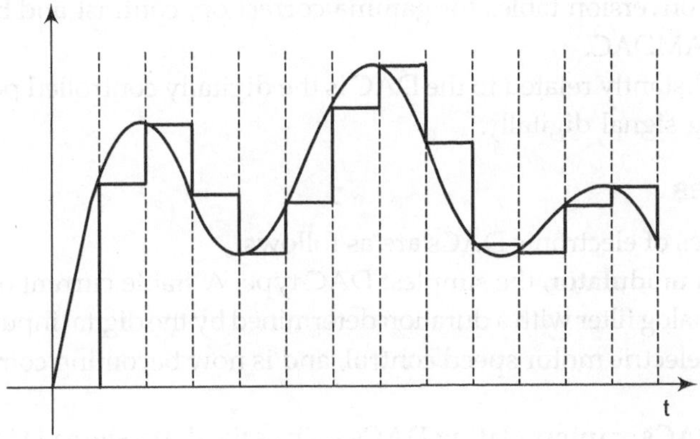

12.2.3 Applications

The following are the applications of Digital to Analog Converters:

(*a*) **Audio:** Most modern audio signals are stored in digital form (for example MP3s and CDs) and in order to be heard through speakers they must be converted into an analog signal. DACs are therefore found in CD players, digital music players, and PC sound cards.
Specialist standalone DACs can also be found in high-end hi-fi systems. These normally take the digital output of a CD player (or dedicated transport) and convert the signal into a line-level output that can then be fed into a pre-amplifier stage.
Similar digital-to-analog converters can be found in digital speakers such as USB speakers, and in sound cards.
Graphically Top-loading CD player and external digital-to-analog converter is represented as follows:

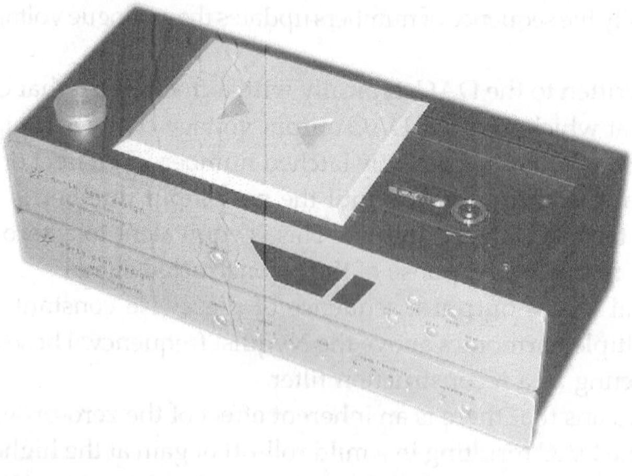

(*b*) **Video:** Video signals from a digital source, such as a computer, must be converted to analog form if they are to be displayed on an analog monitor. As of 2007, analog inputs are more commonly used than digital, but this may change as flat panel displays with DVI and/or HDMI connections become more widespread. A video DAC is, however, incorporated in any digital video player with analog outputs. The DAC is usually integrated with some memory (RAM), which contains conversion tables for gamma correction, contrast and brightness, to make a device called a RAMDAC.
A device that is distantly related to the DAC is the digitally controlled potentiometer, used to control an analog signal digitally.

12.2.4 DAC Types

The most common types of electronic DACs are as follows:

- The **pulse-width modulator,** the simplest DAC type. A stable current or voltage is switched into a low-pass analog filter with a duration determined by the digital input code. This technique is often used for electric motor speed control, and is now becoming common in high-fidelity audio.
- **Oversampling DACs** or **interpolating DACs** such as the **delta-sigma DAC,** use a pulse density conversion technique. The oversampling technique allows for the use of a lower resolution DAC internally. A simple 1-bit DAC is often chosen because the oversampled result is inherently linear. The DAC is driven with a pulse-density modulated signal, created with the use of a

low-pass filter, step non-linearity (the actual 1-bit DAC), and negative feedback loop, in a technique called delta-sigma modulation. This results in an effective high-pass filter acting on the quantization (signal processing) noise, thus steering this noise out of the low frequencies of interest into the high frequencies of little interest, which is called noise shaping (*very* high frequencies because of the oversampling). The quantization noise at these high frequencies are removed or greatly attenuated by use of an analog low-pass filter at the output (sometimes a simple RC low-pass circuit is sufficient). Most very high resolution DACs (greater than 16 bits) are of this type due to its high linearity and low cost. Higher oversampling rates can either relax the specifications of the output low-pass filter and enable further suppression of quantization noise. Speeds of greater than 100 thousand samples per second (for example, 192 kHz) and resolutions of 24 bits are attainable with delta-sigma DACs. A short comparison with pulse-width modulation shows that a 1-bit DAC with a simple first-order integrator would have to run at 3 THz (which is physically unrealizable) to achieve 24 meaningful bits of resolution, requiring a higher-order low-pass filter in the noise-shaping loop. A single integrator is a low-pass filter with a frequency response inversely proportional to frequency and using one such integrator in the noise-shaping loop is a first order delta-sigma modulator. Multiple higher order topologies (such as MASH) are used to achieve higher degrees of noise-shaping with a stable topology.

- The **binary-weighted DAC**, which contains one resistor or current source for each bit of the DAC connected to a summing point. These precise voltages or currents sum to the correct output value. This is one of the fastest conversion methods but suffers from poor accuracy because of the high precision required for each individual voltage or current. Such high-precision resistors and current sources are expensive, so this type of converter is usually limited to 8-bit resolution or less.
- The **R-2R ladder DAC**, which is a binary-weighted DAC that uses a repeating cascaded structure of resistor values R and 2R. This improves the precision due to the relative ease of producing equal valued-matched resistors (or current sources). However, wide converters perform slowly due to increasingly large RC-constants for each added R-2R link.
- The **thermometer-coded DAC**, which contains an equal resistor or current-source segment for each possible value of DAC output. An 8-bit thermometer DAC would have 255 segments, and a 16-bit thermometer DAC would have 65,535 segments. This is perhaps the fastest and highest precision DAC architecture but at the expense of high cost. Conversion speeds of >1 billion samples per second have been reached with this type of DAC.
- **Hybrid DACs**, which use a combination of the above techniques in a single converter. Most DAC integrated circuits are of this type due to the difficulty of getting low cost, high speed and high precision in one device.
    - The **segmented DAC**, which combines the thermometer-coded principle for the most significant bits and the binary-weighted principle for the least significant bits. In this way, a compromise is obtained between precision (by the use of the thermometer-coded principle) and number of resistors or current sources (by the use of the binary-weighted principle). The full binary-weighted design means 0% segmentation; the full thermometer-coded design means 100% segmentation.

12.2.5 Common Test Parameters for DAC's

Common Test Parameters for DAC's are as follows:

Test Parameter	Unit	Typical Description
Resolution	Bits	This is the number of bits that the DAC has. If a DAC has n bits, then its output analog signal span can be divided (or "resolved") into 2^n states, each corresponding to its own digital input code. The higher the number of bits, the better is the resolution of a DAC.
Voltage Reference Input Range	V	This indicates the minimum and maximum voltages that may be used as the DAC's external reference voltage. Example of an Actual Spec: 0V min; Vdd ax
Bipolar Zero Error or Midscale Error	mV	This is the deviation of the actual analog output of the DAC from the ideal output of 0V when the 2's complement input code representing half-scale (all zeros) is loaded into the input register. Example of an Actual Spec: +/- 30 mV typ.
Non-Linearity, Differential (DNL)	LSB or % of FSR	This is the difference between the actual or measured output change and the ideal output change for a digital input change of 1 LSB, *i.e.*, for a transition between two adjacent codes. Examples of Actual Specs: +/- 1 LSB; +/- 0.001% FSR typ.
Non-Linearity, Integral (INL) or Relative Accuracy	LSB	This is the maximum deviation of the output from the straight line between the zero and full-scale points, excluding the effects of zero and full-scale errors. Example of an Actual Spec: +/-1 LSB typ.; +/- 4 LSB's max.
Analog Output Range or Full-Scale Range	V	This is the difference between the maximum and minimum analog output values that the DAC is specified to provide. Example of Actual Spec: -3V to +3V, Bipolar Mode
Logic Input Voltage, Vih (Logic "1")	V	This is the minimum voltage that the digital inputs of the DAC are guaranteed to recognize as a Logic "1". Example of an Actual Spec: 2.4 V min.
Logic Input Voltage, Vil (Logic "0")	V	This is the maximum voltage that the digital inputs of the DAC are guaranteed to recognize as a Logic "0". Example of an Actual Spec: 0.8 V max.
Input Leakage Current, I_{IL}	µA	This is maximum leakage current exhibited by a digital input at logic "0". Example of an Actual Spec: 2 µA max. for Vil = 0.4 V.
Settling Time	µsec	This is the time required by the outputs of a DAC to reach and remain within a specified error band about its final value. This is measured from the digital input transition. Examples of Actual Specs: 1.5 µsec typ. for a 6V step; 1.0 µsec typ. for a 1 LSB step;
Analog Positive Power Supply (+Vs)	V	This is the voltage range that can be used for the analog positive supply of the DAC. Example of an Actual Spec: +4.75V min.; +5.0V typ.; +13.2V max.
Analog Negative Power Supply (-Vs)	V	This is the voltage range that can be used for the analog negative supply of the DAC. Example of an Actual Spec: -13.2V min.; -5V typ.; -4.75V max.
Logic Positive Power Supply (+V_L)	V	This is the voltage range that can be used for the logic positive supply of the DAC. Example of an Actual Spec: +4.75V min.; +5.0V typ.; +13.2V max.

Logic Negative Power Supply ($-V_L$)	V	This is the voltage range that can be used for the logic negative supply of the DAC. Example of an Actual Spec: -13.2V min.; -5V typ.; -4.75V max.
Analog Output Impedance	kW	This is the dynamic load presented by a DAC at its analog output. Example of an Actual Spec: 0.1 kW typ.
Gain Error or Full-Scale Error	mV or %FSR	This indicates how closely the analog output of a DAC matches its ideal value at full scale and may be expressed in mV or as a % of the FSR. For many DAC's, the ideal maximum output voltage is Vref-1 LSB. Example of an Actual Spec: +/-1.22 mV
Signal-to-Noise Ratio (SNR)	dB	This is the ratio of the amplitude of the analog output of the DAC when the output is set to full-scale to the amplitude of the output when it is set to zero. Example of an Actual Spec: 107 dB min.; 110 dB typ. for 20 Hz-30 kHz
Total Harmonic Distortion (THD)	dB	This is the ratio of the rms sum of the first few (2nd to 6th) harmonic components of the DAC output to the DAC output's fundamental value, usually expressed in dB. Example of an Actual Spec: -80 dB typ.

12.2.6 DAC Performance

DACs are very important to system performance. The most important characteristics of these devices are as follows:

(*a*) **Resolution:** This is the number of possible output levels the DAC is designed to reproduce. This is usually stated as the number of bits it uses, which is the base two logarithm of the number of levels. For instance a 1 bit DAC is designed to reproduce 2 (2^1) levels while an 8 bit DAC is designed for 256 (2^8) levels. Resolution is related to the **effective number of bits** (ENOB) which is a measurement of the actual resolution attained by the DAC.

(*b*) **Maximum sampling frequency:** This is a measurement of the maximum speed at which the DACs circuitry can operate and still produce the correct output. As stated in the Nyquist-Shannon sampling theorem, a signal must be sampled at over twice the frequency of the desired signal. For instance, to reproduce signals in the entire audible spectrum, which includes frequencies of up to 20 kHz, it is necessary to use DACs that operate at over 40 kHz. The CD standard samples audio at 44.1 kHz, thus DACs of this frequency are often used. A common frequency in cheap computer sound cards is 48 kHz — many work at only this frequency, offering the use of other sample rates only through (often poor) internal resampling.

(*c*) **Monotonicity:** This refers to the ability of a DAC's analog output to move only in the direction that the digital input moves (*i.e.*, if the input increases, the output doesn't dip before asserting the correct output.) This characteristic is very important for DACs used as a low frequency signal source or as a digitally programmable trim element.

(*d*) **THD+N:** This is a measurement of the distortion and noise introduced to the signal by the DAC. It is expressed as a percentage of the total power of unwanted harmonic distortion and noise that accompany the desired signal. This is a very important DAC characteristic for dynamic and small signal DAC applications.

(*e*) **Dynamic range:** This is a measurement of the difference between the largest and smallest signals the DAC can reproduce expressed in decibels. This is usually related to DAC resolution and noise floor.

12.2.7 Linearity of Digital-to-Analog Converter (DAC)

In selecting a digital-to-analog converter (DAC) one of the most common criteria you look at is linearity. Two tests have evolved as the most popular linearity tests: differential non-linearity (DNL) and integral non-linearity (INL.) If a DAC were ideal, each output step would be exactly the same size; meaning when supplied with equally increasing increments of supply voltage, the output would ramp up in equal increments. DNL is the degree to which each output step (or code width) varies from the ideal step. DNL is generally more critical when outputting small signals. INL measures the deviation of the entire transfer function from the ideal function and is generally more critical for outputting large signals.

To test the linearity of a DAC, you need to generate the digital stimulus and capture the analog response. Usually you would sweep the DAC from 0 to its maximum output step, and compare the response to what it should be at every output step. When most manufacturers specify DNL and INL on a data sheet, they specify the worst cases - the maximum and minimum over the entire transfer functions. You might only be interested in the minimum and maximum, or you might be interested in the measurements for all code widths. LabVIEW provides native array functions to search the data array for the max and min or other points of interest, and you can graph the overall transfer functions for trending information.

To generate the digital stimulus and capture the analog response, you need a high-speed digital source and high-speed digitizer. In hardware, you need to control the timing between the acquisition and generation modules, so you know which acquisition sample should correspond to which generation sample. Platforms such as PXI and VXI are ideal for such applications These platforms are designed with timing and synchronization resources built-in, so modules can share reference clocks, sample clocks, triggers, and events.

Differential Non-linearity of RGB Video DAC is represented as follows:

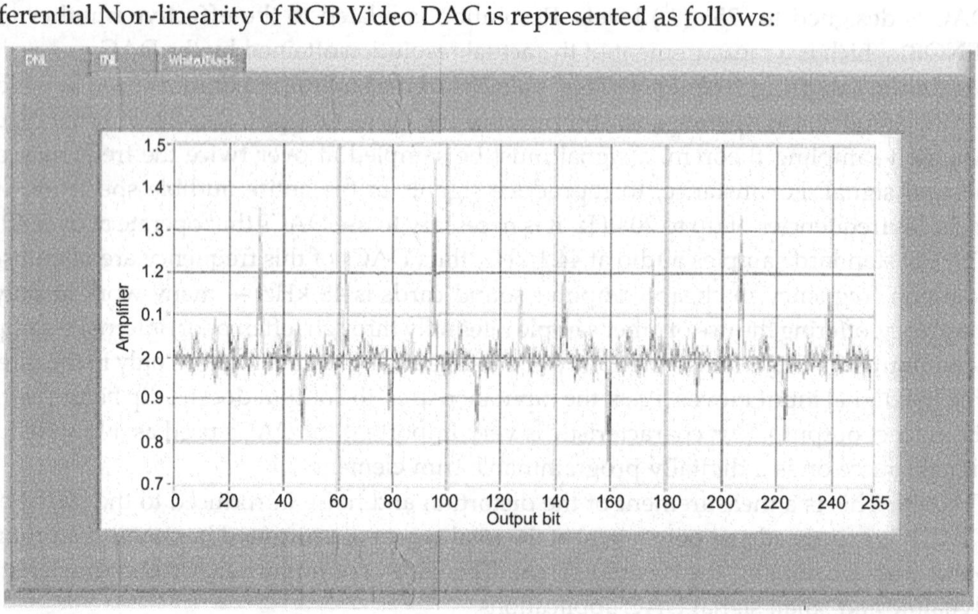

12.2.8 DAC Parameters

In this section we will discuss the various DAC parameters with graphical representation:

Graphically the DAC parameters can be represented as follows:

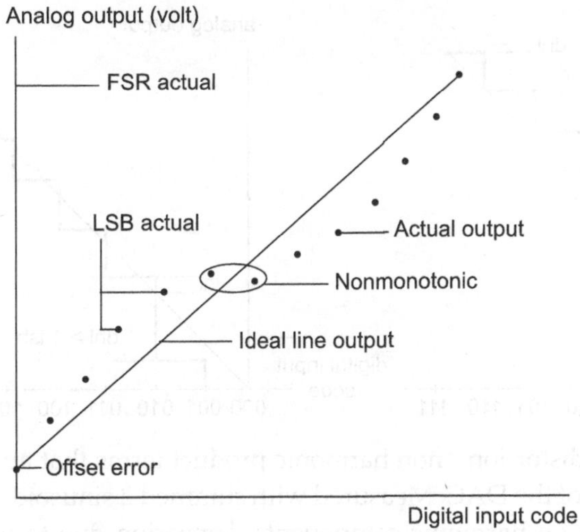

The explanation of the above figure is as follows:

(*a*) **Resolution:** Resolution is normally given in bits. Resolution indicates the smallest increment of its output, corresponding to a 1 LSB input code change. For example for 10 bit DAC, 2 power 10 = 1024 codes, so the resolution is 1/1024 of the output range.

(*b*) **Full scale range (FSR):** Full scale range means maximum output signal for the DAC, specified as current or voltage (ma or V). Can be negative, positive or both.

(*c*) **Offset error:** Difference between an ideal and actual DAC output when zero digital code applied to the input.

(*d*) **Gain error:** Gain error means the difference between an ideal and actual output when full scale digital code applied to the input. Strongly depends on VREF stability.

(*e*) **Differential Nonlinearity:** Differential Nonlinearity is measured within a ramp code applied to the input of DAC. The step between every pair of the adjacent codes should not exceed 1 LSB. 1LSB is calculated from gain and offset measurements:

$$1\text{ LSB} = \text{FSR}/\text{N-1}$$

Differential Nonlinearity represents the error from 1 LSB for every step, Where N=(2 power of bit number).

(*f*) **Integral Nonlinearity:** Shows how the output differs from an ideal line. It is also measured in LSB. Usually 1LSB is an expectable value for nonlinearity.

The following figures show the difference between Integral and Differential Nonlinearity. Offset and Gain are absolute measurements, DNL and INL are referenced to the zero and full scale outputs of the DAC.

(*g*) **SNR:** Signal to noise ratio. Measured with digital code representing sine wave applied to the input. Fundamental and harmonic components of the sine wave are filtered out. Any remaining signal at the output of the DAC is considered as a noise. SNR is a ratio of the full scale sine wave output to noise level. Some companies specify **SINAD** parameter. SINAD - is a signal to noise and distortion ratio. It is similar to SNR, but the harmonic signal components are not removed.

(*h*) **THD:** Total harmonic distortion - measured with digital code representing sinewave,applied to the DAC input continuously. The output is analyzed in the frequency domain to find harmonic components related to the fundamental output signal.

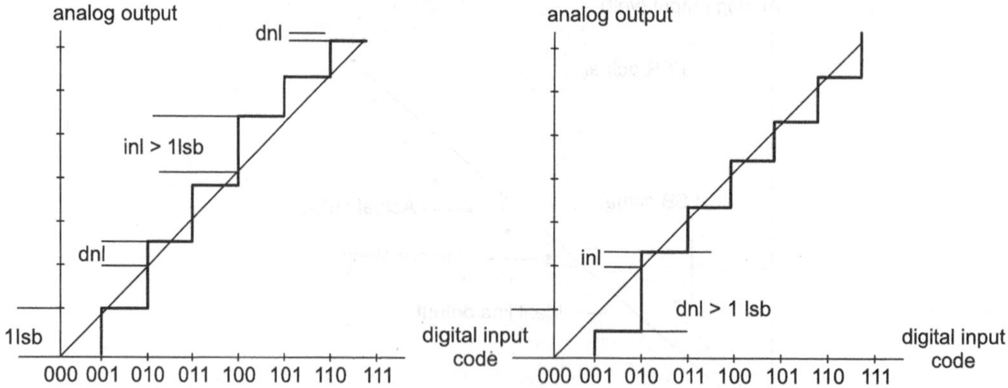

(*i*) **IM:** Intermediation distortion - non harmonic product terms that appear in the output signal due to non-linearity of the DAC. Measured with summed 2 sinusoid tone applied to the input. The output is tested on harmonic components, appearing due to modulation affect on non linear characteristics within DAC. The harmonic components are (F1+F2), (F1-F2), (2F1+/-F2), (F1+/-2F2)...

(*j*) **Max conversion rate:** It is actually max input signal frequency the DAC can handle. The worst case is when input signal changes from zero to max, and the output should reach the max level and settle.

(*k*) **Settling time:** The time required for the output to reach the final value and remain within +- 1 LSB after overshoot.

(*l*) **PSRR:** Power supply rejection ratio .The output signal should remain within limits while power supply voltage changes from Min to Max.

12.2.9 Monotonicity

Monotonicity is a property of certain types of digital-to-analog converter (DAC) circuits. In a monotonic DAC, the analog output always increases or remains constant as the digital input increases. The analog output never decreases during the input sequence. If the analog output decreases at any point during the input sequence, a DAC is said to be non-monotonic.

Monotonicity is an important characteristic in many communications applications in which DACs are used. Such applications can function in the presence of nonlinearity, but not in the presence of non-monotonicity. The term derives from monotonic mathematical functions, also known as non-decreasing functions, in which the value of the dependent variable never decreases as the value of the independent variable increases.

12.3 ADC

An **analog-to-digital converter** (abbreviated **ADC, A/D** or **A to D**) is a device which converts continuous signals to discrete digital numbers. The reverse operation is performed by a digital-to-analog converter (**DAC**).

Typically, an ADC is an electronic device that converts an input analog voltage (or current) to a digital number proportional to the magnitude of the voltage or current. However, some non-electronic or only partially electronic devices, such as rotary encoders, can also be considered ADCs. The digital output may use different coding schemes, such as binary, Gray code or two's complement binary.

4-channel stereo multiplexed analog-to-digital converter WM8775SEDS made by Wolfson Microelectronics placed on X-Fi Fatal1ty Pro sound card is represented as follows:-

12.3.1 Basic Concepts

12.3.1.1 Resolution

The resolution of the converter indicates the number of discrete values it can produce over the range of analog values. The values are usually stored electronically in binary form, so the resolution is usually expressed in bits. In consequence, the number of discrete values available, or "levels", is usually a power of two. For example, an ADC with a resolution of 8 bits can encode an analog input to one in 256 different levels, since $2^8 = 256$. The values can represent the ranges from 0 to 255 (*i.e.*, unsigned integer) or from –128 to 127 (*i.e.*, signed integer), depending on the application.

Resolution can also be defined electrically, and expressed in volts. The voltage resolution of an ADC is equal to its overall voltage measurement range divided by the number of discrete intervals as in the formula:

$$Q = \frac{E_{FSR}}{2^M - 1} = \frac{E_{FSR}}{N}$$

Where:

Q is resolution in volts per step (volts per output codes less one),

E_{FSR} is the full scale voltage range $= V_{Ref\,Hi} - V_{Ref\,Low}$,

M is the ADC's resolution in bits.

N is the number of intervals, (one less than the number of available levels, or output codes), which is:

$$N = 2^M - 1$$

Example 1. In the Full scale measurement range = 0 to 10 volts, ADC resolution is 12 bits: $2^{12} = 4096$ quantization levels (codes). ADC voltage resolution is: (10V – 0V)/4095 steps = 10V/4095 steps ≈ 0.00244 V/step ≈ 2.44 mV/step

Example 2. In the Full scale measurement range = –10 to +10 volts. ADC resolution is 14 bits: 2^{14} = 16384 quantization levels (codes). ADC voltage resolution is: (10V – (–10V))/16383 steps = 20V/16383 steps ≈ 0.00122 V/step ≈ 1.22 mV/step

Example 3. In the Full scale measurement range = 0 to 7 volts. ADC resolution is 3 bits: $2^3 = 8$ quantization levels (codes). ADC voltage resolution is: (7 V – 0 V)/7 steps = 7 V/7 steps = 1 V/ step = 1000 mV/step.

In practice, the smallest output code ("0" in an unsigned system) represents a voltage range which is 0.5Q, that is, half the ADC voltage resolution (Q), as does the largest output code. The other $N-2$ codes are all equal in width and represent the ADC voltage resolution (Q) calculated above. Doing this centers the code on an input voltage that represents the M th division of the input voltage range. For example, in Example 3, with the 3-bit ADC spanning a 7 V range, each of the N divisions would represent 1 V, except the 1st ("0" code) and the last ("7" code) which are 0.5 V wide. Doing this the "1" code spans a voltage range from 0.5 to 1.5 V, the "2" code spans a voltage range from 1.5 to 2.5 V, etc. Thus, if the input signal is at 3/8ths of the full-scale voltage, then the ADC outputs the "3" code, and will do so as long as the voltage stays within the range of 2.5/8ths and 3.5/8ths. This practice is called "mid-tread" operation. This type of ADC can be modeled mathematically as:

$$ADC_{Code} = \text{round}\left(\left(\frac{2^M}{V_{Ref\ Hi} - V_{Ref\ Low}}\right) \cdot \left(V_{In} - V_{Ref\ Low}\right)\right)$$

The exception to this convention seems to be the Microchip PIC processor, where all M steps are equal width. This practice is called "Mid-Rise with Offset" operation.

$$ADC_{Code} = \text{floor}\left(\left(\frac{2^M}{V_{Ref\ Hi} - V_{Ref\ Low}}\right) \cdot \left(V_{In} - V_{Ref\ Low}\right)\right)$$

In practice, the useful resolution of a converter is limited by the best signal-to-noise ratio that can be achieved for a digitized signal. An ADC can resolve a signal to only a certain number of bits of resolution, called the effective number of bits (ENOB). One effective bit of resolution changes the signal-to-noise ratio of the digitized signal by 6 dB, if the resolution is limited by the ADC. If a preamplifier has been used prior to A/D conversion, the noise introduced by the amplifier can be an important contributing factor towards the overall SNR.

12.3.1.2 Response type

(*a*) **Linear ADCs:** Most ADCs are of a type known as linear. The term *linear* as used here means that the range of the input values that map to each output value has a linear relationship with the output value, *i.e.*, that the output value k is used for the range of input values from $m(k+b)$ to $m(k+1+b)$,

Where m and b are constants. Here b is typically 0 or –0.5. When $b = 0$, the ADC is referred to as *mid-rise*, and when $b = -0.5$ it is referred to as *mid-tread*.

(*b*) **Non-linear ADCs:** If the probability density function of a signal being digitized is uniform, then the signal-to-noise ratio relative to the quantization noise is the best possible. Because this is often not the case, it is usual to pass the signal through its cumulative distribution function (CDF) before the quantization. This is good because the regions that are more important get quantized with a better resolution. In the dequantization process, the inverse CDF is needed.

This is the same principle behind the companders used in some tape-recorders and other communication systems, and is related to entropy maximization.

For example, a voice signal has a Laplacian distribution. This means that the region around the lowest levels, near 0, carries more information than the regions with higher amplitudes. Because of this, logarithmic ADCs are very common in voice communication systems to increase the dynamic range of the representable values while retaining fine-granular fidelity in the low-amplitude region.

An eight-bit A-law or the μ-law logarithmic ADC covers the wide dynamic range and has a high resolution in the critical low-amplitude region, that would otherwise require a 12-bit linear ADC.

12.3.1.3 Accuracy

An ADC has several sources of errors. Quantization error and (assuming the ADC is intended to be linear) non-linearity is intrinsic to any analog-to-digital conversion. There is also a so-called *aperture error* which is due to a clock jitter and is revealed when digitizing a time-variant signal (not a constant value).

These errors are measured in a unit called the *LSB*, which is an abbreviation for least significant bit. In the above example of an eight-bit ADC, an error of one LSB is 1/256 of the full signal range, or about 0.4%.

(*a*) **Quantization error:** Quantization error is due to the finite resolution of the ADC, and is an unavoidable imperfection in all types of ADC. The magnitude of the quantization error at the sampling instant is between zero and half of one LSB.
In the general case, the original signal is much larger than one LSB. When this happens, the quantization error is not correlated with the signal, and has a uniform distribution. Its RMS value is the standard deviation of this distribution, given by $\frac{1}{\sqrt{12}}$ LSB $\approx$ 0.289 LSB. In the eight-bit ADC example, this represents 0.113% of the full signal range.
At lower levels the quantizing error becomes dependent of the input signal, resulting in distortion. This distortion is created after the anti-aliasing filter, and if these distortions are above 1/2 the sample rate they will alias back into the audio band. In order to make the quantizing error independent of the input signal, noise with amplitude of 1 quantization step is added to the signal. This slightly reduces signal to noise ratio, but completely eliminates the distortion. It is known as dither.

(*b*) **Non-linearity:** All ADCs suffer from non-linearity errors caused by their physical imperfections, causing their output to deviate from a linear function (or some other function, in the case of a deliberately non-linear ADC) of their input. These errors can sometimes be mitigated by calibration, or prevented by testing.
Important parameters for linearity are integral non-linearity (INL) and differential non-linearity (DNL). These non-linearities reduce the dynamic range of the signals that can be digitized by the ADC, also reducing the effective resolution of the ADC.

(*c*) **Aperture error:** Imagine that we are digitizing a sine wave $x(t) = A \sin(2\pi f_0 t)$. Provided that the actual sampling time *uncertainty* due to the *clock jitter* is Δt, the error caused by this phenomenon can be estimated as $E_{ap} \le |\, x'(t)\, \Delta t \,| \le 2A\pi f_0 \Delta t$.

The error is zero for DC, small at low frequencies, but significant when high frequencies have high amplitudes. This effect can be ignored if it is drowned out by the *quantizing error*. Jitter requirements can be calculated using the following formula: $\Delta t < \frac{1}{2^q \pi f_0}$, where q is a number of ADC bits.

ADC resolution in bit input frequency

	1 Hz	44.1 kHz	192 kHz	1 MHz	10 MHz	100 MHz	1 GHz
8	1243 µs	28.2 ns	6.48 ns	1.24 ns	124 ps	12.4 ps	1.24 ps
10	311 µs	7.05 ns	1.62 ns	311 ps	31.1 ps	3.11 ps	0.31 ps
12	77.7 µs	1.76 ns	405 ps	77.7 ps	7.77 ps	0.78 ps	0.08 ps
14	19.4 µs	441 ps	101 ps	19.4 ps	1.94 ps	0.19 ps	0.02 ps
16	4.86 µs	110 ps	25.3 ps	4.86 ps	0.49 ps	0.05 ps	-
18	1.21 µs	27.5 ps	6.32 ps	1.21 ps	0.12 ps	-	-
20	304 ns	6.88 ps	1.58 ps	0.16 ps	-	-	-
24	19.0 ns	0.43 ps	0.10 ps	-	-	-	-
32	74.1 ps	-	-	-	-	-	-

This table shows, for example, that it is not worth using a precise 24-bit ADC for sound recording if there is not an *ultra low jitter* clock. One should consider taking this phenomenon into account before choosing an ADC.

Clock jitter is caused by phase noise. The resolution of ADCs with a digitization bandwidth between 1 MHz and 1 GHz is limited by jitter.

When sampling audio signals at 44.1 kHz, the anti-aliasing filter should have eliminated all frequencies above 22 kHz. The input frequency (in this case, 22 kHz), not the ADC clock frequency, is the determining factor with respect to jitter performance.

12.3.1.4 Sampling Rate

The analog signal is continuous in time and it is necessary to convert this to a flow of digital values. It is therefore required to define the rate at which new digital values are sampled from the analog signal. The rate of new values is called the *sampling rate* or *sampling frequency* of the converter.

A continuously varying bandlimited signal can be sampled (that is, the signal values at intervals of time T, the sampling time, are measured and stored) and then the original signal can be *exactly* reproduced from the discrete-time values by an interpolation formula. The accuracy is limited by quantization error. However, this faithful reproduction is only possible if the sampling rate is higher than twice the highest frequency of the signal. This is essentially what is embodied in the Shannon-Nyquist sampling theorem.

Since a practical ADC cannot make an instantaneous conversion, the input value must necessarily be held constant during the time that the converter performs a conversion (called the *conversion time*). An input circuit called a sample and hold performs this task-in most cases by using a capacitor to store the analog voltage at the input, and using an electronic switch or gate to disconnect the capacitor from the input. Many ADC integrated circuits include the sample and hold subsystem internally.

12.3.1.5 Aliasing

All ADCs work by sampling their input at discrete intervals of time. Their output is therefore an incomplete picture of the behavior of the input. There is no way of knowing, by looking at the output, what the input was doing between one sampling instant and the next. If the input is known to be changing slowly compared to the sampling rate, then it can be assumed that the value of the signal between two sample instants was somewhere between the two sampled values. If, however, the input signal is changing rapidly compared to the sample rate, then this assumption is not valid.

If the digital values produced by the ADC are, at some later stage in the system, converted back to analog values by a digital to analog converter or DAC, it is desirable that the output of the DAC be a faithful representation of the original signal. If the input signal is changing much faster than the sample rate, then this will not be the case, and spurious signals called *aliases* will be produced at the output of the DAC. The frequency of the aliased signal is the difference between the signal frequency and the sampling rate. For example, a 2 kHz sine wave being sampled at 1.5 kHz would be reconstructed as a 500 Hz sine wave. This problem is called *aliasing*.

To avoid aliasing, the input to an ADC must be low-pass filtered to remove frequencies above half the sampling rate. This filter is called an *anti-aliasing* filter, and is essential for a practical ADC system that is applied to analog signals with higher frequency content.

Although aliasing in most systems is unwanted, it should also be noted that it can be exploited to provide simultaneous down-mixing of a band-limited high frequency signal.

12.3.1.6 Dither

In A to D converters, performance can usually be improved using dither. This is a very small amount of random noise (white noise) which is added to the input before conversion. Its amplitude is set to be about half of the least significant bit. Its effect is to cause the state of the LSB to randomly oscillate between 0 and 1 in the presence of very low levels of input, rather than sticking at a fixed value. Rather than the signal simply getting cut off altogether at this low level (which is only being quantized to a resolution of 1 bit), it extends the effective range of signals that the A to D converter can convert, at the expense of a slight increase in noise – effectively the quantization error is diffused across a series of noise values which is far less objectionable than a hard cutoff. The result is an accurate representation of the signal over time. A suitable filter at the output of the system can thus recover this small signal variation.

An audio signal of very low level (with respect to the bit depth of the ADC) sampled without dither sounds extremely distorted and unpleasant. Without dither the low level always yields a '1' from the A to D. With dithering, the true level of the audio is still recorded as a series of values over time, rather than a series of separate bits at one instant in time.

A virtually identical process, also called dither or dithering, is often used when quantizing photographic images to a fewer number of bits per pixel-the image becomes noisier but to the eye looks far more realistic than the quantized image, which otherwise becomes banded. This analogous process may help to visualize the effect of dither on an analogue audio signal that is converted to digital.

Dithering is also used in integrating systems such as electricity meters. Since the values are added together, the dithering produces results that are more exact than the LSB of the analog-to-digital converter.

Note that dither can only increase the resolution of a sampler, it cannot improve the linearity, and thus accuracy does not necessarily improve.

12.3.1.7 Oversampling

Usually, signals are sampled at the minimum rate required, for economy, with the result that the quantization noise introduced is white noise spread over the whole pass band of the converter. If a signal is sampled at a rate much higher than the Nyquist frequency and then digitally filtered to limit it to the signal bandwidth then there are 3 main advantages:

- digital filters can have better properties (sharper rolloff, phase) than analogue filters, so a sharper anti-aliasing filter can be realized and then the signal can be down sampled giving a better result
- a 20 bit ADC can be made to act as a 24 bit ADC with 256x oversampling

- The signal-to-noise ratio due to quantization noise will be higher than if the whole available band had been used. With this technique, it is possible to obtain an effective resolution larger than that provided by the converter alone
- The improvement in SNR is 3dB (equivalent to 0.5 bits) per octave of oversampling which is not sufficient for many applications. So, oversampling is usually coupled with noise shaping (see /sigma-/delta modulators). With noise shaping, the improvement is 6L+3 dB per octave where L is the order of loop filter used for noise shaping. *e.g.,* - a 2nd order loop filter will provide an improvement of 15dB/octave.

12.3.1.8 Relative Speed and Precision

The speed of an ADC varies by type. The Wilkinson ADC is limited by the clock rate which is processable by current digital circuits. Currently, frequencies up to 300 MHz are possible. The conversion time is directly proportional to the number of channels. For a successive approximation ADC, the conversion time scales with the logarithm of the number of channels. Thus for a large number of channels, it is possible that the successive approximation ADC is faster than the Wilkinson. However, the time consuming steps in the Wilkinson are digital, while those in the successive approximation are analog. Since analog is inherently slower than digital, as the number of channels increases, the time required also increases. Thus there are competing processes at work. Flash ADCs are certainly the fastest type of the three. The conversion is basically performed in a single parallel step. For an 8-bit unit, conversion takes place in a few tens of nanoseconds.

There is, as expected, somewhat of a tradeoff between speed and precision. Flash ADCs have drifts and uncertainties associated with the comparator levels, which lead to poor uniformity in channel width. Flash ADCs have a resulting poor linearity. For successive approximation ADCs, poor linearity is also apparent, but less so than for flash ADCs. Here, non-linearity arises from accumulating errors from the subtraction processes. Wilkinson ADCs are the best of the three. These have the best differential non-linearity. The other types require channel smoothing in order to achieve the level of the Wilkinson.

12.3.1.9 The Sliding Scale Principle

The sliding scale or randomizing method can be employed to greatly improve the channel width uniformity and differential linearity of any type of ADC, but especially flash and successive approximation ADCs. Under normal conditions, a pulse of a particular amplitude is always converted to a certain channel number. The problem lies in that channels are not always of uniform width, and the differential linearity decreases proportionally with the divergence from the average width. The sliding scale principle uses an averaging effect to overcome this phenomenon. A random, but known analog voltage is added to the input pulse. It is then converted to digital form, and the equivalent digital version is subtracted, thus restoring it to its original value. The advantage is that the conversion has taken place at a random point. The statistical distribution of the final channel numbers is decided by a weighted average over a region of the range of the ADC. This in turn desensitizes it to the width of any given channel.

12.3.2 ADC Structures

These are the most common ways of implementing an electronic ADC:

- A **direct conversion ADC** or **flash ADC** has a bank of comparators sampling the input signal in parallel, each firing for their decoded voltage range. The comparator bank feeds a logic circuit that generates a code for each voltage range. Direct conversion is very fast, capable of gigahertz sampling rates, but usually has only 8 bits of resolution or fewer, since the number of comparators needed, $2^N - 1$, doubles with each additional bit, requiring a large expensive circuit. ADCs of this type have a large die size, a high input capacitance, high power dissipation, and are prone to produce glitches on the output (by outputting an out-of-sequence code). Scaling to newer submicrometre technologies does not help as the device mismatch is the dominant design limitation. They are often used for video, wideband communications or other fast signals in optical storage.

- A **successive-approximation ADC** uses a comparator to reject ranges of voltages, eventually settling on a final voltage range. Successive approximation works by constantly comparing the input voltage to the output of an internal digital to analog converter (DAC, fed by the current value of the approximation) until the best approximation is achieved. At each step in this process, a binary value of the approximation is stored in a successive approximation register (SAR). The SAR uses a reference voltage (which is the largest signal the ADC is to convert) for comparisons. For example if the input voltage is 60 V and the reference voltage is 100 V, in the 1st clock cycle, 60 V is compared to 50 V (the reference, divided by two. This is the voltage at the output of the internal DAC when the input is a '1' followed by zeros), and the voltage from the comparator is positive (or '1') (because 60 V is greater than 50 V). At this point the first binary digit (MSB) is set to a '1'. In the 2nd clock cycle the input voltage is compared to 75 V (being halfway between 100 and 50 V: This is the output of the internal DAC when its input is '11' followed by zeros) because 60 V is less than 75 V, the comparator output is now negative (or '0'). The second binary digit is therefore set to a '0'. In the 3rd clock cycle, the input voltage is compared with 62.5 V (halfway between 50 V and 75 V: This is the output of the internal DAC when its input is '101' followed by zeros). The output of the comparator is negative or '0' (because 60 V is less than 62.5 V) so the third binary digit is set to a 0. The fourth clock cycle similarly results in the fourth digit being a '1' (60 V is greater than 56.25 V, the DAC output for '1001' followed by zeros). The result of this would be in the binary form 1001. This is also called *bit-weighting conversion*, and is similar to a binary search. The analogue value is rounded to the nearest binary value below, meaning this converter type is mid-rise (see above). Because the approximations are successive (not simultaneous), the conversion takes one clock-cycle for each bit of resolution desired. The clock frequency must be equal to the sampling frequency multiplied by the number of bits of resolution desired. For example, to sample audio at 44.1 kHz with 32 bit resolution, a clock frequency of over 1.4 MHz would be required. ADCs of this type have good resolutions and quite wide ranges. They are more complex than some other designs.
- A **ramp-compare ADC** produces a saw-tooth signal that ramps up or down then quickly returns to zero. When the ramp starts, a timer starts counting. When the ramp voltage matches the input, a comparator fires, and the timer's value is recorded. Timed ramp converters require the least number of transistors. The ramp time is sensitive to temperature because the circuit generating the ramp is often just some simple oscillator. There are two solutions: use a clocked counter driving a DAC and then use the comparator to preserve the counter's value, or calibrate the timed ramp. A special advantage of the ramp-compare system is that comparing a second signal just requires another comparator, and another register to store the voltage value. A very simple (non-linear) ramp-converter can be implemented with a microcontroller and one resistor and capacitor [6]. Vice versa a filled capacitor can be taken from an integrator, time-to-amplitude converter, phase detector, sample and hold circuit, or peak and hold circuit and discharged. This has the advantage that a slow comparator cannot be disturbed by fast input changes.

- The **Wilkinson ADC** was designed by D. H. Wilkinson in 1950. The Wilkinson ADC is based on the comparison of an input voltage with that produced by a charging capacitor. The capacitor is allowed to charge until its voltage is equal to the amplitude of the input pulse. (A comparator determines when this condition has been reached.) Then, the capacitor is allowed to discharge linearly, which produces a ramp voltage. At the point when the capacitor begins to discharge, a gate pulse is initiated. The gate pulse remains on until the capacitor is completely discharged. Thus the duration of the gate pulse is directly proportional to the amplitude of the input pulse. This gate pulse operates a linear gate which receives pulses from a high-frequency oscillator clock. While the gate is open, a discrete number of clock pulses pass through the linear gate and are counted by the address register. The time the linear gate is open is proportional to the amplitude of the input pulse, thus the number of clock pulses recorded in the address register is proportional also. Alternatively, the charging of the capacitor could be monitored, rather than the discharge. [7] [8]
- An **integrating ADC** (also **dual-slope** or **multi-slope** ADC) applies the unknown input voltage to the input of an integrator and allows the voltage to ramp for a fixed time period (the run-up period). Then a known reference voltage of opposite polarity is applied to the integrator and is allowed to ramp until the integrator output returns to zero (the run-down period). The input voltage is computed as a function of the reference voltage, the constant run-up time period, and the measured run-down time period. The run-down time measurement is usually made in units of the converter's clock, so longer integration times allow for higher resolutions. Likewise, the speed of the converter can be improved by sacrificing resolution. Converters of this type (or variations on the concept) are used in most digital voltmeters for their linearity and flexibility.
- A **delta-encoded ADC** or Counter-ramp has an up-down counter that feeds a digital to analog converter (DAC). The input signal and the DAC both go to a comparator. The comparator controls the counter. The circuit uses negative feedback from the comparator to adjust the counter until the DAC's output is close enough to the input signal. The number is read from the counter. Delta converters have very wide ranges, and high resolution, but the conversion time is dependent on the input signal level, though it will always have a guaranteed worst-case. Delta converters are often very good choices to read real-world signals. Most signals from physical systems do not change abruptly. Some converters combine the delta and successive approximation approaches; this works especially well when high frequencies are known to be small in magnitude.
- A **pipeline ADC** (also called **subranging quantizer**) uses two or more steps of subranging. First, a coarse conversion is done. In a second step, the difference to the input signal is determined with a digital to analog converter (DAC). This difference is then converted finer, and the results are combined in a last step. This can be considered a refinement of the successive approximation ADC wherein the feedback reference signal consists of the interim conversion of a whole range of bits (for example, four bits) rather than just the next-most-significant bit. By combining the merits of the successive approximation and flash ADCs this type is fast, has a high resolution, and only requires a small die size.
- A **Sigma-Delta ADC** (also known as a Delta-Sigma ADC) oversamples the desired signal by a large factor and filters the desired signal band. Generally a smaller number of bits than required are converted using a Flash ADC after the Filter. The resulting signal, along with the error generated by the discrete levels of the Flash, is fed back and subtracted from the input

to the filter. This negative feedback has the effect of noise shaping the error due to the Flash so that it does not appear in the desired signal frequencies. A digital filter (decimation filter) follows the ADC which reduces the sampling rate, filters off unwanted noise signal and increases the resolution of the output. (sigma-delta modulation, also called delta-sigma modulation)

- A **Time-interleaved ADC** uses M parallel ADCs where each ADC sample data every M:th cycle of the effective sample clock. This result in that the sample rate is increased M times compared to what each individual ADC can manage. In practice the individual differences between the M ADCs degrade the overall performance reducing the SFDR. However, technologies exist to correct for these time-interleaving mismatch errors.
- An **ADC with intermediate FM stage** first uses a voltage-to-frequency converter to converts the desired signal into an oscillating signal with a frequency proportional to the voltage of the desired signal, and then uses a frequency counter to convert that frequency into a digital count proportional to the desired signal voltage. Longer integration times allow for higher resolutions. Likewise, the speed of the converter can be improved by sacrificing resolution. The two parts of the ADC may be widely separated, with the frequency signal passed through a opto-isolator or transmitted wirelessly. Some such ADCs use sine wave or square wave frequency modulation; others use pulse-frequency modulation. Such ADCs were once the most popular way to show a digital display of the status of a remote analog sensor.

There can be other ADCs that use a combination of electronics and other technologies:

- A **Time-stretch analog-to-digital converter (TS-ADC)** digitizes a very wide bandwidth analog signal, that cannot be digitized by a conventional electronic ADC, by time-stretching the signal prior to digitization. It commonly uses a photonic preprocessor frontend to time-stretch the signal, which effectively slows the signal down in time and compresses its bandwidth. As a result, an electronic backend ADC, that would have been too slow to capture the original signal, can now capture this slowed down signal. For continuous capture of the signal, the frontend also divides the signal into multiple segments in addition to time-stretching. Each segment is individually digitized by a separate electronic ADC. Finally, a digital signal processor rearranges the samples and removes any distortions added by the frontend to yield the binary data that is the digital representation of the original analog signal.

12.3.3 Commercial Analog-to-digital Converters

These are usually integrated circuits. Most converters sample with 6 to 24 bits of resolution, and produce less than 1 mega sample per second. Thermal noise generated by passive components such as resistors masks the measurement when higher resolution is desired. For audio applications and in room temperatures, such noise is usually a little less than 1 μV (microvolt) of white noise. If the Most Significant Bit corresponds to a standard 2 volts of output signal, this translates to a noise-limited performance that is less than 20~21 bits, and obviates the need for any dithering. Mega- and gigasample per second converters are available, though (Feb 2002). Mega sample converters are required in digital video cameras, video capture cards, and TV tuner cards to convert full-speed analog video to digital video files. Commercial converters usually have ±0.5 to ±1.5 LSB error in their output. In many cases the most expensive part of an integrated circuit is the pins, because they make the package larger, and each pin has to be connected to the integrated circuits silicon. To save pins, it's common for slow ADCs to send their data one bit at a time over a serial interface to the computer, with the next bit coming out when a clock signal changes state, say from zero to 5V. This saves quite a few pins on the ADC package, and in many cases, does not make the overall design any more complexes. (Even microprocessors which use memory-mapped I/O only need a few bits of a port to implement a serial bus to an ADC.)

Commercial ADCs often have several inputs that feed the same converter, usually through an analog multiplexer. Different models of ADC may include sample and hold circuits, instrumentation amplifiers or differential inputs, where the quantity measured is the difference between two voltages.

12.3.4 Applications

The following are the applications of ADC:

(*a*) **Application to music recording:** ADCs are integral to current music reproduction technology. Since much music production is done on computers, when an analog recording is used, an ADC is needed to create the PCM data stream that goes onto a compact disc or digital music file.

The current crop of AD converters utilized in music can sample at rates up to 192 kilohertz. High bandwidth headroom allows the use of cheaper or faster anti-aliasing filters of less severe filtering slopes. The proponents of oversampling assert that such shallower anti-aliasing filters produce less deleterious effects on sound quality, exactly because of their gentler slopes. Others prefer entirely filterless AD conversion, arguing that aliasing is less detrimental to sound perception than pre-conversion brickwall filtering. Considerable literature exists on these matters, but commercial considerations often play a significant role. Most high-profile recording studios record in 24-bit/192-176.4 kHz PCM or in DSD formats, and then down sample or decimate the signal for Red-Book CD production (44.1 kHz or at 48 kHz for commonly used for radio/TV broadcast applications).

(*b*) **Digital signal processing:** AD converters are used virtually everywhere where an analog signal has to be processed, stored, or transported in digital form. Fast video ADCs are used, for example, in TV tuner cards. Slow on-chip 8, 10, 12, or 16 bit ADCs are common in microcontrollers. Very fast ADCs are needed in digital oscilloscopes, and are crucial for new applications like software defined radio.

12.3.5 Electrical Symbol

The following is the electrical symbol of ADC:

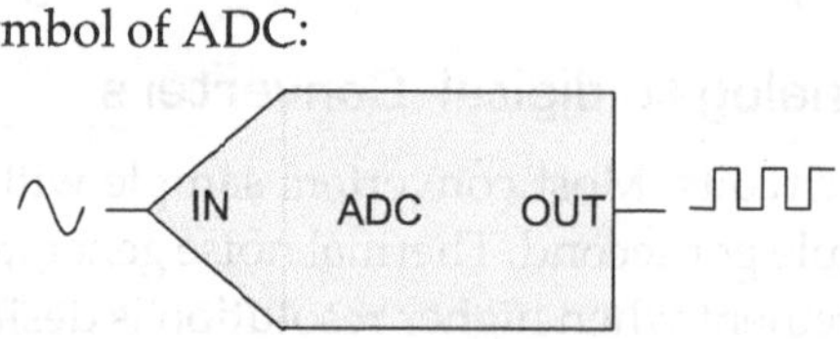

ELECTRICAL SYMBOL FOR ANALOG TO DIGITAL CONVERTER [ADC]

12.4 ADC IC 0808/0809

The ADC0808, ADC0809 data acquisition component is a monolithic CMOS device with an 8-bit analog-to-digital converter, 8-channel multiplexer and microprocessor compatible control logic. The 8-bit A/D converter uses successive approximation as the conversion technique. The converter features a high impedance chopper stabilized comparator, a 256R voltage divider with analog switch tree and a successive approximation register. The 8-channel multiplexer can directly access any of 8-single-ended analog signals. The device eliminates the need for external zero and full-scale adjustments. Easy interfacing to microprocessors is provided by the latched and decoded multiplexer address inputs and latched TTL TRI-STATE outputs. The design of the ADC0808, ADC0809 has been optimized by incorporating the most desirable aspects of several A/D conversion techniques. The ADC0808, ADC0809 offers high speed, high accuracy, minimal temperature dependence, excellent long-term accuracy and repeatability, and consumes minimal power. These features make this device ideally suited to applications from process and machine control to consumer and automotive applications. For 16-channel multiplexer with common output (sample/hold port) see ADC0816 data sheet.

12.4.1 Block Diagram

The following is the block diagram of ADC IC 0808/0809

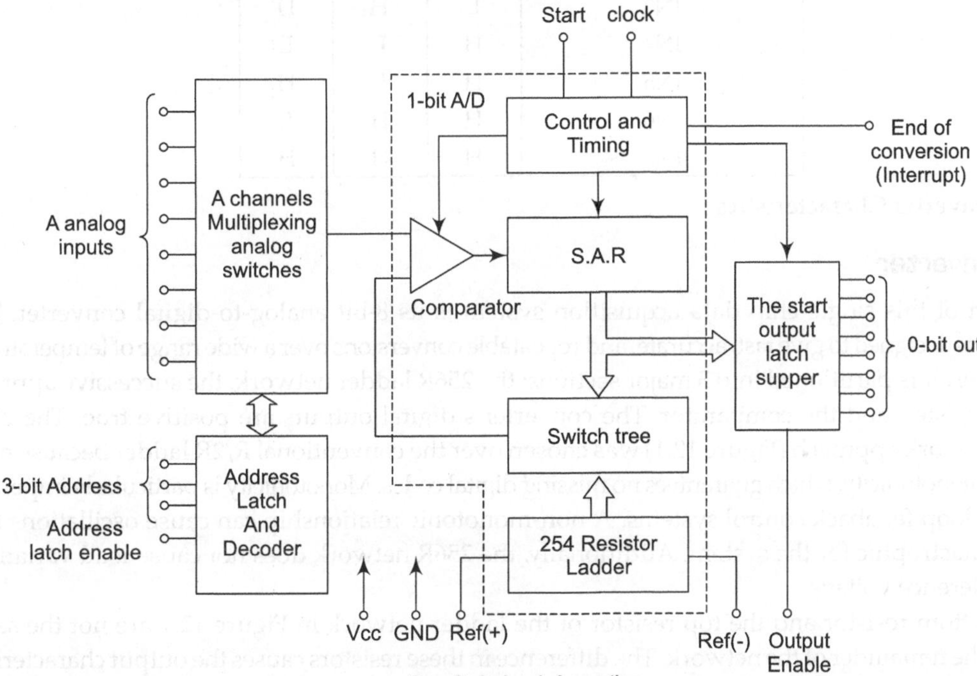

See Ordering Information

12.4.2 Features

The following are the features:

- Easy interface to all microprocessors
- Operates ratio metrically or with 5 VDC or analog span adjusted voltage reference
- No zero or full-scale adjust required
- 8-channel multiplexer with address logic
- 0V to VCC input range
- Outputs meet TTL voltage level specifications
- ADC0808 equivalent to MM74C949
- ADC0809 equivalent to MM74C949-1

12.4.3 Functional Description

(*a*) **Multiplexer:** The device contains an 8-channel single-ended analog signal multiplexer. A particular input channel is selected by using the address decoder. The following *Table shows* the input states for the address lines to select any channel. The address is latched into the decoder on the low-to-high transition of the address latch enable signal.

Table 12.1 Analog Channel Selection

Selected Analog Channel	Address Line		
	C	B	A
IN0	L	L	L
IN1	L	L	H
IN2	L	H	L
IN3	L	H	H
IN4	H	L	L
IN5	H	L	H
IN6	H	H	L
IN7	H	H	H

(*b*) **Converter Characteristics :**

The Converter

The heart of this single chip data acquisition system is its 8-bit analog-to-digital converter. The converter is designed to give fast, accurate, and repeatable conversions over a wide range of temperatures. The converter is partitioned into 3 major sections: the 256R ladder network, the successive approximation register, and the comparator. The converter's digital outputs are positive true. The 256R ladder network approach (Figure 12.1) was chosen over the conventional R/2R ladder because of its inherent monotonicity, which guarantees no missing digital codes. Monotonicity is particularly important in closed loop feedback control systems. A non-monotonic relationship can cause oscillations that will be catastrophic for the system. Additionally, the 256R network does not cause load variations on the reference voltage.

The bottom resistor and the top resistor of the ladder network in Figure 12.1 are not the same value as the remainder of the network. The difference in these resistors causes the output characteristic to be symmetrical with the zero and full-scale points of the transfer curve. The first output transition occurs when the analog signal has reached +½ LSB and succeeding output transitions occur every 1 LSB later up to full-scale. The successive approximation register (SAR) performs 8 iterations to approximate the input voltage. For any SAR type converter, n-iterations are required for an n-bit converter. Figure 12.2 shows a typical example of a 3-bit converter. In the ADC0808, ADC0809, the approximation technique is extended to 8 bits using the 256R network.

The A/D converter's successive approximation register (SAR) is reset on the positive edge of the start conversion start pulse. The conversion is begun on the falling edge of the start conversion pulse. A conversion in process will be interrupted by receipt of a new start conversion pulse. Continuous conversion may be accomplished by tying the end-of-conversion (EOC) output to the SC input. If used in this mode, an external start conversion pulse should be applied after power up. Endof- conversion will go low between 0 and 8 clock pulses after the rising edge of start conversion.

The most important section of the A/D converter is the comparator. It is this section which is responsible for the ultimate accuracy of the entire converter. It is also the comparator drift which has the greatest influence on the repeatability of the device. A chopper-stabilized comparator provides the most effective method of satisfying all the converter requirements. The chopper-stabilized comparator converts the DC input signal into an AC signal. This signal is then fed through a high gain AC amplifier and has the DC level restored. This technique limits the drift component of the amplifier since the drift is a DC component which is not passed by the AC amplifier. This makes the entire A/D converter extremely insensitive to temperature, long term drift and input offset errors. Figure 12.4 shows a typical error curve for the ADC0808 as measured using the procedures outlined in AN-179.

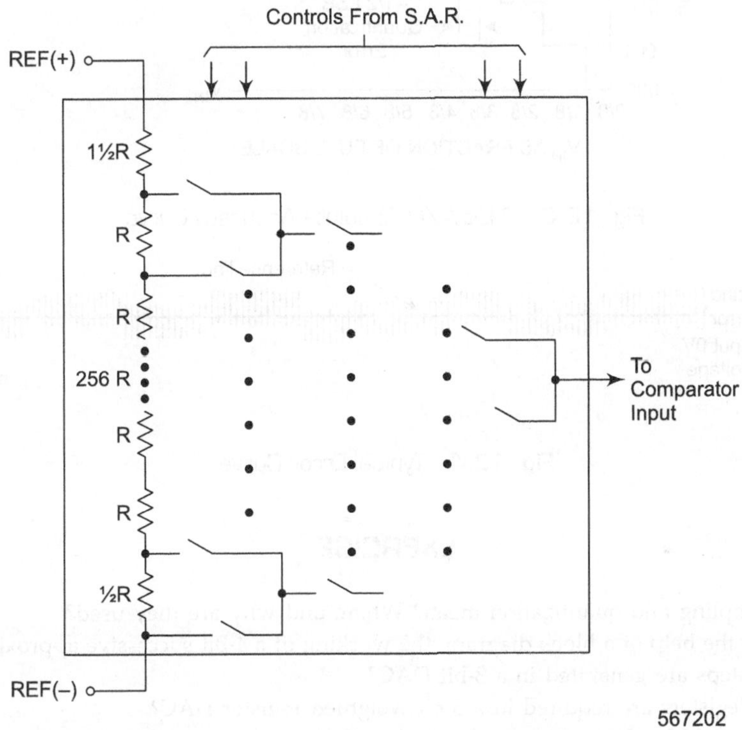

Fig. 12.1 Resistor Ladder and Switch Tree

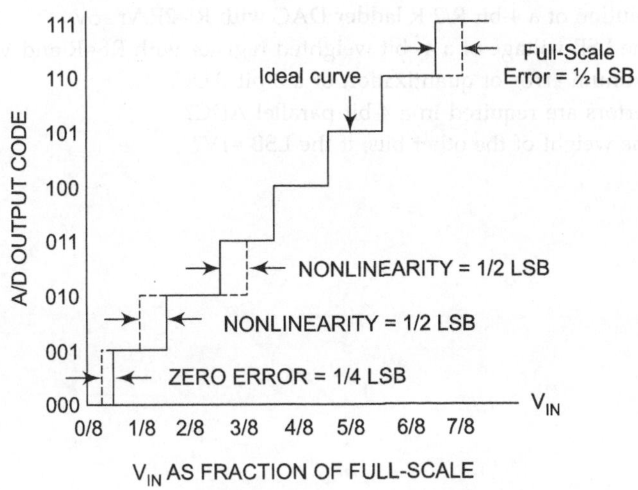

Fig. 12.2 3-Bit A/D Transfer Curve

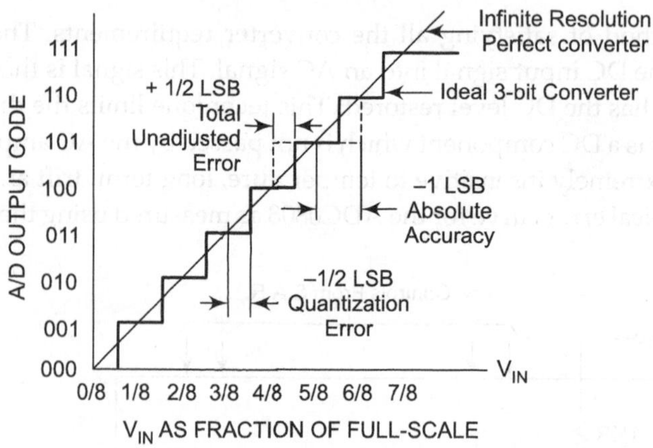

Fig. 12.3 3-Bit A/D Absolute Accuracy Curve

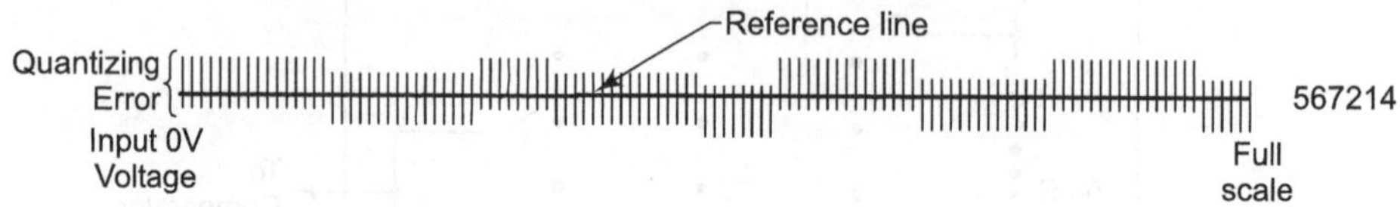

Fig. 12.4 Typical Error Curve

EXERCISE

1. What do sampling and quantization mean? Where and why are they used?
2. Explain with the help of a block diagram, the working of a 4-bit successive approximation ADC.
3. How many steps are generated in a 3-bit DAC?
4. How many registers are required in a 3-bit weighted register DAC?
5. How much would be the conversion time of an 8-bit successive approximation ADC with 1 MHz Clock?
6. Calculate the resolution of a 4-bit R/2 R ladder DAC with Rf=2R Vr =5v.
7. What would be the LSB voltage of a 4- bit weighted register with Rf =R and Vr=5v?
8. Calculate the maximum error of quantization of a 3-bit ADC.
9. How many convertors are required in a 4-bit parallel ADC?
10. What would be the weight of the other bits, if the LSB =1V?

13

ADVANCED MICROPROCESSORS

In this chapter, we will discuss the advanced microprocessor just like 80286, 80386, 80486 and Pentium family.

This chapter provides detailed technical information about all the major processor families used on PC-compatible computers. It covers every major x86 processor on the market, from the first Intel 8088 used in the original IBM PC, to the latest released hot chips. The processors here are grouped by "families", where I consider a family of processors to be a group of processors that vary only in clock speed, not in architecture (with a couple of possible minor exceptions to keep the section from getting too large).

I organize this section by processor "generation". Which generation a processor belongs to used to be clear when it was just Intel making processors and they were coming out at a slow rate and with very obvious performance differences. Now the situation is much cloudier, with many more chips on the market and many types of technology being used to implement them. I have made the decision of which processor to include in which generation based on technology and performance considerations, along with my interpretations of how the market perceives the chip. Pentium Overdrives use fifth generation technology so they are listed as fifth-generation chips.

Most of the information about each processor is presented in tabular format for easy reference. The rows of these tables summarize the statistics for each processor, and correlate to the other sections in the processor reference on this site where the terms and characteristics are explained in detail. The first section contains explanations for each of the rows in the tables, and links to where each of the characteristics is described in more detail.

Each section contains a brief discussion of the processor and its more interesting points. Included is an illustration of the processor's improvements and changes from the previous generation processor it most directly replaced.

Note: The information on the newer processors, especially fifth generation and later, is more likely to be complete and accurate than that for the earlier CPUs. This is due to a combination of factors, including my greater experience with the newer processors, more information generally being available on them, and my perception of there being much more interest in the newer chips.

Note: There are many variants and licensed versions of different processors on the market. Due to time constraints I do not have all or even most of these covered in this section; they were primarily on older processors in which there is currently little interest.

13.1 INTEL 8088

The first chip used in PCs was Intel's 8088. This was not, at the time it was chosen, the best available CPU, in fact Intel's own 8086 was more powerful and had been released earlier. The 8088 was chosen for reasons of economics: its 8-bit data bus required less costly motherboards than the 16-bit 8086. Also, at the time that the original PC was designed, most of the interface chips available were intended for use in 8-bit designs. It's ironic, isn't it, that Intel's first production chip was in a way, the "8086SX"? :^) It originally shipped at 4.77 MHz and a "turbo" version was later produced that ran at 8 MHz's.

This original chip used what would be considered today to be archaic technology. The 8088 offers performance less than one-thousandth that of a modern processor, showing just how far we have come in 15 years.

Note: The NEC V20 is an 8088-compatible but actually delivers slightly better performance due to a slightly more efficient internal design.

Note: The original design of this chip was in NMOS, with newer ones in CMOS (80C88) for lower power consumption.

13.2 INTEL 8086

The Intel 8086 chip is almost identical to the 8088, except that it uses a 16-bit-wide data bus instead of the 8-bit bus of the 8088. It was actually invented before the 8088 but was not chosen for the original PC because IBM wanted to keep costs down by going with an 8-bit bus design.

The 8086 was used in some PC clones and later on in early IBM PS/2 models, but was never a popular choice for PCs. This is probably because by the time it started to be used, the 80286 was introduced, offering much improved performance over the 8086. The 8086 is architecturally identical to the 8088 aside from the wider data bus.

Note: The NEC V30 is an 8086-compatible but actually delivers slightly better performance due to a slightly more efficient internal design.

Note: Original designs were made in NMOS, newer ones in CMOS (80C86) for lower power consumption.

13.3 INTEL 80286

The 80286 was the chip used in IBM's AT (advanced technology) system. The 286 was the first major step up in PC processors, providing significant performance increases over the 8088 and 8086—double or more performance at the same clock speed. The 286 also widened the address bus to allow access to 16 MB of memory, and introduced protected mode operation. It was originally available in 6 MHz and 8 MHz versions, but was later expanded to faster versions, all the way up to 20 MHz.

The 286 opened up the PC world to many users, but still was used mainly as the equivalent of a "turbo-charged 8088". At this time DOS was still the virtually exclusive operating system, and the protected mode the 286 offered was largely ignored.

13.4 INTEL 80386DX

The Intel 80386DX was the first true 32-bit processor used on the PC platform. Its internal register size was increased to 32 bits, and its data and address buses were as well, doubling data path width to the processor and increasing addressable memory to 4 GB theoretical. The 80386 family of chips offered more performance than the 80286s they replaced, largely through processor speed increases. The 386 did not offer the very large improvements over the 286 that the 286 did over the 8088.

The 386DX's increased power and the improved processor modes it offered (including full protected mode and virtual real modes) spurred the introduction of GUI-based operating systems on the PC, such as Microsoft Windows (although they are still quite slow on 386 chips). The instruction set of the 386 has set the standard for what is now called "x86" and hasn't changed very much since it was introduced. The 80386DX was the first to use pipelining to allow much improved processor performance through the use of much higher clock frequencies.

Invented by Intel, the 386 was also "cloned" by AMD and Cyrix. These are very good copies with no real compatibility problems. Intel only produced the 386DX up to 33 MHz (presumably to prevent overlapping into the 486's performance range) while AMD and Cyrix produced a 40 MHz version as well. This latter chip produced very good performance (for the time), comparable to many lower-end 486s. They are still however obsolete by today's standards, of course.

13.5 INTEL 80386SX

The 80386SX is a "lite" version of the 80386DX chip. It uses only a 16-bit data bus, the same external bus width as the 80286. It also only can address 16 MB of memory, like the 80286. The SX version of the 386 chip was in fact released well after the DX, and in some ways was intended to move the market away from the 286 since it had roughly the same interfaces but better performance. It was actually introduced several *years* after the DX.

Note: It is a common myth that the 80386SX can be substituted in place of an 80286 in a 286 motherboard. While the chips are compatible in terms of their external interfaces they use different packaging. A 386SX could be put into a 286 motherboard if a proper adapter were used.

The narrower data bus width of the 386SX creates a reduction in performance of about 20-25% compared to an equivalent-speed 386DX; a significant difference but not excessive. The 80386SX chip was a popular choice for the first small notebook computers, especially the 386SL variant that introduced the SMM power management features to the Intel line. The 386SX is still a 32-bit processor internally and will run 32-bit software (really slowly :^)). It is available in speeds from 16 to 33 MHz; a 40 MHz version was not produced for the SX, and is also available in cloned versions from AMD and Cyrix.

The 386SX chip is today considered obsolete.

13.6 INTEL 80486DX

The Intel 80486DX was the first member of the 486 family (which has many more members than the 386 family did). It provided a very significant increase in power over the 80386DX processor, in fact, far more proportionately than the 386 did over the 286. A 486DX processor provides approximately 100 to 150% more performance than a 386DX of the same clock speed. The 80486 brought GUIs to the mainstream on PCs; it is the minimum processor that most people consider "usable" for running an operating system like Microsoft Windows.

Interestingly, the 486 does not provide its performance improvements by widening any of the buses, as had been the case in the previous two generations: it is still a 32-bit processor with 32-bit data and address buses, just like the 386DX. However, internally, the 486 incorporates several significant improvements over the 386:

- **Faster Overall Instruction Execution time:** The core of the chip can execute instructions in less time than earlier processors.
- **Deeper Pipeline:** The execution pipeline was increased by one step.
- **Primary Cache:** The 486 processor was the first to incorporate level 1 cache on the chip, to reduce the number of required accesses to memory.
- **Integrated Floating Point Unit:** The chip includes an integrated math coprocessor (not on the SX version however). In addition, the coprocessor provides much more performance than the optional 80387 used with 386 chips, in part because it is integrated into the chip.
- **Burst Mode:** The 486 introduced the use of burst mode to reduce wait time on memory accesses.
- **Power Management:** SL power management enhancements as an option (instead of a specialized SL chip as was the case for the 386SL).
- **Improved Support Architecture:** In general, 486 motherboards were more efficient than 386 ones, and began to use secondary cache as well. This improves processor performance significantly.

Interestingly, the 80487SX coprocessor that is intended for use with the SX version of the 80486, is in fact a full-blown *80486DX* processor. This is part of how Intel structured its fourth generation family's upgrade path. See the description of the 80486SX for more details on this.

AMD and Cyrix did produce clones of the original Intel 80486 processors, but they were not a big player in the 486 clone market until the higher-speed DX2 and DX4 processors.

The 486DX is considered obsolete, although the chip still has fairly good power for performing a wide variety of light tasks, such as word processing and some older games, and light Internet access. The 486DX-50, which runs on a 50 MHz system bus, provides performance comparable to the 486DX2-66 in many ways, because the latter uses only a 33 MHz system bus. The 486DX-50 was not used in nearly as many systems as the other processor speeds were. It should not be confused with the 486DX2-50, which runs at the same processor clock speed but is clock-doubled relative to the system bus (which runs at 25 MHz).

Note: The 486DX processor normally was purchased as part of a new system only, not as part of an upgrade. Most early 486 systems used a 168-pin socket for the chip, which predates the numbered standardized socket system that Intel created. The 486DX will fit into a Socket 1, Socket 2 or Socket 3 however.

13.7 INTEL 80486SX

The Intel 80486SX is the same chip as the 80486DX with one exception: the lack of an integrated math coprocessor (floating point unit). Note that this is a different kind of difference between the SX and DX versions than is the case with the 386; the 386SX and 386DX both had no coprocessor and the 386SX had narrower data and address buses than the 386DX. The 486DX and 486SX have the same bus widths. Since it is the same chip except for the floating point processor, the 486SX has the same advantages over the 386 that the 486DX does. Note that the 486SX was made available in slower clock speeds than the 486DX; the SX comes in 16, 20, 25 and 33 MHz versions, while the DX is 25, 33 and 50 MHz.

Intel made some rather interesting (some would say *bizarre*) decisions about how to market the 486 line. The 486SX was in many ways a marketing gimmick only; it was in fact a 486DX with the math coprocessor disabled! Why did Intel bother to sell a full 486DX and a "crippled" one (the 486SX)? Mainly for market targeting reasons: they wanted to be able to fill the niche of low-end buyers without dropping the price on their "top of the line" 486DX. The 486SX was a popular choice for laptops due to its lower cost and lower power consumption.

The 80486SX uses a "math coprocessor" called the 80487SX (there is no 80487DX at all so don't be confused by that). The 80487SX is, in fact, a fully functional 80486*DX* chip! What Intel wanted people to think was that (like with its earlier coprocessors) you would put the 80487SX in and it would handle the math functions. In fact, when inserted, the 80487SX *shuts down* the 80486SX and handles both integer and floating point operations (since it is internally a 80486DX, which does both). This makes no difference from a performance standpoint but is kind of a technical curiosity.

With the large number of very cheap 486DX and faster chips and systems around, the 486SX is really quite obsolete. In addition, it does not come in a 50 MHz version. If you are not doing any floating point operations at all it is comparable to a 486DX of the same speed, but there is usually no need to have to worry about this given that even the 486DX is quite obsolete at this time anyway.

Note: The 486SX processor normally was purchased as part of a new system only, not as part of an upgrade. Most early 486 systems used a 168-pin socket for the chip, which predates the numbered standardized socket system that Intel created. The 486SX will fit into a Socket 1, Socket 2 or Socket 3 however.

13.8 INTEL 80486DX2 AND 80486DX2 OVERDRIVE

The 80486DX2 was the first chip to use "clock doubling" technology, where the processor runs at a faster speed than the memory bus it talks to. This was done to allow the processor speed to be increased without having to deal with the much more difficult task of increasing motherboard speed. Chips that run at faster than memory bus speed improve performance but at a diminishing rate as the multiplier increases, due to the processor waiting for data from memory.

Intel produced 50 and 66 MHz DX2 chips, intended for use in 25 and 33 MHz system bus systems. These chips have been sold as regular chips intended for use in new systems, which generally come in 168 pin packages to go in the original 168 pin socket used in 486 systems. They have also been made in 169 pin OverDrive versions to go in Socket 1 (the original OverDrive socket). These can be used to upgrade older 486DX or 486SX systems.

AMD and Cyrix not only cloned the 66 MHz DX2 processor, they took Intel one step further with the 80486DX2-80, running at 80 MHz. This uses a 40 MHz system bus, which isn't a speed that is normally used by Intel systems but that became more popular late in the 486 life cycle due to the performance increase it gives over 33 MHz bus systems. In addition, the AMD (enhanced version) and Cyrix chips have several advantages over the Intel chips (they had the benefit of developing them well after Intel):

- **Power Management:** They have SMM (power management) built in.
- **Write-Back Cache:** They have write-back capability for the primary cache which provides a small boost in performance.
- **Lower Power:** They use less power because they run at 3 volts, but will still work in standard 5 volt motherboard sockets without a converter or regulator (they are said to be "5 volt tolerant").

Other than clock speed, the 80486DX2 is virtually identical to the 80486DX. They are obsolete due to the availability of faster, very inexpensive processors such as the 5x86-133 that go in the same sockets, but 486DX2 systems are perfectly viable for many uses, including routine office word processing and spreadsheet work under DOS and Windows 3.x. The 80486DX2-66 is by far the most common version of this chip; a great number of these systems were produced and many are still in use today, especially in small businesses.

Note: The 486DX2 was the first processor to really *require* a heat sink in order to operate reliably. The increased speed of this chip means that it runs very hot (at least the Intel 5 volt versions).

13.9 INTEL 80486DX4 AND 80486DX4 OVERDRIVE

The 80486DX4 continued the trend started by the 80486DX2 toward faster clock speed processors on slower motherboards. The DX4 uses "clock tripling", where the processor runs at three times the speed of the memory bus (motherboard). Note that despite the name the DX4 does not run at four times the memory bus speed ("DX3" refers to 2.5 times the memory bus, but never became a shipping product for the 486.)

Intel's DX4 runs at two speeds: 75 MHz (for the 25 MHz bus) and 100 MHz (for the 33 MHz), with the 100 being by far the most popular. In order to keep power and heat to a manageable level the voltage of these chips is reduced to 3.3 volts. The AMD and Cyrix versions are very similar to their DX2 cousins, in that they are 5 volt tolerant, meaning they can handle being put in a 5 volt motherboard; the Intel chip is not 5 volt tolerant and cannot be put into a 5 volt board (well, you can physically put them into the board, but you shouldn't do it! :)). AMD and Cyrix also make a 120 MHz version that runs on a 40 MHz bus; this is relatively rare. The Intel, AMD and Cyrix versions of the DX4 chip are quite similar. One difference is that the AMD and Cyrix support write-back cache while the Intel does not; however the Intel has its level 1 cache doubled to 16 KB. All three chips support power management.

The 486DX4 was most commonly put into systems as OverDrive processors for older, slower systems. Intel did not originally make the DX4 available as a stand-alone chip, presumably since it shipped after the Pentium had already been introduced. However, they did make the chip available stand-alone later on. New 486DX4-100 systems were often made with the AMD version of the chip. The OverDrive version is generally required for older motherboards that don't support 3.3 volt power.

Like the 80486DX2, the 80486DX4 is technically obsolete, but the 100 MHz version especially has good processing power for routine applications. It is also acceptable for some other uses, but is not found in modern systems.

13.10 AMD 5X86 (80486DX5)

The 486DX4-100 was the fastest 486 processor made by Intel before they decided to leave the fourth generation and concentrate on the Pentium. AMD took clock multiplying one increment further with what it calls the "5x86" chip. Despite the implication that it is a fifth-generation chip, it is not—it is a high speed, clock-quadrupled 486 processor that runs in 486 motherboards. It does use a rather advanced 0.35 micron CMOS process (advanced for a fourth-generation chip that is).

The AMD 5x86 was made available in one speed only, 133 MHz, for use in 33 MHz motherboards. The processor runs at four times the system clock and fits into a Socket 3. Since having a clock multiplier of four was not part of the original Socket 3 design, AMD made the 5x86 look for a *two times* setting from the motherboard and interpret that as four times instead. In other words, to use the 5x86 you want to set the motherboard to the 2x setting. This will actually cause the 5x86 to run at 4x. The chip will actually physically fit into an older 486 socket such as a socket 1 or 2 or the original 168-pin 486 socket, but doing this requires a voltage regulator since the AMD chip runs at 3.3 volts.

AMD calls this chip the "5x86-P75" because it offers performance comparable to low-end fifth-generation chips. In fact, it is comparable in integer performance to a 75 MHz Pentium. It is also called the 80486DX5-133, which is a more accurate reflection of what the chip really is. Architecturally, the chip is virtually identical to the 80486DX4, except that it matches Intel's 16 KB level 1 cache (the AMD 80486DX4 has only 8 KB). Note that this is not the case with Cyrix's 5x86 chip, which is quite different.

The 5x86-133 is the most powerful 486-class chip available (the Cyrix 5x86-120 is actually faster but was discontinued by Cyrix many months before AMD stopped mass-producing the 5x86-133). The 5x86-133 has traditionally been an excellent choice for an economy PC, especially for home-builders, because not only is the chip itself very inexpensive, so are the motherboards that use it. With 200+ MHz Pentium and Pentium-compatible chips coming well down in price, the 5x86-133 is following the rest of the fourth-generation chips to obsolescence, quickly. However, since it is the "king of the hill" for 486 motherboards, it remains a good choice for those who want to upgrade their 486 systems without replacing the motherboard. The chips are still out there, though they are getting much more difficult to find now.

13.11 CYRIX 5X86 ("M1SC")

Despite having the same name as AMD's 5x86 processor, the Cyrix 5x86 is a totally different animal. While AMD designed its 5x86 by further increasing the clock on the 486DX4, Cyrix took the opposite approach by modifying its M1 processor core (used for the 6x86 processor) to make a "lite" version to work on 486 motherboards. As such, the Cyrix 5x86 in some ways resembles a Pentium OverDrive (which is a Pentium core modified to work in a 486 motherboard) internally more than it resembles the AMD 5x86. This chip is probably the hardest to classify as either fourth or fifth generation.

The 5x86 employs several architectural features that are normally found only in fifth-generation designs. The pipeline is extended to six stages, and the internal architecture is 64 bits wide. It has a larger (16 KB) primary cache than the 486DX4 chip. It uses branch prediction to improve performance.

The 5x86 was available in two speeds, 100 and 120 MHz. The 5x86-120 is the most powerful chip that will run in a 486 motherboard—it offers performance comparable to a Pentium 90 or 100. The 5x86 is still a clock-tripled design, so it runs in 33 and 40 MHz motherboards. (The 100 MHz version will actually run at 50x2 as well, but normally was run at 33 MHz.) It is a 3 volt design and is intended for a Socket 3 motherboard. It will run in an earlier 486 socket if a voltage regulator is used. I have heard that some motherboards will not run this chip properly so you may need to check with Cyrix if trying to use this chip in an older board. These chips have been discontinued by Cyrix but are still good performers, and for those with a compatible motherboard, as good as you can get. Unfortunately, they are extremely difficult to find now.

13.12 INTEL PENTIUM ("P5" / "P54C")

Intel's new fifth-generation chip was expected to be called the 586, following their earlier naming conventions. However, with the rise of AMD and Cyrix, Intel wanted to be able to register as a trademark the name of their new CPU, and numbers can't be trademarked. Thus, the Pentium was born. It is now one of the most recognized trademarks in the computer world, one reason why Intel doesn't seem to ever want to make another processor whose name doesn't have "Pentium" in it somewhere. :^)

The Pentium is the defining processor of the fifth generation. It has in fact had several generations itself; the first Pentiums are different in many ways from the latest ones. It has been the target for compatibility for AMD's K5 and Cyrix's 6x86 chips, as well as generations that have followed. The chip itself is instruction set compatible with earlier x86 CPUs, although it does include a few new (rarely used) instructions.

The Pentium provides greatly increased performance over the 486 chips that precede it, due to several architectural changes. Roughly speaking, a Pentium chip is double the speed of a 486 chip of the same clock speed. In addition, the Pentium goes to much higher clock speeds than the 486 ever did. The following are the key architectural enhancements made in the Pentium over the 486-class chips (note that some of these are present in Cyrix's 5x86 processor, but that chip was developed *after* the Pentium):

- **Superscalar Architecture:** The Pentium is the first superscalar processor; it uses two parallel execution units. Some people have likened the Pentium to being a pair of 486s in the same chip for this reason, though this really isn't totally accurate. It is really only partially superscalar because the second execution unit doesn't have all the capabilities of the first; some instructions won't run in the second pipeline. In order to take advantage of the dual pipelines, code must be optimized to arrange the instructions in a way that will let both pipelines run at the same time. This is why you sometimes see reference to "Pentium optimization". Regardless, the performance is much higher than the single pipeline of the 486.
- **Wider Data Bus:** The Pentium's data bus is doubled to 64 bits, providing double the bandwidth for transfers to and from memory.
- **Much Faster Memory Bus:** Most Pentiums run on 60 or 66 MHz system buses; most 486s run on 33 MHz system buses. This greatly improves performance. Pentium motherboards also incorporate other performance-enhancing features, such as pipelined burst cache. The Pentium processor was also the first specifically designed to work with the (then new) PCI bus.
- **Branch Prediction:** The Pentium uses branch prediction to prevent pipeline stalls when branches are encountered.
- **Integrated Power Management:** All Pentiums have built in SMM power management (optional on most of the 486s).
- **Split Level 1 Cache:** The Pentium uses a split level 1 cache, 8 KB each for data and instructions. The cache was split so that the data and instruction caches could be individually tuned for their specific use.
- **Improved Floating Point Unit:** The floating point unit of the Pentium is significantly faster than that of the 486.

The Pentium is available in a wide variety of speeds, and in regular and OverDrive versions. It is also available in several packaging styles, although the pin grid array (PGA) is still the most prevalent. The original Pentiums, the 60 and 66 MHz versions, were very different than the later versions that are used in most PCs; they used older, 5 volt technology and significant problems with heat. Intel solved this with later (75-200 MHz) versions by going to a smaller circuit size and 3.3 volt power.

Pentiums use three different sockets. The original Pentium 60 and 66 use Socket 4. Pentiums from 75 to 133 will fit in either socket 5 or socket 7; Pentium 150s, 166s and 200s require Socket 7. Intel makes Pentium Overdrives that allow the use of faster Pentiums in older Pentium sockets (in addition to Overdrives that go in 486 motherboards).

The Pentium processor achieved a certain level of "fame" as a result of the bug that was discovered in its floating point unit not long after it was released. This is commonly known as the "FDIV" bug after the instruction (floating point divide) that it most commonly turns up in. While bugs in processors are relatively common, they usually are minor and don't have a direct impact on computation results. This one did, and achieved great notoriety in part because Intel didn't own up to the problem and offer to correct it immediately. Intel does offer a replacement on affected processors, which were only found in early versions (60 to 100) sold in 1994 and earlier.

If you suspect your Pentium of having the FDIV bug, try this computation test using a spreadsheet or calculator program: take the number 4,195,835 and divide it by 3,145,727. Then take the result and multiply it by the same number again (3,145,727). You should of course get the same 4,195,835 back that you started with. On a PC with the FDIV bug you will get 4,195,579 (an error of 256), but beware that some operating systems and applications have been *patched* to compensate for this bug, so a simple math test isn't necessarily conclusive.

For many years, the Pentium processor was the mainstream processor of choice, but finally the Pentium with MMX has driven it to the economy market. With the regular Pentium maxing out at 200 MHz and the Pentium with MMX 166 dropping well below $200, the "Pentium Classic" doesn't make nearly as much sense as it used to for new PCs. The 60 and 66 are obsolete due to their slow speed and older technology, and the 75 to 150 are obsolete because their performance is much lower than the 166 and 200, for almost the same amount of money.

The entire classic Pentium line is now technically obsolete, due to the availability of inexpensive, faster Pentium with MMX chips (as well as comparable offerings from AMD and Cyrix). The non-MMX Pentium is no longer generally used in new systems. However, since the Pentium with MMX requires split rail voltage, the classic Pentium 200 remains a great chip for those who have socket 7 motherboards and want to upgrade, but who do not have split rail voltage support.

13.13 INTEL PENTIUM OVERDRIVE

In order to allow the easy processor upgrades of earlier motherboards, Intel has made available Pentium-class OverDrive chips. These are internally Pentium chips, but have subtle changes made to them to accommodate the unusual sockets and motherboards they are designed to work in. This allows them to be used where a regular Pentium would not work, which is the whole idea behind the OverDrive product—their interfaces appear to the motherboard like an older chip, so you don't need to worry about whether or not the motherboard supports the higher speed of what would be a native chip, in most cases.

Note: In general, if your motherboard can support a higher-speed regular (non-OverDrive) Pentium, that is the better way to go as opposed to an upgrade, because the regular chips are cheaper than the OverDrives. Your motherboard manual will tell you what your system's limitations are.

OverDrive processors have several changes made so that they can be used in older machines. First, when necessary they include integrated voltage regulators (sandwiched between the chip and the heat sink) so they will work with the voltage of the socket they are intended to go into. Second, they of course have the right pin configuration for the socket. Third, they are hard-wired to a specific clock multiplier; they do not set their multiplier based on motherboard jumpers like regular Pentiums do. This saves the user from worrying about what jumpers to change when replacing the processor.

The Pentium OverDrive comes in three basic flavors:

- **OverDrive for 486 Systems:** This Pentium OverDrive is specially modified to fit the 32-bit data bus of a 486 system. As you know a Pentium normally uses a 64-bit data bus, but this is not consistent with a 486 motherboard's design. Therefore, this OverDrive is modified to use an external 32 bit bus. This reduces performance compared to a real Pentium, which is one reason why these chips score below what would be expected for a real Pentium of the same speed. Interestingly, this chip has a 32 KB primary cache, double the size of regular Pentiums. Presumably this was done to help mitigate the effects of this chip running on a 32-bit motherboard. (It's still slow, and in fact, slower than the top-end 486 and 5x86 chips that run in the same motherboards and cost much less).
  The Pentium OverDrive for 486s is available in two speeds: 63 MHz for 25 MHz bus systems and 83 MHz for 33 MHz systems. They obviously use a clock multiplier of 2.5. These chips are designed to work in a 5 volt system, and go into a Socket 2 or Socket 3 (the socket 3 must be set on 5 volts).
- **OverDrive for Pentium 60/66:** The original Pentium chips were different than later versions, in terms of voltage, socket size and power consumption. Intel created a clock-doubling OverDrive for these chips, which is sold as one "120/133" chip: when replacing a Pentium 60 it runs at 120 MHz, and when replacing a 66 it runs at 133. This is a true Pentium chip since it is on a Pentium motherboard, although it still benchmarks below the real Pentium 120/133, most likely due to the older design of the Pentium 60/66 motherboards.
- **OverDrives for the Pentium 75, 90 and 100:** These three chips run with a multiplier of 1.5 on system buses of 50, 60 and 66 MHz respectively. Intel has made for these OverDrives running at 125, 150 and 166 MHz (clock multiplier of 2.5). The 125 is an oddity because Intel never made a Pentium 125 as a stand-alone processor. Note that these three OverDrives have been replaced by versions of the Pentium with MMX OverDrive running at the same speed.

Except where noted above, the Pentium OverDrive has the same design and internal functioning as the regular Pentium. Refer to the section on the Pentium for more description of the Pentium's functions and features, and improvements over 486-class chips.

For many systems, OverDrive processors are the simplest way to get a performance increase at a reasonable price. Since those who need to use OverDrives are somewhat of a "captive market", Intel has traditionally priced OverDrives significantly higher than equivalent regular Pentiums. In some cases, the price difference can be exorbitant, with OverDrives selling for double what the street price is for the regular chip. The best time to buy OverDrives is after they have been on the market for a while and the prices have begun to decrease. Still, an upgrade of the motherboard and processor is often less expensive than an OverDrive processor, and yields better performance (at the cost of more work and risk, of course).

It's important to remember that increasing the performance of the processor is only part of the solution to increasing overall system performance. Many other factors impact on system performance, and increasing the processor clock while leaving the system bus speed the same is an exercise in diminishing returns, because the processor is increasingly stuck waiting for the rest of the system.

13.14 INTEL PENTIUM WITH MMX TECHNOLOGY ("P55C")

In January of 1997 Intel released its newest, and presumably last, fifth generation processor, the Intel Pentium with MMX Technology. It is an evolutionary design, adding some enhancements and new capabilities to the "classic" Pentium. In most ways however, it is the same chip.

The chief (well-publicized) difference between the Pentium with MMX and the Pentium Classic is the inclusion of the MMX instruction set extensions. Running software that is designed for MMX on a Pentium MMX will result in a significant increase in performance over the regular Pentium; some claim up to 70%. Even running non-MMX software on the MMX chip results in an improvement of about 20% over a regular Pentium of the same clock speed, due to the following other improvements over the older chip:

- **Doubled Primary Cache:** The Pentium with MMX has 16 KB for each of the level 1 data and instruction caches, as opposed to 8 KB each for the regular Pentium.
- **Improved Cache Mapping:** The primary cache is now 4-way set associative instead of 2-way.
- **Deepening of Internal Pipelines:** Both of the internal integer executions units are increased from 5 to 6 stages.
- **Better Use of Internal Pipelines:** More types of instructions can be run in parallel down the two execution pipes than on the older Pentium, so more use is made of the second pipe.
- **Improved Branch Prediction Unit:** The branch prediction unit's accuracy is enhanced over the classic Pentium.
- **Improved Instruction Decoder:** The instruction decoder is more efficient than the Pentium's.

The Pentium with MMX is pin-compatible with the regular Pentium and goes in a Socket 7 just like the old chip. However, there is one important difference between the two: the Pentium with MMX uses split-rail voltage: 3.3 volts for the external voltage and 2.8 volts for the core. Not all motherboards with a socket 7 support this dual voltage setting, because before the Pentium with MMX came out no processors needed it, so some manufacturers skimped by not planning for the future. Intel is making OverDrives for motherboards that don't support the lower core voltage.

Warning: Some unscrupulous vendors are misleading buyers by telling them that the Pentium with MMX will run with 3.3 volts applied to both external *and* internal voltage. In many cases this will not instantly fry the chip; it may even work for a little while. Eventually however it will fail and you could void the warranty on the chip.

The MMX capability is implemented on the Pentium with MMX by sharing the registers used by the floating point unit. This has led to silly rumors that using MMX would cause a "major" performance penalty when switching between applications using MMX and those using floating point applications. This is a myth; aside from the fact that very few applications make extensive use of the floating point unit, a switch between MMX and non-MMX applications takes no more time than a switch between floating point and non-floating-point applications. And the amount of time for a switch is less than 100 cycles on a processor that is usually running at 200,000,000 cycles per second or more.

Originally, the Pentium with MMX was supposed to be available only in 166 and 200 MHz versions, after which Intel was going to stop production on Socket 7 chips entirely. Intel surprised the PC world a bit a few months after the introduction of the original 166 and 200 chips by announcing a Pentium with MMX 233, running at 66 MHz and 3.5x multiplier. They now say that *this* will be the last Pentium, but who knows. :^)

Since socket 7 boards don't support a 3.5x multiplier, Intel made the chip respond to a 1.5x signal as if it were 3.5x. This is becoming a common trick, which I believe AMD used first in creating its 5x86-133 (which runs in 486 motherboards as 4x when configured as 2x). So to set up a Pentium with MMX 233, you jumper its system bus speed and multiplier (but *not* core voltage!) as if it were a Pentium 100.

Overall, the Pentium with MMX is a very good chip, and with Intel aggressively cutting prices it has become pretty much the mainstream processor of choice over the last year or so. In recent months, however, with new sixth-generation chips by AMD (the K6) and Cyrix (the 6x86MX) that will run in fifth-generation motherboards becoming commonplace, the high-end performance picture on fifth-generation motherboards has become much more cloudy. For its part, Intel has moved on to the Pentium II, and as a result the Pentium with MMX is being used on fewer and fewer new systems, in favor of the newer chip.

13.15 INTEL PENTIUM WITH MMX TECHNOLOGY OVERDRIVE

With the introduction of the Intel Pentium with MMX Technology, Intel also created OverDrive processors to upgrade existing Pentium motherboards to the new MMX chip. Older Pentium motherboards cannot handle the new Pentium with MMX because of its requirement for a 2.8V core. Keeping with the tradition of Intel's OverDrive line, the Pentium with MMX OverDrive includes a converter that lets it run in Socket 5 motherboards (except for the 200) and Socket 7s that do not have 2.8V support. Otherwise the chip is identical to the standard Pentium with MMX. Refer to the section on the Pentium with MMX for more description of its improvements over the Pentium.

Note: In general, if your motherboard can support the MMX chip directly, it is almost always better to buy a real Pentium with MMX as opposed to an OverDrive, because they are typically less expensive. Your motherboard manual will tell you what your system's limitations are, but generally if your Pentium motherboard supports split-rail voltage, you should be able to run a standard Pentium with MMX.

The initial Pentium with MMX OverDrives run at 125 MHz, 150 MHz and 166 MHz, for use in Pentium 75 (1.5x50), Pentium 90 (1.5x60) and Pentium 100 (1.5x66) systems respectively. These chips replaced identical non-MMX OverDrives which Intel quickly phased out when MMX hit the market. In each case, the chip has a fixed clock multiplier of 2.5, intended for use in place of the older 1.5 multiplier chips on these three system speeds.

Intel later introduced two new Pentium with MMX OverDrives, running at 180 MHz and 200 MHz These chips run with a 3x clock multiplier; the 180 is intended for upgrading 90, 120 and 150 MHz systems running at 60 MHz system bus speeds, and the 200 for upgrading 100, 133 and 166 systems running with 66 MHz system buses. No 3x clock multiplier OverDrive was created specifically for 50 MHz Pentium 75 systems, but the 180 will function in Pentium 75 systems, running at 150 MHz (Note that the older 150 MHz OverDrive has a 2.5x multiplier and will only run at 125 in a Pentium 75 PC).

Note: Unlike the other OverDrives, which will work in either socket 5 or socket 7 motherboards, the Pentium with MMX OverDrive 200 is intended for use only in socket 7 systems.

These OverDrives provide significant performance improvements over the original chips and the ability to execute MMX software as well. It's still important to remember that increasing the performance of the processor is only part of the solution to increasing overall system performance. Many other factors impact on system performance, and increasing the processor clock while leaving the system bus speed the same is an exercise in diminishing returns, because the processor is increasingly stuck waiting for the rest of the system.

13.16 CYRIX 6X86 ("M1")

Cyrix entered the fifth generation processor market in a big way with its (IMO poorly-named) 6x86 processor, formerly called the M1. This is an advanced processor that is an alternative to the Intel Pentium, and is pin- and voltage-compatible with it. Cyrix apparently gave it the "6x86" moniker in reference to some of its more advanced features, which it calls "sixth generation". In reality, the processor is comparable in power (and in many ways, architecturally as well) to the fifth-generation Pentium. It is, like the Pentium, a superscalar, native x86 execution CPU.

The 6x86, like the AMD K5, is *not* a Pentium "clone", in the traditional sense of the word. Clones are exact or near-exact copies, usually being reversed engineered or based on licensed code. The 6x86 is based on an original design of Cyrix's and is not 100% Pentium compatible. (It is 100% *x86* compatible, but the Pentium uses some extensions to the base x86 instruction set that the 6x86 doesn't support).

The 6x86 incorporates several advanced architectural features that allow it to outperform a Pentium of equal clock speed. For example, a Cyrix running at 133 MHz will typically equal or outperform a Pentium 166 (not Pentium with MMX 166). For this reason, and because the public seems overly obsessed with thinking that clock speed is all that matters, Cyrix helped invent the "P rating" system. This was a good idea, to help people make more valid comparisons. Unfortunately however, it has caused a great deal of confusion of its own, because Cyrix named the chips with their P rating and not their clock speed. Many users (and even vendors!) confuse the P rating with the clock speed and try to set their motherboards to run, say, a 6x86-PR166+ to run at 166 MHz This is over clocking the chip by 25% and the usual result is that the chip fails. It is important to make sure that the chip is set to its correct clock speed.

Compared to the Pentium (classic), the Cyrix 6x86 has the following enhancements or improvements that contribute to its greater performance to clock speed ratio (it also has all the advantages over the 486 that are mentioned in the section discussing the original Pentium):

- **Deepening of Internal Pipelines:** Both of the internal integer executions units are increased from 5 to 7 stages.
- **Out of Order Completion:** Instructions flowing down the two integer pipelines can complete out of order.
- **Superior Branch Prediction Unit:** The branch prediction unit of the 6x86 can handle multiple branches and has better performance than the Pentium's.
- **Improved Cache Mapping:** The primary cache in the 6x86 is 4-way set associative instead of the Pentium's 2-way. (It is also unified instead of being split like the Pentium's, though I don't think this is really an enhancement).
- **Register Renaming:** This feature improves parallel performance of the pipelines.
- **Speculative Execution:** The 6x86 uses speculative execution to reduce pipeline stall time.

Of course the 6x86 is also cheaper than a Pentium of the same performance rating. Often, it is less than half the cost. This is in fact the main reason that the 6x86 is purchased instead of the Pentium. Compared to the Pentium (classic), the 6x86 has the following disadvantages:

- **Software Compatibility Problems:** This is probably the biggest problem with the 6x86 and one reason that many people won't use it, even though it provides excellent value for many applications. Some rather parochial software has been written by programmers who think Intel is the only company making CPUs, and therefore they have used Pentium-specific instructions (recall that the Pentium has extensions to the standard x86 instruction set). This causes problems with the 6x86 because the 6x86 is not 100% Pentium compatible. Some companies have made "patches" available to allow their software to run on the 6x86. This problem is most pronounced with games and high-end graphical utilities that have been heavily optimized for the Pentium. One of the most publicized problems with the 6x86 is that its early versions had a bug that causes Windows NT to run more slowly than it should have. Cyrix has addressed this issue and is offering chip replacements to affected users. New systems shouldn't have this problem.

- **Processor Identification Problems:** In addition to programs that won't work with the 6x86, there are those that would work with the 6x86 if only they knew what it was. There is a special instruction called "CPUID" that is used to identify the make and model of the processor. Some programs use this to detect if the computer is running on a Pentium, and will refuse to run (or run in a reduced mode) if they don't find one. These tests were originally used by software requiring a lot of power, to distinguish between Pentiums and 486s (which the software would be too slow to run on). The 6x86 by default does not respond to the "CPUID" instruction so some programs think it is a 486. Again, this problem is most pronounced with games. Also, early versions of Windows 95 will not identify the 6x86 properly since they predate it—they think it is a 486. This doesn't cause any performance problems however, other than programs (games) running under Windows 95 looking at the information that Windows 95 detects to decide if the machine is a Pentium or not. This problem is also fixed with the OEM SR2 release, which detects the 6x86 properly. Various utilities are available to "trick" the system into identifying the chip as a Pentium, if needed. This can solve many of these problems, although not all of them.
- **Motherboard Compatibility Problems:** Some motherboards (especially cheapo ones) have quirks that cause them not to work correctly with the 6x86. *Cyrix* maintains an approved motherboard list to help 6x86 buyers choose boards that are compatible with the chip. There are some boards that will work with it despite not being on the list, but the ones on it are known to work.
- **Heat and Power Problems:** The 6x86 uses significantly more power and generates more heat than the Pentium. This causes problems in some boxes and with some motherboards. Later versions address this issue and the low-power 6x86L eliminates it almost completely (but requires a motherboard with 2.8 volt core power). The best way to avoid this problem is to use an approved motherboard (as mentioned just above) since often the heat problems are caused by motherboards with insufficient voltage regulators. You should also use an active heat sink specifically meant for use on the 6x86; special cooling units are designed for the 6x86 and cool better than generic Pentium fans.
- **No Multi-Processor Support:** The implementation of SMP (symmetric multiprocessing) supported by today's motherboards is customized to Intel processors and will not work with the 6x86.
- **Lower-Power Floating Point Unit:** While the 6x86 equals or exceeds a Pentium of its "P rating" at integer operations, it lags far behind in floating-point performance. Cyrix intentionally decided not to spend time optimizing the FPU due to its lack of importance for most applications (which is true). A 6x86 generally has floating point performance slightly below a Pentium of its *clock speed* (as opposed to P rating). So a 6x86-PR166+ has integer performance equal to a Pentium 166 or higher, but floating point performance more comparable to a Pentium 120 or even 100. The most famous problem this causes is with Quake, the popular game that uses floating point instructions. Most games actually do not use the FPU much at all.

Many of the compatibility problems with the 6x86 are a result of the fact that most software was written when Intel was the "only game in town". With AMD and Cyrix establishing themselves as serious players, it is likely that in the future programmers will be more comprehensive in their choice of coding methods. However, the 6x86 really isn't totally Pentium compatible, which the buyer needs to take into consideration.

The 6x86 chip is designed for Socket 7, and is available in several different clock speeds, some of them rather unusual. All of the chips use a 2x multiplier, and they support bus speeds of 50, 55, 60, 66 and 75 MHz. Of course only 3 of those will be readily recognizable to those familiar with Pentium processors. The 55 MHz bus speed of the PR133+ is in particular an oddball; many (if not most) motherboards will not support this speed.

The 75 MHz bus speed is an interesting advance in many ways, as it pushes the traditional 66 MHz memory speed limit forward. The system bus speed is for most systems a bottleneck that severely limits overall performance; increasing it can increase performance greatly. This in fact is one major reason that the 150 MHz 6x86-PR200+ performs at the level of a Pentium 200. However, there are several additional caveats associated with the 75 MHz bus speed used by the 6x86-PR200+:

- **Chipset Support:** Intel is the major producer of Pentium-class chipsets, which are rated to 66 MHz, the fastest memory bus speed that Intel processors use. Not surprisingly :^) Intel is not in a big hurry to support a higher bus speed used only by a Cyrix chip. This leaves two choices for motherboard manufacturers: overclock an Intel chipset (which some do) or use an alternative.
- **Motherboard Support:** Largely because of the chipset problem, motherboards that support the 75 MHz bus speed of the PR200+ are harder to find than regular motherboards.
- **PCI Device Overclocking:** Since the PCI bus runs at half of the memory bus speed, when you increase the memory bus to 75 MHz you are increasing the PCI bus to 37.5 MHz. When you do this, you are effectively overclocking all of your PCI cards by 12.5%. This is not usually a problem, but you should bear in mind that *many* peripheral makers do not do extensive testing at the higher bus speed because of the very small market of the Cyrix chips. This will hopefully change in the future as higher bus speeds become more prevalent.

The Cyrix 6x86 is manufactured by IBM, which in fact also sells the 6x86 under its own name. (Cyrix doesn't have its own manufacturing facilities). These chips are identical; they are manufactured in the same IBM fabrication plant, and independently packaged and tested. The rumor is that the IBM version is superior due to better testing, but I've found nothing reliable to prove this.

Responding to complaints about the heat and power problems of the 6x86, Cyrix came out with a lower powered version called the 6x86L. This chip reduces power consumption by 25% or more compared to the 6x86. It uses a smaller, 0.35 micron circuit size, and split rail voltage much like the Pentium with MMX (3.3V external, 2.8V internal). This version is preferred for those seeking to reduce the power problems of earlier models.

Overall, the 6x86 is a decent processor that provides a lot of value, but at the cost of some headaches that users will have to deal with. When it first came out, it was an excellent alternative to the Intel Pentium CPU, because it was selling for several hundred dollars less. As Pentium prices dropped, so did the spread in cost between it and the 6x86, and there was increasingly little reason to use the Cyrix chip. Today the 6x86 is essentially obsolete; the main reason is the inexpensive availability of Cyrix's own 6x86MX MMX chip.

13.17 AMD K5 ("K5" / "5K86")

AMD's entry in the fifth generation processor sweepstakes is called the K5. This processor was eagerly awaited and it was hoped that it would provide a viable alternative to the Pentium early in the Pentium's life cycle. Unfortunately, AMD delivered the processor over a year late and at much lower clock speeds than had been originally anticipated. As a result instead of being the "Pentium killer" AMD had hoped for, the K5 was positioned as a low-cost Pentium alternative, much like the Cyrix 6x86.

The K5 is, internally, a very advanced processor, the most advanced of the fifth-generation chips. Internally it is more comparable to the Pentium Pro. It is an x86 translation/emulation processor, decoding x86 instructions into RISC-like microinstructions and executing them on a 6-pipeline internal core. This allows the K5 to achieve higher performance than a Pentium of the same speed. In many ways, the K5 is sixth-generation, but its performance level is held back to fifth-generation levels due its low clock speeds.

The K5 is not a Pentium "clone" in the traditional sense of the word, because it is a totally different internal design. It is also not 100% Pentium compatible because it is not an exact clone. However, it is a very close approximation, and suffers from fewer compatibility problems than the 6x86.

Due to its advanced internal architecture, the K5 (like the 6x86) is labeled according to its "P rating", which attempts to label the processor according to what speed Pentium the chip's performance approximates. This is a good way for AMD to position its products in a marketplace that overvalues the clock speed number. The K5 is available in P rating speeds from 75 to 166 and goes in a Socket 5 or Socket 7 motherboard. It is produced on a 0.35 micron CMOS process, until recently state of the art. AMD manufactures its own chips.

Compared to the Pentium (classic), the K5 has the following architectural improvements that contribute to its greater performance to clock speed ratio (it also has all the advantages over the 486 that are mentioned in the section discussing the Pentium):

- **RISC-Based Internal Architecture:** The K5 is internally a highly-parallel RISC processor using an x86 decoding front-end.
- **Six Execution Units:** The K5 has five integer units and one floating point unit for processing micro-instructions, while the Pentium has only two integer units.
- **Out of Order Completion:** Instructions flowing down the two integer pipelines can complete out of order.
- **Superior Branch Prediction Unit:** The branch target buffer is four times the size of the Pentium's (although K5 claims a lower accuracy with it than Intel does with the Pentium. Seems odd to me.)
- **Larger Primary Cache:** The K5 has a 16 KB instruction cache instead of the Pentium's 8 KB. The data cache is unchanged at 8 KB.
- **Improved Cache Mapping:** The primary cache in the K5 is 4-way set associative instead of the Pentium's 2-way.
- **Register Renaming:** This feature improves parallel performance of the pipelines.
- **Speculative Execution:** The K5 uses speculative execution to reduce pipeline stall time in its RISC core.

While these improvements are impressive, they are the reason that the K5 has P ratings well above its internal clock speeds; if it didn't have the higher P ratings the K5 could not compete with the Pentium at all. Like the Cyrix 6x86, the K5's main's attraction is significantly lower price than the Pentium. Unlike the 6x86, the K5 has a reputation for being reasonably compatible and has far fewer problems than the Cyrix chip does. There aren't the heat and motherboard compatibility problems of the 6x86 associated with the K5. However, it (arguably) does not offer as high a level of performance as the 6x86, and there is no K5 chip near the power level of the 6x86-PR200+.

The K5 can have some compatibility problems due to the fact that it is not a Pentium. Some Pentium-specific software can cause problems with the K5, but these programs are relatively rare. The "CPUID" problem associated with the 6x86 under Windows 95 is not an issue for the K5.

While the K5 exceeds or beats a Pentium of its "P rating" at integer operations, it lags behind in floating-point performance. The K5 falls somewhere between the 6x86 and the Pentium in terms of its FPU, with performance roughly equal to a Pentium 25% slower than the K5's P rating. This really isn't much of an issue for most people since few applications make extensive use of the FPU (although Quake is one high-profile game that does.)

The 6x86 suffers from the problem where it is occasionally overclocked by accident because users set the motherboard to the clock speed of the P rating instead of the actual clock speed of the chip itself. The K5 avoids this by making two design decisions. First, all of the K5 chips run on the same memory bus speed as the equivalent Pentium of its P rating. So a K5-PR133 runs on a 66 MHz bus speed, like the Pentium 133, and unlike the 6x86-PR133+. Second, all K5 chips are designed so that if you configure them with the multiplier that an equivalent-rated Pentium would use, the chip will run with the correct internal multiplier. The combination of these two characteristics means that to set up a K5 you jumper your motherboard as if it were a Pentium and it will work. This avoids a lot of confusion.

The K5 also has an undocumented, generally unused multipler setting: if you set it to 3x, it runs at 2x. The reason for this is generally believed to be that AMD intended to release a K5-PR200, but never did. This chip was supposed to run at 133 MHz and have performance equivalent to a Pentium 200. Of course the Pentium 200 has a multiplier of 3x, so this would have been consistent with the other K5s' ability to be jumpered as if they were a Pentium. The full table of AMD K5 multipliers is:

Motherboard Multiplier Setting	CPU Interpreted Multiplier	Processors
1.5x	1.5x	K5-PR75 (75 MHz), K5-PR90 (90 MHz), K5-PR100 (100 MHz)
2.0x	1.5x	K5-PR120 (90 MHz), K5-PR133 (100 MHz)
2.5x	1.75x	K5-PR166 (116 MHz)
3.0x	2.0x	K5-PR200 (133 MHz, not released)

As you can see above, the K5 has its own confusions to compensate for its better handling of jumper settings. :^) The main one is that the PR90 and PR120 both run at 90 MHz processor speed, and the PR100 and PR133 both run at 100 MHz How is it possible for two processors with the same design to have different performance ratings running at the same speed? It isn't—and that's the confusion, because the PR75, PR90 and PR100, have a different internal design than the PR120, PR133 and PR166. The first three chips were created earlier, and were actually at one point called the "5k86".

The K5 is a good alternative to the classic Pentium, with performance equal to a mid-level Pentium. It does not have many of the problems that the Cyrix 6x86 does, but it also doesn't offer performance as high as the top-end 6x86s. However, it is now as obsolete as the 6x86, and for that matter the classic Pentium that both compete with. For the home builder on a very tight budget the K5-PR166 is a decent choice, but for the power user it is already too weak to be considered. The price of the K6 has come down so far, and the K6 is so much better than the k5, that the latter is really only a good choice for a motherboard that does not support split voltage.

13.18 INTEL PENTIUM PRO ("P6")

The Pentium Pro was introduced in 1995 as the successor to the Pentium. It introduced several unique architectural features that had never been seen in a PC processor before. The Pentium Pro was the first mainstream CPU to radically change how it executes instructions, by translating them into RISC-like microinstructions and executing these on a highly advanced internal core. (The Nexgen Nx586 processor was actually the first x86 CPU to use this design, but this chip was used in very few systems.)

The Pentium Pro achieves performance approximately 50% higher than a Pentium of the same clock speed. In addition to its new way of processing instructions, the Pentium Pro incorporates several other technical advances that contribute to this increased performance:

- **Super pipelining:** The Pentium Pro dramatically increases the number of execution steps, to 14, from the Pentium's 5.
- **Integrated Level 2 Cache:** The Pentium Pro features a dramatically higher-performance secondary cache compared to all earlier processors. Instead of using motherboard-based cache running at the speed of the memory bus, it uses an integrated level 2 cache with its own bus, running at full processor speed, typically three times the speed that the cache runs at on the Pentium. The Pentium Pro's cache is also non-blocking, which allows the processor to continue without waiting on a cache miss.
- **32-Bit Optimization:** The Pentium Pro is optimized for running 32-bit code (which most modern operating systems and applications use) and so gives a greater performance improvement over the Pentium when using the latest software.
- **Wider Address Bus:** The address bus on the Pentium Pro is widened to 36 bits, giving it a maximum addressability of 64 GB of memory.
- **Greater Multiprocessing:** Quad processor configurations are supported with the Pentium Pro compared to only dual with the Pentium.
- **Out of Order Completion:** Instructions flowing down the execution pipelines can complete out of order.
- **Superior Branch Prediction Unit:** The branch target buffer is double the size of the Pentium's and its accuracy is increased.
- **Register Renaming:** This feature improves parallel performance of the pipelines.
- **Speculative Execution:** The Pro uses speculative execution to reduce pipeline stall time in its RISC core.

For a few reasons, the Pentium Pro is *still*, despite its age, an ideal choice for servers. First, it is a fast chip in general. Second, its integrated level 2 cache makes it ideal for multiprocessing; instead of having a single motherboard-based level 2 cache that all the processors must share, each has its own. Third, the Pentium Pro has chipsets available for it that are designed for high-end server use, moreso than the Pentium.

The most widely-publicized advanced feature of the Pentium Pro is of course the integrated level 2 cache. The Pentium Pro is shipped in a special *dual cavity* SPGA package that includes the chip itself and the integrated cache. It goes into a special Socket 8 interface unique to the Pentium Pro. One disadvantage of this arrangement is that the cache is not upgradable without also replacing the processor.

The integrated-cache design has been both a blessing and a curse for Intel. The blessing is that it greatly improves the performance of the chip. The curse is that it has been very difficult for Intel to manufacture the Pentium Pro at the volumes and cost levels necessary for it to become a mainstream processor. There are two main reasons for this. First, the cache itself is highly miniaturized and therefore much more expensive to produce than the typical SRAM chips used on a Pentium motherboard for level 2 cache. Second, some problems with the cache are not found until after it has been mated with the processor and installed in their shared package; when this happens the whole package (including the processor) must be thrown away, reducing yields and increasing costs. Due to the problems with its design, Intel has abandoned the integrated-cache concept and it is unlikely that any future PC processors will use it in the same way that the Pentium Pro does.

The Pentium Pro is usually found in either 180 MHz or 200 MHz versions. Older Pentium Pros ran at 150 and 166 MHz; these are far less common and the 166 MHz chip is in particular rarely seen. The 150 and 180 chips ship only with 256 KB level 2 cache, while the 200 is available with 256 KB, 512 KB or 1 MB of level 2 cache. The cost of the 200 MHz chip with 512 KB or 1 MB of cache is very high due to production costs and demand. The 166 MHz chip is unusual in that it was available with 512 KB of cache only.

Despite being almost two years old, the Pentium Pro processor is still commonly used in high-end systems, although the Pentium II is now starting to take some of this market. Until Intel comes out with a proper Pentium II chip set for servers, demand for the 200 MHz version (especially with 512 KB or 1 MB of cache) will continue to be high. In addition, multiple-Pentium-Pro servers are quite common and provide good performance at a reasonable price. The Pentium Pro often competes against non-Intel server processors such as DEC's Alpha.

13.19 8051 MICROCONTROLLER

The **Intel 8051** is a Harvard architecture, single chip microcontroller (μC) which was developed by Intel in 1980 for use in embedded systems. Intel's original versions were popular in the 1980s and early 1990s, but has today largely been superseded by a vast range of faster and/or functionally enhanced 8051-compatible devices manufactured by more than 20 independent manufacturers including Atmel, Infineon Technologies (formerly Siemens AG), Maxim Integrated Products (via its Dallas Semiconductor subsidiary), NXP (formerly Philips Semiconductor), Nuvoton (formerly Winbond), ST Microelectronics, Silicon Laboratories (formerly Cygnal), Texas Instruments and Cypress Semiconductor. Intel's official designation for the 8051 family of μCs is **MCS 51.**

Intel's original 8051 family was developed using NMOS technology, but later versions, identified by a letter C in their name (e.g., 80C51) used CMOS technology and were less power-hungry than their NMOS predecessors. This made them more suitable for battery-powered devices.

13.19.1 8051 Derivatives

Along the way, this 8031 architecture gained enviable market acceptance. Many semiconductor manufacturers started either manufacturing the 8031 devices as such (Intel was liberal in giving away license to whoever asked) or developing a new kind of microcontrollers based on 8031 core architecture.

Manufacturers modified the basic 8031 architecture and added many new peripheral functions to make them attractive to the designers.

Because of the rush, electronic community started getting a variety of 8031 based devices with range of options. To beat the competition, manufacturers developed different microcontrollers with many unique features.

These parts are popularly known as '8031 Derivatives'. Almost every decent manufacturer boasted of having an 8031 based microcontroller in the line card.

First major manufacturer was the Philips who brought out more than 40-50 derivatives with a variety of I/O options, memory combinations, and peripheral functions. Devices became available in regular DIP and SMD packages. With the basic 8031 core, Philips ported high capacity Program Memory (upto 32K/64K), its patented I2C interface bus, 8/10 bit Analog to Digital Converters, CAN Bus, Capture and Compare registers, Watch dog timer, PWM facilities and etc. More I/O ports (as many as eight ports), additional timer/counter, second serial port was also made available in Philips devices.

Apart from all these, Philips developed many consumer devices meant for telecom, computer and TV applications. A smart card controller was also developed by incorporating a cryptographic engine. So Philips clearly established itself as the market leader in 8031 derivatives and still caters to this segment.

Then came Dallas semiconductor. Dallas redesigned the 8031 architecture and eliminated waste clock cycles of original core and made all instructions executed in less clock cycles (maximum of 4) which has traditionally taken upto 12 clock cycles. So, came the birth of High speed 8031 Derivatives.

Dallas also maintained the same device pin out configurations to enable the user get upto 3X performance by replacing slower parts with a Dallas device. So, existing compiled code started running faster without any modification. These days, you can find Dallas devices giving upto 50 MIPS (Million Instructions Per Second).

Apart from this, Dallas introduced additional Serial port, Watch Dog Timer, Precision Reset Circuitry, Real Time Clock, Power Fail Monitor in the 8031 devices. Also a second data pointer, more onchip RAM space and more interrupt lines were also made available.

Dallas semiconductor also has got a range of secure microcontrollers based on 8031 core. This microcontroller family uses non volatile RAM to keep both program and data. Because of this RAM, the controller gives the In System Reprogrammability. Dallas has combined this microcontroller, SRAM and lithium cell in a single pack. This device guarantees 10+ years of data retention in the RAM area. This 8031 also boasts the tamper proof security features like Real Time Memory Encryption, user selected 48 bit Encryption key, memory contents, security lock and the facility to hide interrupt vector table. As you can agree, this particular 8031 device has found a niche market in banking and security related applications.

Atmel Corporation is the another major semiconductor manufacturer who introduced many flash memory based 8031 derivatives at a competitive cost. Atmel used its expertise in flash memory technology into the basic 8031 core and brought out microcontrollers with a variety of flash memory options and few devices also carry In System Reprogramming facility. You can program/reprogram this microcontroller after soldering the device in the target board. If this programming facility is embedded in the system software, then the tasks like remote calibration, onsite system upgradation become as easy as sending your data/program in a floppy disk or by internet. Atmel devices sport security lock to its flash memory to protect the contents from the prying eyes.

Meantime, Intel itself tried to cash in the popularity of this 8031 architecture and introduced improved versions of microcontrollers: 80151 and 80251 families. These devices sport 16 bit architecture using 8031 core and unfortunately these devices have not become as popular as 8031.Even after many years of introduction, 8031 core is still going strong in 8 bit arena.

13.19.2 Memory Organization

All 80C51 devices have separate address spaces for program and data memory, as shown in Figures 1 and 2. The logical separation of program and data memory allows the data memory to be accessed by 8-bit addresses, which can be quickly stored and manipulated by an 8-bit CPU. Nevertheless, 16-bit data memory addresses can also be generated through the DPTR register.

Program memory (ROM, EPROM) can only be read, not written to. There can be up to 64k bytes of program memory. In the 80C51, the lowest 4k bytes of program are on-chip. In the ROMless versions, all program memory is external. The read strobe for external program memory is the PSEN (program store enable). Data Memory (RAM) occupies a separate address space from Program Memory. In the 80C51, the lowest 128 bytes of data memory are on-chip. Up to 64k bytes of external RAM can be addressed in the external Data Memory space. In the ROMless version, the lowest 128 bytes are on-chip. The CPU generates read and write signals, RD and WR, as needed during external Data Memory accesses.

External Program Memory and external Data Memory may be combined if desired by applying the RD and PSEN signals to the inputs of an AND gate and using the output of the gate as the read strobe to the external Program/Data memory.

13.19.3 Instruction Sets

The instruction set is divided in to 5 categories. They are as follows:

1. Arithmetic instructions
2. Logic instructions.
3. Data transfer instructions.
4. Boolean variable manipulation instruction.
5. Program and machine control instruction.

We have listed all the instructions of microcontroller.

Instruction Set Summary

Mnemonic		Description	Byte	Cycle
Arithmetic Operations				
ADD	A.Rn	Add register to accumulator	1	1
ADD	A.direct	Add direct byte to accumulator	2	1
ADD	A, @Ri	Add indirect RAM to accumulator	1	1
ADD	A,#data	Add immediate data to accumulator	2	1
ADDC	A.Rn	Add register to accumulator with carry flag	1	1
ADDC	A.direct	Add direct byte to A with carry flag	2	1
ADDC	A. @Ri	Add indirect RAM to A with carry flag	1	1
ADDC	A,#data	Add immediate data to A with carry flag	2	1
SUBB	A.Rn	Subtract register from A with borrow	1	1
SUBB	A.direct	Subtract direct byte from A with borrow	2	1
SUBB	A.@Ri	Subtract indirect RAM from A with borrow	1	1
SUBB	A.#data	Subtract immediate data from A with borrow	2	1
INC	A	Increment accumulator	1	1
INC	Rn	Increment register	1	1
INC	direct	Increment direct byte	2	1
INC	@Ri	Increment indirect RAM	1	1
DEC	A	Decrement accumulator	1	1
DEC	Rn	Decrement register	1	1

DEC	direct	Decrement direct byte	2	1
DEC	@Ri	Decrement indirect RAM	1	1
INC	DPTR	Increment data pointer	1	2
MUL	AB	Multiply A and B	1	4
DIV	AB	Divide A by B	1	4
DA	A	Decimal adjust accumulator	1	1

Logic Operations

ANL	A,Rn	AND register to accumulator	1	1
ANL	A.direct	AND direct byte to accumulator	2	1
ANL	A,@Ri	AND indirect RAM to accumulator	1	1
ANL	A,#data	AND immediate data to accumulator	2	1
ANL	direct.A	AND accumulator to direct byte	2	1
ANL	direct.#data	AND immediate data to direct byte	3	2
ORL	A,Rn	OR register to accumulator	1	1
ORL	A,direct	OR direct byte to accumulator	2	1
ORL	A,@Ri	OR indirect RAM to accumulator	1	1
ORL	A,#data	OR immediate data to accumulator	2	1
ORL	direct,A	OR accumulator to direct byte	2	1
ORL	direct,#data	OR immediate data to direct byte	3	2
XRL	A,Rn	Exclusive OR register to accumulator	1	1
XRL	A direct	Exclusive OR direct byte to accumulator	2	1
XRL	A,@Ri	Exclusive OR indirect RAM to accumulator	1	1
XRL	A,#data	Exclusive OR immediate data to accumulator	2	1
XRL	direct.A	Exclusive OR accumulator to direct byte	2	1
XRL	direct,#data	Exclusive OR immediate data to direct byte	3	2
CLR	A	Clear accumulator	1	1
CPL	A	Complement accumulator	1	1
RL	A	Rotate accumulator left	1	1
RLC	A	Rotate accumulator left through carry	1	1
RR	A	Rotate accumulator right	1	1
RRC	A	Rotate accumulator right through carry	1	1
SWAP	A	Swap nibbles within the accumulator	1	1

Data Transfer

MOV	A,Rn	Move register to accumulator	1	1
MOV	A,direct"	Move direct byte to accumulator	2	1
MOV	A,@Ri	Move indirect RAM to accumulator	1	1
MOV	A,#data	Move immediate data to accumulator	2	1
MOV	Rn,A	Move accumulator to register	1	1
MOV	Rn.direct	Move direct byte to register	2	2

MOV	Rn,#data	Move immediate data to register	2	1
MOV	direct.A	Move accumulator to direct byte	2	1
MOV	direct.Rn	Move register to direct byte	2	2
MOV	direct.direct	Move direct byte to direct byte	3	2
MOV	direct,@Ri	Move indirect RAM to direct byte	2	2
MOV	direct,#data	Move immediate data to direct byte	3	2
MOV	@Ri,A	Move accumulator to indirect RAM	1	1
MOV	@Ri,direct	Move direct byte to indirect RAM	2	2
MOV	@Ri, #data	Move immediate data to indirect RAM	2	1
MOV	DPTR. #data16	Load data pointer with a 16-bit constant	3	2
MOVC	A,@A + DPTR	Move code byte relative to DPTR to accumulator	1	2
MOVC	A,@A + PC	Move code byte relative to PC to accumulator	1	2
MOVX	A,@Ri	Move external RAM (8-bit addr.) to A	1	2
MOVX	A,@DPTR	Move external RAM (16-bit addr.) to A	1	2
MOVX	@Ri,A	Move A to external RAM (8-bit addr.)	1	2
MOVX	@DPTR,A	Move A to external RAM (16-bit addr.)	1	2
PUSH	direct	Push direct byte onto stack	2	2
POP	direct	Pop direct byte from stacK	2	2
XCH	A,Rn	Exchange register with accumulator	1	1
XCH	A,direct	Exchange direct byte with accumulator	2	1
XCH	A,@Ri	Exchange indirect RAM with accumulator	1	1
XCHD	A.@Ri	Exchange low-order nibble indir. RAM with A	1	1

Boolean Variable Manipulation

CLR	C	Clear carry flag	1	1
CLR	bit	Clear direct bit	2	1
SETB	C	Set carry flag	1	1
SETB	bit	Set direct bit	2	1
CPL	C	Complement carry flag	1	1
CPL	bit	Complement direct bit	2	1
ANL	C,bit	AND direct bit to carry flag	2	2
ANL	C./bit	AND complement of direct bit to carry	2	2
ORL	C,bit	OR direct bit to carry flag	2	2
ORL	C,/bit	OR complement of direct bit to carry	2	2
MOV	Cbit	Move direct bit to carry flag	2	1
MOV	bit.C	Move carry flag to direct bit	2	2

Program and Machine Control

ACALL	addr11	Absolute subroutine call	2	2
LCALL	addr16	Long subroutine call	3	2
RET		Return from subroutine	1	2

RETI		Return from interrupt	1	2
AJMP	addrl 1	Absolute jump	2	2
LJMP	addrl 6	Long iump	3	2
SJMP	rel	Short jump (relative addr.)	2	2
JMP	@A + DPTR	Jump indirect relative to the DPTR	1	2
JZ	rel	Jump if accumulator is zero	2	2
JNZ	rel	Jump if accumulator is not zero	2	2
JC	rel	Jump if carry flag is set	2	2
JNC	rel	Jump if carry flag is not set	2	2
JB	bit,rel	Jump if direct bit is set	3	2
JNB	bit,rel	Jump if direct bit is not set	3	2
JBC	bit,rel	Jump if direct bit is set and clear bit	3	2
CJNE	A,direct,rel	Compare direct byte to A and jump if not equal	3	2

Program and Machine Control (cont'd)

CJNE	A,#data,rel	Compare immediate to A and jump if not equal	3	2
CJNE	Rn,#data rel	Compare immed. to reg. and jump if not equal	3	2
CJNE	@Ri.#data,rel	Compare immed. to ind. and jump if not equal	3	2
DJNZ	Rn.rel	Decrement register and jump if not zero	2	2
DJNZ	direct.rel	Decrement direct byte and jump if not zero	3	2
NOP		No operation	1	1

EXERCISE

1. What are the differences between 8086 and 80186 microprocessor?
2. Explain the architecture of 80186 microprocessor.
3. On the 80186 which pin is used to refresh the DRAM memory?
4. What is virtual memory machine?
5. What would be the state of 80286 microprocessor following a system reset?
6. Explain the architecture of 80186 microprocessor.
7. How the address bus and data bus of 80186 differ from 80286?
8. Explain the features of Pentium processor.
9. Give the difference between 80186 and 80286.
10. Compare the features of 80386 with 80486.

INDEX

E

F

G

H

I

T

U

W